Handbook of the Effects of Landfill Emissions on Sustainable Development

Abdelkader Anouzla · Salah Souabi ·
Tonni Agustiono Kurniawan
Editors

Handbook of the Effects of Landfill Emissions on Sustainable Development

Springer

Editors
Abdelkader Anouzla
Faculty of Science and Engineering
University of Hassan II Casablanca
Mohammedia, Morocco

Tonni Agustiono Kurniawan
College of Environment and Ecology
Xiamen University
Xiamen, Fujian, China

Salah Souabi
Department of GPE, Faculty of Science and Techniques
University of Hassan II Casablanca
Mohammedia, Morocco

ISBN 978-3-032-01950-9 ISBN 978-3-032-01951-6 (eBook)
https://doi.org/10.1007/978-3-032-01951-6

This Springer imprint is published by the registered company Springer Nature Switzerland AG
The registered company address is: Gewerbestrasse 11, 6330 Cham, Switzerland

Contents

Chapter 1
Atmospheric Pollution: Current Knowledge Causes, Effects, and Relevance

Wafa Hassen, Bilel Hassen, Marwa El Ouaer, and Abdennaceur Hassen

Abstract Air pollution contributes to the degraded quality of life and reduces the life expectancy of the population. The World Health Organization estimates that air pollution is responsible for 7 million deaths per year worldwide. It contributes to worsening respiratory diseases and causes lung cancer and heart attacks. Air pollution thus has important medical consequences for human life and biodiversity. This past year, there were considerable signs of progress in the control of this atmospheric pollution to sensitize the public authorities and the population about the need to reduce the intensity of this pollution and its harmful effects. In this chapter, we reviewed this pollution, which is mostly ignored by governments for economic reasons and ignored by the overall public out of ignorance. Technological advances and public awareness campaigns of the overall public thus allowed and contributed to the design and production of the tools within an academic framework, making it possible to better evaluate and assess the quality of the air and its potential effects on humans and their environment. The sector of heavy industries associated with the services of transport and road traffic was the factor showing and inducing the greatest number of harmful environmental effects, with earlier occupants being more specialists, decision-makers, and the general population. The aim of this chapter book consists of summarizing this pollution, a long time ignored by decision-makers and the overall public, and synthetically and transparently evaluating the environmental effects of atmospheric pollutants with their collective and individual components,

W. Hassen · M. El Ouaer · A. Hassen (✉)
Laboratory of Treatment and Wastewater Valorization, Center of Research and Water Technologies (CERTE), Soliman, Tunisia
e-mail: abdohas@gmail.com

W. Hassen
e-mail: hassen.wafa@gmail.com

M. El Ouaer
e-mail: maroua_el@yahoo.fr

B. Hassen
Laboratories of Microbiology and Pathology of Aquatic Organisms, National Institute of Navy Sciences and Technologies (INSTM), Salammbo, Tunisia
e-mail: bilel.hassen@fst.utm.tn

A. Anouzla et al. (eds.), *Handbook of the Effects of Landfill Emissions on Sustainable Development*, https://doi.org/10.1007/978-3-032-01951-6_1

parts of which are professional or deprived and emitted via transport, industries, agriculture, and varied human activity.

Keywords Air pollution · Principal pollutants · Environmental effect · Medical effects · Current state

1.1 Introduction

Atmospheric pollution results from the change in the chemical composition of the air under the effects of human activities and the cosmic. It reproduces from the urban and local scales to the planetary scales. On the planetary scale, the changes in the chemical composition of the air appear in particular as an increase in the contents of gases with a greenhouse effect, such as carbon dioxide or methane, and condition a major change in the planet's climate. However, local pollution in the industrial or urban environment, for example, presents risks for human health because of pollutant air contents such as nitrogen and sulfur oxides, various other oxides, and oxidizing particle acids (Chen et al., 2020; Tong, 2005).

Atmospheric pollution has been recognized in recent years as pollution with an equivalent whole share or more dangerous than that of the aquatic environment because of its diffuse capacity and nature and because it does not control gases and air emissions. Thus, in 2013, the World Health Organization recognized that atmospheric pollution represents an unquestionable carcinogenic factor; many pathologies are related to the human, animal, and plant respiratory tracts. Since air is a constituent of the atmosphere and is fundamental for the lives of humans, animals, plants, and even microbes, it is critical to protection against solar and cosmic radiation by generating the ozone layer and regulating climate through the greenhouse effect. These properties weaken with the evolution of human civilization in relation to air pollution that flows out of cities and the countryside, moves, transforms, and ultimately affects human health, the environment, and especially the pressures of thermal balance,... exchanges, and components of the atmosphere.

Atmospheric pollution results from the change in the chemical composition of the air because of human and cosmic activities. It reproduces from the urban and local scales to the planetary ones. On the planetary scale, the changes in the chemical composition of the air appear in particular as an increase in the contents of gases with a greenhouse effect, such as carbon dioxide or methane, and conditions mainly at risk of major changes in the planet's climate. However, local pollution in the industrial or urban environment, for example, presents risks for human health because of the contents of pollutants such as sulfur oxides, oxides, and oxidizing particle acids (Chen et al., 2020; Tong, 2005).

Atmospheric pollution has been recognized in recent years as pollution with an equivalent whole share or more dangerous than that of the aquatic environment because of its diffuse nature capacity and lack of controlled gas emissions in the air. Thus, in 2013, the World Health Organization recognized that atmospheric pollution represents a carcinogenic factor. Many pathologies are related to the respiratory tracts of humans, animals, and plants. Since air is a constituent of the atmosphere and fundamental for the life of humans, animals, plants, and even microbes, it is critical to protection against radiation from the sun and cosmic harm by generating a layer of ozone and in climatic regulation by the greenhouse effect. These last properties weaken with the evolution of human civilization and are connected to the atmospheric pollution that escapes from cities and campaigns, which is transformed and affected in final human health, the environment, and especially the thermal balance of pressure, … of exchanges, the components of the atmosphere.

Considering the importance of this pollution within the framework of the life of the man and his environment, the regulations at the national and international levels, and multiple international conventions aiming at the improvement of the air quality, the main thing that ties together the man's life and his environment became a world concern. This concern is conceived on the scale of individuals, members, and governments. Each one must act in their daily lives to reduce the many and considerable emanations with a harmful tendency.

Various technical disciplines and scientists are concerned with atmospheric pollution problems, and this file wants to be a general introduction to air pollution. It is centered on the problems of air quality at the local and regional scales and will be followed by other files treating more of the approached topics such as regulation, state of knowledge, toxicology and ecotoxicology methods of air pollution investigation, pollutant monitoring, processing technologies of liquid waste, tools for air pollution simulation, etc.

1.2 Independent Factors of Atmospheric Pollution Causes

These independent causes of atmospheric pollution are of two types, according to their source. Thus, we could divide his factors into two categories: those caused by and coming directly from the universe and those related to man's destroying and devastating action. The natural origins are represented by the earth's geological activities, such as volcanism, where volcanoes in eruptions send sulfur into the air and contribute to considerable atmospheric pollution. The weather events are represented by abrupt falls in the pressure generating the storms and by the flashes that reveal nitrogen dioxide and other potentially harmful gases, especially following the frequent summer period and involuntary fires of the forests. The other great natural cause of atmospheric pollution is combustion, whether caused by humans or not, which ensures a more or less massive release of CO_2 and its principal derivatives, primarily CO.

The main source of human pollution is breathing, which involves oxygen consumption and carbon dioxide exhalation. Thus, with the man's installation and the cities' and villages' appearance, atmospheric pollution would come primarily from heating, waste incineration operations, the transport sector, and the circulation sector. The latter has reached its peak in the last few years with uncontrolled development and industry growth. The emergence of new energy sources such as oil contributed to this degradation of the atmosphere via the increase in marine, air, and especially terrestrial traffic. Thus, car pollution was and is the leading human cause of the generation of combustion gases. Transport handles the rejection of carbon monoxide (87%) because of incomplete combustion and the rejection of nitrogen oxide (70%). The combustion of other energy sources, such as oil, natural gas, and coal, releases large quantities of CO2, CO, and SO_2. Agricultural activities are presented as the oldest source of air pollution, with methane release (CH_4) into the atmosphere via poultry, cattle and sheep farming, and rice plantations. Unintentional deforestation, whether brought on by forest fires or started intentionally, remains a significant source of carbon dioxide.

The consequences of air pollution on the environment mainly affect the evolution processes and plant development via the generation of acid rain, the reduction in the ozone layer thickness, climate warming, greenhouse effects, etc. The need for reinforcing the monitoring and prevention of air pollution met the planetary scales, leading governments to issue a law on air and rational energy use. These laws established the right of everyone to breathe clean air and the right to information about the state of the air and its potential consequent effects.

Additionally, they mandate the deployment of a variety of planning instruments, including regional and national plans for air quality to safeguard the atmosphere, relocation plans for cities, and emergency measures to prevent atmospheric pollution and its negative and undesirable effects.

1.3 Determination of Air Quality

Air is the most fundamental element for all living beings. Every day, an average adult consumes 14 kg of air, 1.5 kg of food, and 2 kg of water. This consumption plays an important role in regulating the climate and maintaining the biosphere's living conditions. However, air quality is changing in time and space since its composition comes from complex balances between matter contributions that permanently come from natural sources or are related to human activities. The elimination and continuous recycling of this matter via various chemical phenomena, such as the dry and wet deposits of gases and particles that are primarily the result of man's activity, the universe's transformations, the assimilation by vegetation, and the wholesale general exchange of ocean, biosphere, and lithosphere, etc. This concept of complex atmospheric balances has become increasingly changing with the development of modern life and the dangers or results of unfavorable consequences, sometimes dreadful for health and the environment, emerging in front of humans and their life framework. The zone of the terrestrial atmosphere concerned with pollution problems extends up to 50 km from an altitude, and it is classically divided into two large parts: the stratosphere between 10 and 50 km above the soil and the troposphere at an altitude of 10 km.

A transition zone separates these two principal zones, called the tropopause. Their rate of variation characterizes these three areas, particularly in the air's temperature; consequently, the average air temperature decreases with altitude in the troposphere but increases in the stratosphere.

As with any definition, air pollution is primarily the result of a biophysical-chemical imbalance caused by various human activities, such as introducing various gaseous, liquid, and solid compounds that pose health risks and interfere with the proper development of any living being, including plants, animals, and microbes, as well as the evil operation of natural ecosystems, often involving degradation and harmful effects, directly or indirectly, into a closed atmosphere.

1.4 Consequences of Air Pollution on Human Health, Animal, and Plants.

All the undesirable compounds in the air are disturbing elements of good performance in the atmosphere, such as these principal living beings. According to the definition provided above, "atmospheric pollution" can refer to any physical, chemical, or biological alteration of the air that results in an imbalance of good performance. When we think about air pollution, we think especially of the damage caused by human activities. We often overlook how air pollution affects the surrounding ecosystem without taking plants and animals into account.

1.4.1 Effects of Atmospheric Pollution on Plants

The plants will mostly absorb the pollutants delivered by the air through their leaves, which serve as both seats and receivers of the needed oxygen for their existence. A minor penetration rate, lower than 10% and variable according to the plant, would be recorded through the stems and trunk. The "layer boundary" must first be crossed by the pollutant. The size and structure of the sheet are determined by the thickness of the foliar external layer and the wind speed, which influence oxygen penetration. The thickness of the external layer of foliar tissue is approximately a tenth of the thickness of the outer layer of leaf tissue. The characteristics of the plant, in particular the size, the form, and the foliar density, control the penetration of gases, particularly oxygen, in the plant system. The nature of the pollutant, its physicochemical characteristics such as pH and density, its electrophile character, its condition, its level of penetration, and its insertion into the plant system. Thus, in the zones of heavy industries related to production or raw material transformation, such as mines, metallurgy, the iron and steel industry, petrochemistry, the paper mill, and the chemistry of the first transformation, we often record deserted zones deprived of almost any vegetative form, regardless of the country, climate, and geography. After penetration, the physiological reaction of plants to air pollution appears as a progressive drying of the sheets, vital systems, and bodies, leading to a slow death. We would see a slow, ill-fated, suffering plant (Verma et al., 2010).

1.4.2 Effects of Atmospheric Pollution on Humans

As we have mentioned, air pollution is a complex pollutant mix in constant evolution in space and time. This pollution is mostly represented by three categories of contaminants:

Chemical Mainly represented by ozone, dioxides, nitrogen oxides, sulfur dioxides, metals (arsenic, lead), some volatile organic compounds (VOCs), such as butane, ethanol, or benzene, or the hydrocarbons (polycyclic aromatic hydrocarbons—HAP) present in petroleum, oil, or coming from fuels or wood combustion. In this way, COVs are a large family of compounds frequently produced out of total hydrocarbons such as methane or propane. They can be emitted by anthropogenic factors via the production of gasoline, the emanation of solvents, and by vegetation following various compound emanations, including hydrocarbons, among which isoprene and monoterpenes are considered the most abundant and reactive in photochemical processes (Gao et al., 2014). The other COVs emitted include sesquiterpenes and especially terpene families of oxygenated compounds (such as aldehydes, ketones, alcohols, etc.). These compounds are largely found in the ambient air of domesticated areas, and they can cause many respiratory and skin diseases. Fuel, paint, adhesives, solvents, insecticides, perfume interiors, and cleaning products are strongly carcinogenic and sources of VOCs that are responsible for some breathing difficulties and reproduction problems (Grange et al., 2007).

Physics and Biology This category included fine particles, dust made up of large chemical components; and finally, biology, assembling the general pollen, molds, and microbial aerosols, grouping various microbial and viral species. These air pollutants, which are in the minority, originate from frequent natural phenomena, such as the Sahara sand winds, soil erosion, and volcanic eruptions, and from most activities caused by the man following his various interventions, particularly those linked to industries and transport. We are all too familiar with the fact that air pollution is the primary cause of cardiovascular problems, allergies, asthma attacks, conjunctivitis, diseases of the bronchi, lung cancer, skin, vision problems, blood diseases, and children's problems related to mental development. The most vulnerable people are children, elderly individuals, expectant mothers, and patients affected by diseases such as tension and diabetes. Table 1.1 summarizes the major pollutants likely to be present in polluted air and the pathologies exposed to these pollutants for a short or long time. A study not yet published and carried out in 2018 by our research team on the site of two wastewater treatment plants installed in the suburbs of Tunis, Tunisia, revealed that diversified biological agents, and in particular the microorganisms with their toxins normally conveyed by wastewater, the pasty or dry sewage sludge, are found in the ambient air at the site as aerosols. This bioaerosol is a suspension of droplets or particles generated by water jets with high pressure, operations blowing compressed air into the wastewater treated in the activation and clarification basins of the activated sludge, etc. The operational data obtained on the scale of forced aspiration revealed that the operators' exposure to the bioaerosol varied according to the zones of the plant and their activities. Thus, in a wastewater plant, aerosols charged with liquid or dry particles containing microorganisms and their components, in particular endotoxins, will often be inhaled by operators. Endotoxins are complex molecules that are components of the wall of gram-negative bacteria. These molecules, released during the multiplication or destruction of the bacteria, persist for a long time in the environment and resist many chemical or physical agents. These latter are often specific allergens for sensitive people and responsible for a strong excitation of the immune system, triggering a hypersensitivity phenomenon.

These bioaerosols can be displaced by moving objects (convoying, shoveling sludge, etc.) or by falling water or using water (water jets, retro-washing of the fabrics and grids of the processes of treating sewage sludge, tables of draining, band filters, etc.). They can also be suspended in the air by ventilation systems (ventilation of biological basins, use of compressed air, suffragettes, etc.). The studies of Lidwien et al. (2005) and Shang et al. (2023) cover this subject well for the specific cases of bioaérosols with endotoxin and virus, respectively.

Recent research has shown that airborne particles are among the most hazardous and cancer-causing substances for humans. The toxicity of these particles is correlated with both their size and composition. The finer the particles are, the more able

Table 1.1 Major air pollutants likely causing disease effects in humans

Pollutants	Principal health risks
Ammonia	Exposure to very strong concentrations causes irritations, even ocular and respiratory burns
Sulfur dioxide	Irritations of the respiratory tracts (cough, respiratory embarrassment, asthma)
Volatile hydrocarbons and compounds	Frequent irritations, breathing difficulties, smell pollution Benzene is classified as carcinogenic for the Man
Nitrogen oxides	In the short run, irritations and aggravations of respiratory diseases (asthma) Eventually, the development of respiratory diseases or weak, cardiovascular weight of the infant, and increased risk of death
Ozone	Respiratory embarrassment, cough, irritations of the eyes, asthma attacks, the appearance of respiratory diseases
Particulate particles or matters	Eventually, the development of cancers (lung, bladder), cardiovascular diseases, respiratory diseases, developed children, diabetes,…

they are to penetrate deeply into the organism and pass via blood circulation toward other bodies. In the same way, the exposure time to polluted air is fundamental for the appearance of more or less serious symptoms in humans; thus, exposure of a few hours to a few days (exposure acute, known as in the short run) can lead to ocular irritations or respiratory tracts, asthma attacks, exacerbation of cardiovascular and respiratory diseases sometimes leading to hospitalization, and in the most serious cases, death (Boraschi et al., 2023). Rather than short-term incidents, the main medical effects of air pollution are associated with prolonged exposure and can last for a few years. As opposed to the peaks, exposure to air pollution throughout the year at the average level showed a more significant medical impact.

1.5 Medical and Economic Effects of Air Pollution

Thus, on a worldwide level and according to the WHO, air pollution, in addition to that of water, is the principal environmental risk for human health worldwide, and air pollution leads each year to the death of approximately 4.2 million people worldwide. In 53 European countries, the WHO estimates that approximately 600,000 deaths per year are related to air pollution. For example, the Ministry of French Public Health estimates that exposure to fine particles among people older than 30 would be responsible for approximately 40,000 mortalities yearly in France. Air pollution is an important risk factor for health in France since it accounts for 7% of the total mortality of the French population attributed to exposure to fine particles [particles of diameter lower than 2.5 μm ($PM_{2.5}$)]. In terms of economic impact, the Board of Inquiry into the Economic Costs and Financing of Air Pollution of the Senate estimates that the total price of air pollution (external and interior) is established at between 68 and 97 billion euros per year for France, of which a large part is correlated to the medical effect.

1.6 Independent Sources of Atmospheric Pollution

The independent sources of pollution in the atmosphere are well known, and they include road traffic, the services of general transport, heating, and various industrial activities. Road traffic generates many emissions associated with large quantities of pollutants released near the road infrastructure. These quantities of pollutants come not only from the emissions from the exhaust of the vehicles but also from other causes such as the wear of the tires and the brakes, the technologies of air-conditioning the vehicle, the surface of the roadways, and the maintenance of their access (use of plant health products, etc.). Vehicles are the origin of two types of emissions distinct from those in the atmosphere: pollutants of the air known as local pollutants, composed of pollutant gases (e.g., carbon monoxide, nitrogen oxides (NO_x)) or harmful particles that directly affect health (respiratory tracts and cardiovascular diseases), and carbon dioxide (CO_2), considered the principal gas with the greenhouse effect responsible for climate warming, which has little direct effect on health.

1.6.1 Anthropic Independent Sources in France

In the annual report of the French government on the quality of the air in 2014 (appeared on September 30th, 2015), the principal pollutants are reported in Table 1.2.

Air pollution thus results mainly from heating installations, thermal plants, and industrial facilities, the means of transportation for which are motor vehicles (except electric vehicles) and agriculture.

These pollutants, known as "primary" because they are emitted directly by sources of pollution, are added by pollutants known as "secondary," since they are often generated by very complex and random interactions with the various pollutants likely to be in the atmosphere. Thus, according to the data from recent studies carried out in 2019, road transportation accounted for approximately 12, 54, and 40% of the average emissions of fine particles ($PM_{2.5}$), nitrogen oxides (NO_x), and carbon soot, respectively; these proportions can be more important locally near axes with strong road traffic. These pollutant emissions usually occur near inhabited areas and at the soil level, which involves a strong potential for population exposure to road traffic

Table 1.2 Universal air pollutants: primary sources and regulation status in France as a developed country

Pollutants	Independent primary sources	Regulations respected
Sulfur dioxide (SO_2)	Industry	Yes
Nitrogen oxides (NO_x) of which nitrogen dioxide (NO_2)	Road transport	Not
Ozone (O_3)	No direct sources	Not
Particles of diameter lower than 10 μm (PM_{10})	Residential, industry, agriculture, road transport	Not
Particles of diameter lower than 2.5 μm ($PM_{2.5}$)	Residential, heating with wood in particular	Not
Carbon monoxide	Residential, industry	Yes
Benzene (C_6H_6)	Residential, transport	Not
Arsenic	Industry	Not
Cadmium	Industry	Yes
Nickel	Industry	Yes
Lead	Road transport, industry, aviation slight 67, shot-firing fume	Yes
Polycyclic aromatic hydrocarbons, mainly the benzopyrene	Residential, heating with wood mainly	Not

polluting emissions. A research study conducted by our team between 2013 and 2018 on a sizable parcel in Cap-Bon, Tunisia, that was irrigated with municipally treated wastewater and planted with orange trees, revealed high lead soil contents in the parcel bordering a major road with heavy traffic, but the lead soil content in the areas farther from this road seemed normal. This result showed the effect of car traffic on the level of lead pollution in soil, considering that lead is a particularly dangerous substance.

According to epidemiological studies, a relationship exists between the distance from the dwelling compared to the main roadways and various medical manifestations, such as asthma in children, respiratory symptoms in no asthmatics, pulmonary function disorders, cardiovascular pathologies, leukemia, etc. Thus, the dwelling's proximity to the main roads of circulation would be responsible for approximately 15 to 30% of the new cases of asthma in children, according to a study relating to 10 European cities. Thus, a French study highlighted a frequency of leukemia of the myeloblastic type higher than 30% for children living less than 150 m from a road in strong traffic. In 2012, the French International Research Center on Cancer (CIRC) classified the effluents of the exhaust of diesel engines (vehicles, boats, trains, machines on the construction site, etc.) as carcinogens for humans. The effluents of the exhaust of petrol engines are "perhaps carcinogenic" for humans. Diesel engines are also the major source of nitrogen dioxide (NO2), known as a strongly irritating substance in the respiratory tract.

In addition to road traffic, residential heating with wood in a report published in 2015 by the WHO showed that residential heating with wood or coal is a significant source of air pollution, both outside and inside buildings, except in powerful heating appliances with regard to the reduction of polluting emissions (Jaramillo et al., 2007). Fine and ultrafine particles, therefore easily transportable over long distances, can reach several hundred kilometers. Therefore, various particle compounds, such as carbon soot or "black carbon" and organic carbon, originate from biomass combustion, particularly wood combustion, and are involved in various toxic gas emissions in the air, represented in particular by carbon monoxide (CO) and nitrogen oxides (NO_x). In addition, some other volatile organic compounds (VOCs), such as benzene, formaldehyde, acrolein…, some polycyclic aromatic hydrocarbons, such as aldehydes and phenols…, some heavy metals, such as mercury, arsenic, and lead…, and dioxins and furans are often released by household and industrial activities. These last residual compounds, known as toxic elements and mainly as microparticles, are confirmed as coming from the combustion of various kinds of products, and according to the WHO, they are associated with an exacerbation of respiratory pathologies, in particular asthma and chronic obstructive bronchoneuropathy, as well as general bronchitis and otitis.

According to the WHO statistics, each day, an adult breathes between 10,000 and 20,000 L of air according to its morphology, size, activities, etc., and in addition to oxygen (O_2) and diazoxide (N_2), which represent on average 99% of its composition, the air also carries pollutants of natural origin, such as marine spray, dust, pollen, etc., and/or those resulting from human activities, such as road traffic, energy production, industry, agriculture, etc. The varied penetration of these pollutants in humans could, in the short- and long-term, have negative health consequences.

As evoked above, exposure to airborne pollutants can worsen chronic respiratory conditions such as asthma and bronchiolitis, support cardiovascular system issues, and even result in death. Among the immediate risks and with short-term, even weak effects, exposure to pollutants can cause itchy symptoms at the level of the eyes, nose, and throat on the same day or in the days that follow. French Public Health has undertaken a study over the past year on the short-term impact of suspended particles on mortality in 17 French cities. The results showed that an increase of 10 $\mu g/m^3$ suspended particles of the day and the 5 days prior increased 0.5% of no accidental mortality. The excess risk is higher in 75-year-old people and more + 1.04%, and the effects on mortality seem more important in the summer. A one-second study on the short-term impact of nitrogen dioxide (NO_2) on mortality in 18 French agglomerations over 2010 and 2014 showed and confirmed that exposure to NO_2, as a tracer of pollution related to traffic, is associated with short-term mortality, even with an average annual concentration in conformity with the European regulation and the WHO guidelines.

Even at low levels of pollutant content, exposure to air pollution over several years can lead to significant health effects, sometimes considerably more important than in the short run. Many studies have shown the role of atmospheric pollution in reducing life expectancy and mortality, as well as in the development of cardiovascular diseases, respiratory diseases, and lung cancer. Following direct toxicity on the cells, the pollutant inhalation by the air can involve genetic deteriorations, through an indirect action and via a pro-inflammatory reaction and oxidative stress, and by weakening the defense mechanisms of the organism; thus, pollutants can mainly induce (Table 1.3):

- At the cardiovascular level: a reduction in cardiac rhythm variability; an increase in arterial pressure and blood coagulability; and a progression of atherosclerosis leading to the development of coronary diseases such as myocardial infarction and cerebral vascular accidents.
- At the respiratory level: a reduction in breathing capacity; an increase in bronchial reactivity; abnormal cell multiplication leading to the development of chronic obstructive broncho neuropathy; asthma; lower respiratory infections; and, in certain cases, lung cancer (Günter et al., 2005).

New studies have also shown the role of air pollution in disorders of reproduction, disorders of child development, neurologic affections, and type 2 diabetes. The general recent evaluation showed that, on average, in developed and industrialized countries, nearly 40,000 people die each year from air pollution, corresponding to a 7.6-month loss of life expectancy (Blanchard et al., 2014).

Table 1.3 Main disorders and pathological manifestations caused by the prevalent air pollution (Dr. J. Harkema cited by Günter et al., 2005)

Disorders and bodies	Pathological demonstrations
Systemic ignition and oxidative stress	An increase in C reactivates proteins; an increase in the pro-inflammatory mediators; activation of the leucocytes and the plates
Disorders of reproduction and the development of the child	Fertility; miscarriages; growth of the fetus; premature birth; weak weight with the birth
Blood	Rheological disorders; an increase in coagulability; diffusion of the particles through the capillary wall; Peripheral thrombosis; reduction in oxygen saturation
Brain	Increase in cerebral ischemia; Cognitive disorders; neurodegenerative diseases
Heart	Deterioration of the cardiac function; oxidative stress; increase in the frequency of disorders of the rate/rhythm; disorders of the repolarization of the cardiac fabric; increase in myocardial ischemia
Vascular system	Atherosclerosis, acceleration of the evolution of the plates of atheroma, and instability of the plates; degradation of endothelial, Vasoconstriction and hypertension
Lungs	Ignition; oxidative stress; aggravation of the chronic obstructive broncho-neuropathy and progression faster of the disease; increase in the respiratory symptoms; degradation of the pulmonary function

Source Monitoring program air and health, NSSI, Blanchard et al. (2014)

1.7 Role and Effects of Ozone in the Atmosphere

Ozone is known as a gas with a natural intensive oxidant capacity, and its molecule is made of 3 oxygen atoms (O_3). The ozone molecule has bent molecular symmetry of the C2V type (like the water molecule). The angle between the oxygen atoms is 116.48°. It is generated naturally following the reaction of natural excitation (of the flash electric shock) or caused (by the electric shock) by two oxygen molecules. It is known as allotropic oxygen formation. Air pollution is caused by ozone existing in the low layers of the atmosphere between 0 and 12 km above the soil. Ozone is naturally present in trace amounts in the atmosphere, but it has potentially toxic effects on living systems when these concentrations in the low layers become too high. Its polluting atmosphere effect is called secondary because of its low content in the lower layers of the atmosphere and its unstable character related to its strong chemical and photochemical reactivity and its high oxidation capacity. The strong oxidant effect of ozone involves a complex whole of chemical and photochemical reactions, which imply the generation of some residual byproducts known as “precursors” such as nitrogen oxides (NO_x), volatile organic compounds (VOCs), and mainly the generation of methane (CH_4) and carbon monoxide (CO).

Photochemical pollution shows a complex mixture of pollutants chemically formed in the air under the effect of solar radiation, starting from some precursory compounds emitted by natural sources and varied human activities, commonly known as volatile organic compounds (VOCs), as cited above. Therefore, the principal photochemical air pollutant is ozone (O_3), and this ozone-rich air pollution is called “smog.”

The photochemical production of O_3 resulted from the dissociation of nitrogen dioxide (NO_2) by ultraviolet radiation, with this reaction making up the independent atomic source of oxygen (O) in the lower layers of the atmosphere. The production of ozone depends on the sun and prevalent temperature, so the production of photochemical pollutants is favored in the spring or summer.

Ozone is included in the category of “secondary” atmospheric pollutants since its formation processes involve slow chemical reactions that produce secondary byproducts at low concentrations and that can be carried in clouds and are likely to move over long distances.

Thus, often, for ozone and other secondary pollutants such as certain air particles, the highest levels of concentration are always far away from the precursory pollutant sources. A part of ozone ensures protection from the sun’s harmful rays, in particular UVA, UVB, and UVC, via the atmosphere’s ozone layer. Additionally, ozone is usually used to disinfect and purify water, deodorize the air, and treat industrial wastes. As mentioned above, ozone contributes to protecting living beings from UV rays since stratospheric ozone is viewed as a beneficial compound. The strong ozone concentration found between 15 and 30 km above the Earth’s surface is commonly called the ozone layer. Thus, by absorbing notably harmful UV-B sun rays, ozone contributes to protecting our planet, and stratospheric ozone in the high layer is essential to prevent harmful UV-B sun rays from reaching the Earth’s surface. Considerable efforts allowed by political and communal power are made at a worldwide scale toward protecting this ozone layer, ensuring life on our planet.

Ozone, also known as tropospheric, “basic altitude,” or “bad ozone,” is a typical smog pollutant that is photochemically charged with fine particles and generated following complex chemical reactions between various residual pollutants such as NO_x and VOCs. Under the solar radiation effect and air stagnation conditions, the contribution of ozone to mortality could be evaluated by several studies. France’s Public Health data showed that chronic exposure to ozone would be responsible for almost 500 deaths from respiratory causes each year, with a marked north–south gradient expanding because of the influence of weather on ozone formation. Therefore, studies carried out in 18 French cities revealed that the death risk brought on by ozone and fine particles outweighed the risk brought on by hot days.

There is always synergy between the pollutant-negative effects and the temperature. High temperatures support the production of ozone, and this pollutant will be largely present in the summer. Thus, the ozone content of the air is known to be more important during hot and sunny days (Jhun et al., 2014). People can spend more time outside when the weather is nice and will be more exposed to ozone. Last, co-exposure to ozone and temperature can cause amplified negative reactions in an organism compared to single agent exposure by intensifying inflammatory reactions. These factors can contribute to worsening the ozone effect on health when the weather is hot. In the peak event of pollution from concomitant ozone and a heat wave, the results of the analyses carried out in 2003 have shown that the risks related to high temperatures are much more important than those related to ozone. Therefore, it will be a necessary priority to be protected from heat (Lacour et al., 2006).

1.8 Biological Air Pollution or Bioaerosols

We define an aerosol more commonly as fine particles, solids, or liquids of a chemical substance or a mixture of substances in a gaseous suspension. Emitted following varied human or natural activities such as volcanoes and forest fires, aerosols contribute on planetary and local scales to the phenomena of air pollution and allergies. The microbial quality of the interior air represents an important medical stake, particularly in the professional sector. Inhabitants of industrialized countries spend 80%–90% of their time in semiclosed spaces, such as residences, offices, transportation, schools, etc. (Gustavsson et al., 2024).

Exposure to aerosols can be associated with some important infectious diseases and hypersensitivity episodes following toxin, allergen, and carcinogen compound inhalation. There are several measuring techniques available, enabling the examination and measurement of biophysical-chemical particles in the air at workplaces. The aerosols correspond to an important category of the air's interior parts.

They are solid or liquid particles in suspension in the air, whose falling speed is negligible. The spectrum of their sizes varies from a few nanometers to approximately 100 μm (Harrison, 2020). A subclass of aerosols is distinguished by bioaerosols, which are airborne particles of microbial, animal, or vegetable origin (Douwes et al., 2003). They are present in the air in several physical forms that can be individualized as bacterial spores, cells, etc., and grouped as an aggregate made up of several individualized elements or other biological materials. They can also result from subproducts, such as mycotoxins, endotoxins, etc., or fragments of individualized elements (Duquenne & Greff-Mirguet, 2005). The bioaerosol particles whose size is lower than 5 μm showed an important time of flight and could remain suspended in the air for a long time, while the coarser particles (>10 μm) form a deposit more quickly to settle on surfaces (Mohr et al., 2003).

The microorganisms in the air are mainly found as aggregates (Fig. 1.1). These aggregates allowed microbial protection from desiccation, oxidation, and light. Lighthart and Kim (1989) and Lighthart and Shaffer (1994) specified that the bacteria located inside the large particles (> 9.2 μm) formed by the aggregates would survive better over long distances, which explains why most of the cultivable bacteria are associated with particles.

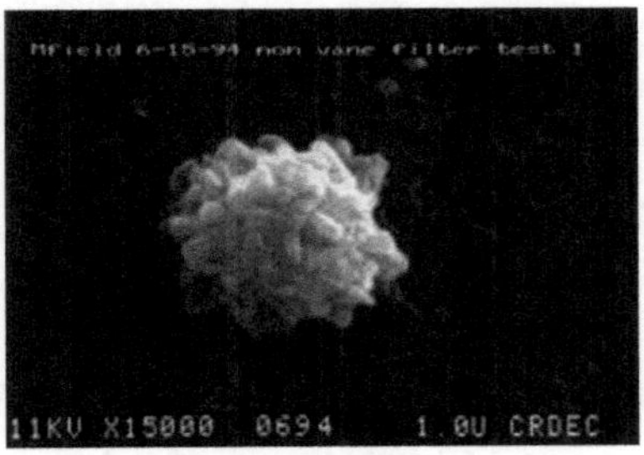

Fig. 1.1 Scanning electron micrograph of atmospheric microbial aggregates found near the ocean (Lighthart, 1997)

A scanning electron micrograph identified the cluster as a natural microbial cluster found in the air near the ocean. The clump showed a bacterial aggregate of approximately 2.8 μm in diameter (Lighthartl, 1997).

Bioaerosols are mainly derived from living organisms, mainly microorganisms, including bacteria, fungi, protozoa, algae, viruses, and some acarians. Microbial aerosols depend on several physical factors that also influence their development. The drafts, relative humidity, and temperature are environmental physical parameters that can significantly change the size, density, and shape of particles (Mohr et al., 2017, Pedgley, 1991). The airborne microbes mainly included bacteria with sizes ranging between 0.2 and 50 μm, fungi of variable size ranging between 1 and 100 μm, and other various microflora of variable size averaging between 10 and 400 nm.

1.9 Main Microbial Residual Components

Among the bioaerosols, it is also advisable to be interested in the biological components resulting from the microorganisms likely to be harmful to human health. Endotoxins are components of the external cellular membrane of gram-negative bacteria and include proteins, lipids, and lipopolysaccharides known to be responsible for the toxic effect. The term "endotoxin" refers to a toxin present either in the bacterial cell or in the fragments of the cell walls released during bacterial lysis. In the environment, the endotoxins of the air are often associated with particles of dust or liquid aerosols. Significant endotoxin exposure has usually been linked to respiratory issues, including asthma, and low doses of endotoxins can play a positive role by ensuring protection against allergies and asthma (Douwes & Pearce, 2014). These endotoxins are also suspected to play a role in rheumatic diseases in old and unmaintained buildings affected by moisture (Lorenz et al., 2006). Their effects on health could vary according to the species, the people, the amount, and the way of entry. The symptoms usually reported according to respiratory exposure to endotoxins are cough, breath runs, fever, obstruction, and the ignition of the lungs, as well as gastrointestinal problems (Goyer et al., 2001a, 2001b, 2002; Hinds et al., 1982). Figure 1.2 illustrates the relationship between the size of particulate matter (PM) and its ability to penetrate the human respiratory tract (Costa et al., 2016).

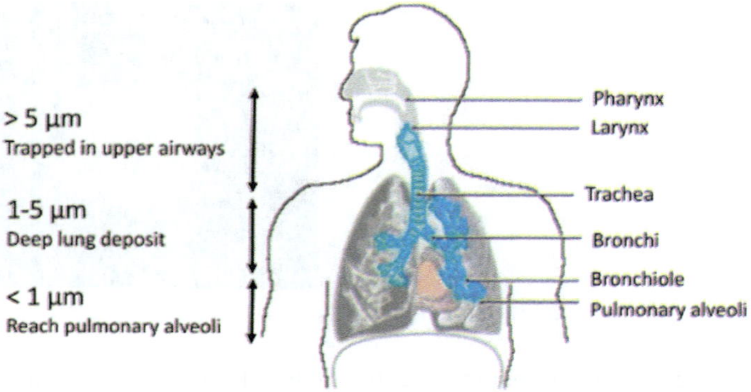

Fig. 1.2 Relation between the size of particulate matter (PM) and its ability to penetrate the human respiratory tract. Adapted from Costa et al. (2016)

Mycotoxins, also called fungal toxins, are low-weight biomolecules that can be toxic to animals and humans. These specific secondary metabolites of the fungi are released during the processes of food degradation and fermentation as residual compounds. Mycotoxins are a defense tool for fungi against other microorganisms, including other fungi. Volatile compounds are present in the air only in the case of medium agitation and disturbance (Straus, 2009). According to Edwards (2004), Edwards et al. (2012), the trichothecene produced by the fungal strains of *Stachybotrys chartarum* was an immune system stimulant even if used in low doses. Nevertheless, mycotoxins interfere with the synthesis of ARN and can cause damage to DNA. Many of these mycotoxins are deleterious to the immune system. Several mycotoxins, such as aflatoxin, produced by *Aspergillus flavus* and *Aspergillus parasiticus*, are potentially carcinogenic. Mycotoxins can be found in high quantities in buildings affected by moisture and fungal problems, but it is not settled that the increased fungal development in a building affected by moisture problems systematically induces enough mycotoxin quantities to affect residents' health (Al Hallak et al., 2023; Bloom et al., 2009; Brasel et al., 2004, 2005a, 2005b; WHO, 2009).

The WHO guidelines for airborne contamination measurement by fungal particles consider that airborne fungal spore contents above 500 colony-forming units (CFU)/m^3 are hazardous for occupants, and those above 1000 CFU/m^3 are extremely hazardous (ANSES, 2020). Several important studies have reported a serious comparison of airborne fungal spore measurements in normal dwellings, moldy dwellings, and some hospitals based on quantitative measurements.

1.10 Airborne Measurement of Fungal Particle Contamination

Fungal organisms are in strong regression in zones of intensive agriculture and urbanization, such as certain lichens used as biological indicators of air quality. It was confirmed that the fungicides present in the air and washed away by the rain caused the most significant regression of the fungal species. In addition, several no known pollutants may not have intentional fungicidal properties. Fungi can also bioaccumulate several compounds, especially some micropollutants such as certain heavy metals such as mercury and radionuclides, which can be indicators of old pollution.

Sampling of the bioaerosol can be carried out by various techniques, such as impaction on an agar medium or air filtration through a filter system, but other methods allowing microbial collection in a solution are available (Mayol et al., 2014). Microbial detection and measurement in the air can be carried out using methods of analysis used in microbiology. We define biological aerosols or bioaerosols as particles of microbial, animal, or vegetable origin suspended in the air. Interest in the study of these aerosols has shown a net increase in the last few years, and we recognize that exposure to aerosols can be associated with some medical infectious diseases, allergic reactions, and carcinogenic cases (Bellanger et al., 2009; International Agency for Research against Cancer, 1994).

Sampling and biological agent analysis are necessary to identify sources of contamination and check the effectiveness of corrective measures. Several standards exist for the microbes measured in the workplace's air. In practice, there are many methods and apparatuses usable for aerosol studies, but most are not standardized, and their characterization is often incomplete.

1.11 Acid Rains

Acid rains are defined as abnormally acidic precipitation with a pH slightly lower than 5, considering that rainwater has an average pH of 5.6. The expression and its definition are related to the initiative of the Scottish chemist and specialist in atmospheric pollution, Robert Angus Smith (1872). These rains have been regarded as a main issue since the emergence of coal- and steel-heavy industries, particularly at the beginning of the Industrial Revolution. This pollution was recognized only in the eighties because of its visible effects and the general conscience toward general environmental problems and its acute pollution, particularly in countries and industrialized areas in Europe and North America and later in Asia. The continuation of this general awakening on an international scale has been associated with the regulations flowing to remedy and improve this hazardous situation. As mentioned above, the main cause of acid rain is atmospheric pollution, which is essentially related to high industrial activities, and the harmful gases and derivatives that are present in abundance are compounds such as CO_2, NO_2, and SO_2 (Fig. 1.3).

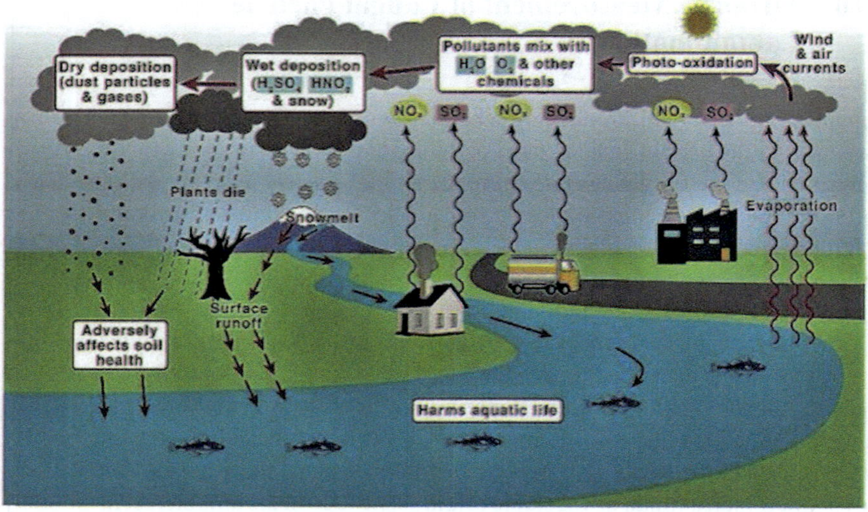

Fig. 1.3 Origin and main causes of acid rains (Sivaramanan, 2015)

A large part of the gases present in the air will be dissolved in rainwater, involving a light fall into pH and producing acid; the pH value of this slightly acidified rainwater would be closely related to its load of toxic pollutants and their dissolving power. Therefore, CO_2, NO_2, and SO_2 induce the formation of hydrodioxides as $H_n\ (CO_2)_n$, $H_n\ (NO_2)_n$, and $H_n\ (SO_2)_n$, and their presence in rainwater is natural and explains the slight rainwater acidity. This rainwater acidity is related to various atmospheric flows that are constantly changing and conditioning the pH values of rainwater over time, especially according to the seasons, human activities, and air gas releases. These gas releases have drastically worsened air pollution during the last century, in addition to acid rain (Sivaramanan, 2015).

Different human actions and other various elements may contribute to the rise of acid rain. In addition, other varied human activities and biophysical-chemical factors can cause acid rain. Some sporadic natural events, such as volcanic eruptions or forest fires, might occur, leading to acid rains of brief duration in contrast to those caused by or relying on human activity and long gas rejections.

As we mentioned several times above, the sectors of coal, transportation, and industry, as well as power plants, are the principal sources of the atmosphere's polluting gases. The establishment of legislation on local and international scales for atmospheric pollution fighting has allowed from the year 80 to drastically reduce the emissions of carbon dioxide, nitrogen and sulfur, and other oxides by half. These various dioxides are mainly derived from sulfur dioxide (SO_2), emitted by volcanic activities, fires, and industrial activities; nitrogen oxides (NO_x), coming from storms and diesel engines; and finally, ammonia (NH3), derived from industrial phosphate refining and intensive agriculture. The risks and dangers of acid rain on living beings and their environments are variable according to the rain's intensity and duration.

Thus, acid rain contributes to the killing, elimination, and/or changing of many animal, plant and microbial species, making them sterile, fragile, and weakened by insect and fungal attacks. Thus, this change involved weak disturbances in the restorative and durable development of plants, animals, and even microbes. These negative changes contribute to the generation of new animal, bacterial, and plant species. Species of vegetable, bacterial, fungal, and algal origin include a whole group of organisms that prefer and require acidic soils or compositions to develop and grow in an optional, preferential, or obligatory way in an acidic environment. These organisms are known as extremophiles, grouping primarily some important bacterial and fungal species. These are selected as excellent operators for bioremediation processes at polluted sites (Hassen et al., 2021; Werheni Ammeri et al., 2021a, 2021b, 2022a, 2022b, 2022c). These acidophilus organisms grow under very acidic conditions, with a pH often lower than 2.0. These acidophiles include archaebacteria, bacteria, and even some eukaryotes. Therefore, these species are often found in food or pharmaceutical products as probiotics, such as *Lactobacillus acidophilus, Lactobacillus casei, Lactobacillus bulgaricus, Bifidobacterium bifidum,* and *Streptococcus thermophilus,* to help with digestion. The other acidophilus present in the natural spring water is bacteria such as Acidithiobacillus, Acetobacter, and Acidobacterium. Thus, for example, the microorganism *Picrophilus oshimae* can survive in hot sources with a negative pH up to -0.06.

While microbes are important, acidophilic microbes depend particularly on biogeochemical cycles and human and environmental development. Examples include the regulation of the basic sulfate and iron cycles, the management of water and soil contaminated by acid effluents, and the use of bio-extraction, including bio-lixiviation and bio-extraction remediation technologies.

The varieties of plant acidophilus include azaleas, camellias, heathers, digital ones, ferns, bilberries, and rhododendrons. Chestnuts usually develop in acidic soil rich in silica. As species and acidophilic varieties are found, we also encounter species and varieties of organisms known as neutrophils, basophils, or alcalinophils, which are adapted to grow in a neutral medium for neutrophils or an alkaline-basic medium for basophilic or alcalinophilic ones. As mentioned above, these organisms are known as extremophiles. Natural and artificial acid habitats are widely distributed worldwide in terrestrial and oceanic ecosystems, such as sulfur-rich acid thermal springs, marine volcanic chimneys, and acid mining drainage.

Thus, intermittent acid rain contributes to polluting water, particularly lakes and rivers, and disturbs them by unbalancing the aquatic ecosystem, which includes fish, trees, and animals. Certain lakes saw the disappearance of all their populated aquatic species. In the United States, acid rain erodes rocks and limestone by solubilizing them and releasing them into the natural water and rivers often used for drinking water provision. This phenomenon of limestone acid dissolution has importantly contributed to the increase in problems related to drinking water treatment. Likewise, the aquatic fauna suffered from this water acidification; the soils will be subjected to various impoverishments, mainly through flora and fertility weakening. If the watery species are directly affected, the others can face a reduction in their food resources in quality or quantity via the transformation of their habitat or their ecosystem. Historical buildings are also affected and degraded by acid rain, which erodes limestone and corrodes metals. Many famous monuments, such as the Taj Mahal Coliseum, had to undergo the effects of atmospheric acidity.

Separately from the wet nature of the acid rains, atmospheric pollution simultaneously generates dry deposits represented by small acid particles and gases in the acid matter, such as sulfur dioxide and nitrogen oxide. These acidophilic particles settle on the surface of all the alive or inert components of the planet, such as water, vegetation, and structures, and transform them into acid via contact with the naturally prevailing humidity. These particles will often have an average pH value lower than 5.6, damaging natural ecosystems. As we have mentioned, the deposits of atmospheric particles contribute to damaging lakes, rivers, forests, soils, fish and wild population species, buildings, and even people's health. Thus, before falling on the soil, the acidifying emissions as acid particles, nitrogen dioxide, sulfur dioxide, and various other oxides contribute to putting the formation of fog and smog at risk and harming public health. Thus, acid deposits pose a problem in many areas of Canada and the USA since the atmospheric emissions responsible for acid rain can travel thousands of kilometers from their source. This is all about acid rain, its sources, and its effects. Therefore, acid rain is a major natural problem that has become progressively worse because of pollution. Hence, everyone should take these natural disasters seriously.

1.12 Management of the Atmospheric Environment

This management operation has moved forward at the regional, national, and international levels via establishment and public awareness campaigns for legislation and regulation implementation. Prevention is shown to be the best solution against air pollution in addition to other collective or individual solutions. These solutions can include the use of better apparatus, industrial or personal, with low air pollution generation, a judicious choice of fuels with less polluting gases in road traffic and household heating, a measured and preliminary study of the construction sites of factories and industrial complexes, and the maintenance and safeguarding of forests and green spaces surrounding cities. Several reports have indicated that many natural zones at a worldwide scale are deserted and given up by humans following bad programming and implementation of industrial parks (De Ridder et al., 2004).

The anthropic emissions of aero-contaminants such as carbon monoxide, sulfur, soot, and various particles related to the combustion of coal were probably the first major source of air pollution at the beginning of the industrial era. Rail-bound transport, road traffic, and air reinforced the release of air-polluting emanations. For phosphate-heavy industries, steel and oil-primarily industries also reinforced and brought strong air pollution during the second half of the twentieth century.

The United Nations (UNO), the World Health Organization (WHO), and many other governmental organizations acted to draw the attention of governments and populations to the great danger related to atmospheric pollution problems. The WHO reported that pollutants of natural or anthropic origins in the world over the period 1990–2016 were the fifth highest risk factor for health after malnutrition, food perils, arterial hypertension, and nicotine use. This pollution causes the premature death of seven million people each year. The data on this part are multiple and have varied over the last 20 years. Thus, in July 2020, the Energy Institute of the University of Chicago will publish a report on the loss of life expectancy due to air pollution. It estimates that the life expectancy of humans would increase by almost 2 years if all countries conformed to the air quality directives endorsed by the World Health Organization. This loss is unequally distributed: In 2018, the largest losers, with 5 years of potential longevity on average, are the inhabitants of Bangladesh (6.2 years), India (5.2 years), Nepal, and Pakistan. In this part of southern Asia, the levels of air pollution increased by 44% in 20 years. Contrary to expectations, China reduced air pollution by almost 40% in 5 years, but the United States and Europe spent several decades reaching the same rates of depollution (Greenstone & Hanna, 2014; Greenstone et al., 2015; Greenstone & Fan, 2018; Greenstone et al., 2022; Greenstone, 2020).

1.13 Atmospheric Pollutants and Gas with the Greenhouse Effect

The greenhouse effect is a natural phenomenon related to the absorption of infrared radiation at large wavelengths, reflected by the terrestrial surface, by compounds present in the atmosphere called the "gas greenhouse effect," mainly carbon dioxide (CO_2), methane (CH_4), steam (H_2O), ozone (O_3), nitrogen protoxide (N2O), sulfur hexafluoride (SF6), and halation carbides (CFC). Part of the infrared radiation is not returned to space, and it transforms absorptive energy into heat. The gases associated with the greenhouse effects of anthropic origin are mainly carbon dioxide, methane, protoxide of nitrogen, sulfur hexafluoride, hydrofluorocarbons (HFC), and perfluorocarbons. Air pollution and the gases associated with the greenhouse effect should not be a priority, but they are sometimes the same ones, and a climatically disordered state could often have consequences that are worse and synergistic with the air pollution majority. There are many polluting species emitted or transformed in the atmosphere (Ko et al. 2015). Even if their concentrations are weak (measured in micrograms per cubic meter), they can have particular effects on health. The gases associated with the greenhouse effect are not, strictly speaking, atmospheric pollutants. Road pollution, for example, is at the origin of two distinct types of atmosphere emissions: air pollutants known as local, composed of pollutant gases such as carbon monoxide, nitrogen oxides, or harmful particles directly affecting health via respiratory tracts and cardiovascular diseases, and carbon dioxide (CO_2), considered the principal gas with the greenhouse effect and being in charge of climate warming, whose effects directly on health are benign (Dentener et al., 2011).

1.14 Evolution of Atmospheric Pollution Over Time

From 1990 to 2021, we saw urban pollution decrease overall in very many European cities following actions of fight, prevention, and sensitization by the public authorities. For example, nitrogen dioxide (NO_2) dropped by 54–30 g/m^3, nitrogen oxides (NO_x) dropped by 105 to 47 μg of equivalent NO_2/m^3, nitric oxide (NO) dropped by 36–11 g/m^3, and carbon monoxide dropped by 500–300 g/m^3, according to values recorded in 2007 in Paris. The most spectacular fall of 97% from 1992 is recorded for the case of lead and is explained by the lead prohibition in unleaded gas. We see a rise in the tropospheric ozone rate in the air from 20 to 50 μg/m^3 and the emergence of new air pollution components (HANDLES, 2018). Among these emergent pollution parameters, we can quote the chlorinated products of dioxanes or furans resulting from waste incineration, spray residues of crystallized microdroplets or microagglomerates of pesticides from agricultural sprayers carried by wind drift or separated from the soil, or in absorbent form on soil particles transformed into dust.

Additionally, new biological pollutants or abnormalities are present, such as the fine particles coming from the dehydrated droppings of birds, the excrement of dogs and cats, and settings suspended in the air by the wind, along with fungal or microbial spores and metals, including three metals of the platinum group lost by the catalytic pots. The new pollens introduced by the plants have allergens such as ragweed or certain pollens containing lipids, which adsorb pollutants, making these pollens allergenic or better known as "lipid mediators" or PALM for pollen-associated lipid mediators (Kopferschmitt-Kubler & Pauli, 1999) and to radioactive atmospheric elements related to nuclear experiments and operations of nuclear installations or accidents such as those of Chernobyl, Fukushima, Tokaimura, etc. Heavy metals are well known for posing problems for the environment and human health. As newly formed nanoparticles, they are usually associated with aerosols of small size. When they are present in the air following industrial pollution, fuel combustion, waste incineration, etc., they will be mainly evacuated by wet deposits. Then, they are found building up in the soils, sediments, microorganisms, and interstitial water where they can pose ecosystem problems (Marinussen et al., 1997; Strong et al., 1986).

1.15 Conclusion

Because of air pollution, homeowners and professionals, policymakers, and the public are all affected by several environmental effects. Given the diversity and complexity of these effects, each of us moves forward in its own way to put them into perspective. The statistics of the World Health Organization (WHO) attribute air pollution to over 7 million deaths in 2012, including 600,000 children younger than 5 years old. The most touched areas are the Far East: 2.8 million (172 deaths for every 100,000 inhabitants) and Southeast Asia: 2.3 million (124 deaths for every

100,000 inhabitants); Europe would count 582,000 deaths per year (106 deaths for every 100,000 inhabitants in the countries with weak and average incomes and 47 in the countries with high incomes) (WHO, 2018).

Because of the lower presence of human activity, agriculture, and industry, the Southern Hemisphere is touched more often than the Northern Hemisphere. In addition, scientific research has made human activities less polluting. Because of this research, significant progress and projections have been made in this field of atmospheric pollution. Thus, better-quality fuel containing less sulfur or without lead and vehicles equipped with catalytic converters that produce fewer emissions contributed to the reduction in lead, nitrogen, and sulfur oxide emissions.

The political establishment and community have reacted to this major air pollution problem, which has been contributing to it for 20 years. This is because (1) they started a comprehensive policy to make human activities cleaner, and (2) they sensitized the public via ceaseless specific conferences, treating the reduction of gas emissions with the greenhouse effect and other general pollution parts.

All these operations go toward safeguarding our planet from many dangers building and providing a solid foundation without major problems for future generations.

References

Al Hallak, M., Verdier, T., Bertron, A., Roques, C., & Bailly, J.-D. (2023). Fungal contamination of building materials and the aerosolization of particles and toxins in indoor air and their associated risks to health: A review. *Toxins, 15*, 175. https://doi.org/10.3390/toxins15030175

ANSES, Agence Nationale de Sécurité Sanitaire et Environnement. (2020). Impact sanitaire de l'exposition aux moisissures présentes dans l'air ambient Avis de l'ANSES.- Rapport d'expertise collective. Juillet 2020 - Édition scientifique.

Bellanger, A.-P., Reboux, G., Roussel, S., Grenouillet, F., Didier-Scherer, E., Dalphin, J.-C., & Millon, L. (2009). Indoor fungal contamination of moisture-damaged and allergic patient housing analyzed using real-time PCR. *Letters in Applied Microbiology, 49*, 260–266.

Blanchard, B., Borrelli, D., Chardon, B, Chatignoux, E., Declercq, C., Fabre, P., Host, S., Jusot, J.-F., Larrieu, S., Lefranc, A., Medina, S., Pascal, L., Prouvost, H., Saoudi, A., & Wagner, V. (2014). *Monitoring the program's air and health. Analyzes short-term bonds between urban atmospheric pollution and mortality in nine French cities* (p. 44). National Sanitary Surveillance Institute—Monitoring Program Air and Santé.

Bloom, E., Nyman, E., Must, A., Pehrson, C., & Larsson, L. (2009). Molds and mycotoxins in indoor environments: A survey in water-damaged buildings. *Journal of Occupational and Environmental Hygiene, 6*, 671–678.

Boraschi, D., Duschl, A., Lynch, I., & Stoeger, T. (2023). Editorial: Exploring impacts of combined exposures to particles and chemicals on immune reactions across living organisms. *Frontiers in Toxicology, 5*, 1148374. https://doi.org/10.3389/ftox.2023.1148374

Brasel, T. L., Campbell, A. W., Demers, R. E., et al. (2004). Detection of trichothecene mycotoxins in sera from individuals exposed to *Stachybotrys chartarum* in indoor environments. *Archives of Environmental Health, 59*, 317–323.

Brasel, T. L., Douglas, D. R., Wilson, S. C., & Straus, D. C. (2005a). Detection of airborne *Stachybotrys chartarum* macrocyclic trichothecene mycotoxins on particulates smaller than conidia. *Applied and Environmental Microbiology, 71*, 114–122.

Brasel, T. L., Martin, J. M., Carriker, C. G., Wilson, S. C., & Straus, D. C. (2005b). Detection of airborne *Stachybotrys chartarum* macrocyclic trichothecene mycotoxins in the indoor environment. *Applied and Environmental Microbiology, 71*, 7376–7388.

Chen, Y., Cai, J., Wang, Z., Peng, C., Yao, X., Tian, M., Han, Y., Shi, G., Shi, Z., Liu, Y., Yang, X., Zheng, M., Zhu, T., He, K., Zhang, Q., & Yang, F. (2020). Simultaneous measurements of urban and rural particles in Beijing, Part 1: Chemical composition and mixing state. *Atmospheric Chemistry and Physics, 20*, 9231–9247, https://doi.org/10.5194/acp-20-9231-2020

Costa, A., Pinheiro, M., Magalhaes, J., Ribeiro, R., Seabra, V., Reis, S., & Sarmento, B. (2016). The formulation of nano medicines for treating tuberculosis. *Advanced Drug Delivery Reviews, 2016*(102), 102–115.

De Ridder, K., Adamec, V., Bañuelos, A., Bruse, Mr., Bürger, Damsgaard, Mr. O., Dufek, J., Hirsch, J., Lefebre, F., Pérez-Lacorzana, J., Thierry, A., & Weber, C. (2004). Year integrated methodology integrated methods to assess the benefits of urban green space. *Science of Total the Environment, 334–335*, 489–497.

Dentener, F., Derwent, R., Dlugokencky, E., Holland, E., Isaksen, I., Katima, J., Kirchhoff, V., Matson, P., Midgley, P., Wang, M., Berntsen, T., Bey, I., Brasseur, G., Buja, L., Collins, W.J., Daniel, J., De More, W.B., Derek, N., Dickerson, R., Etheridge, D., …, Ren, C. (2014). Photochemical properties and sources of pollutants during continuous pollution episodes in Beijing, October 2011. *Journal of Environment of Sciences—China, 26*, 44–53. https://doi.org/10.1016/s1001-0742 [13] 60379-4

Douwes, J., Pearce, N. (2014). Epidemiology of respiratory allergies and asthma. In *Handbook of epidemiology* (pp. 2263—2319). https://doi.org/10.1007/978-0-387-09834-0_50. PMCID: PMC7121502

Douwes, J., Thorne, P., Pearce, N., & Heederik, D. (2003). Bioaerosol health effects and exposure assessment: Progress and prospects. *Annals of Occupational Hygiene, 47*(3), 187–200. https://doi.org/10.1093/annhyg/meg032

Duquenne, P., & Greff-Mirguet, G. (2005). L'échantillonnage et l'analyse des aérosols microbiens. *Hygiène et Sécurité du Travail-Cahiers de notes documentaires*, (198), 23–28.

Edwards, S. G., Imathiu, S. M., Ray, R. V., Back, M., & Hare, M. C. (2012). Molecular studies to identify the Fusarium species responsible for HT-2 and T-2 mycotoxins in UK oats. *International Journal of Food Microbiology, 156*, 168–175. https://doi.org/10.1016/j.ijfoodmicro.2012.03.020

Edwards, S. G. (2004). Influence of agricultural practices on fusarium infection of cereals and subsequent contamination of grain by trichothecene mycotoxins. *Toxicology Letters, 153*, 29–35.

El Fuladi, A. (2002). Macro pollution: Acid rain. In *Causes, affects, mechanism of control, Canadian strategy* (45 p). The University of Montreal. www.lemig.umontreal.ca/cours/geo1312/

Jhun, I., Fann, N., Zanobetti, A., & Hubbell, B. (2014). Effect modification of ozone-related mortality risks by temperature in 97 US cities. *Environment International, 73*, 128–134. https://doi.org/10.1016/j.envint.2014.07.009

Gao, P., Postiglione, M. P., Krieger, T. G., Hernandez, L., Wang, C., Han, Z., Streicher, C., Papusheva, E., Insolera, R., Chugh, K., Kodish, O., Huang, K., Simons, B. D., Luo, L., Hippenmeyer, S., & Shi, S. H. (2014). Deterministic progenitor behavior and unitary production of neurons in the neocortex. *Cell, 159*(4), 775–788. https://doi.org/10.1016/j.cell.2014.10.027

Goyer, N., & Lavoie, J. (2001a). Emissions of chemical compounds and bioaerosols during the secondary Treatment of Paper Mill Effluents. *AIHAJ—American Industrial Hygiene Association, 62*(3), 330–341. https://doi.org/10.1080/15298660108984635

Goyer, N., & Lavoie, J. (2002). Identification of sources of chemical and bioaerosols emissions into the work environment during secondary treatment of pulp mill effluents. *Tappi Journal, 84*(2), 395–407.

Goyer, N., Lavoie, J., Lazure, L., & Marchand, G. (2001b). Les bioaérosols en milieu de travail: guide d'évaluation, de contrôle, et de prévention. In *Études et recherches, guide technique T-23* (63 p). Institut de recherche Robert-Sauvé en santé et en sécurité du travail.

Grange, D., Host S., & Gremy. I. (2007). Les composés organiques volatils [COV]: état des lieux: définition, sources d'émissions, exposition, effets sur la santé. Rapport Décembre 2007, Observatoire Régional de Santé Rapport Décembre 2007, Observatoire Régional de Santé d'Ile-de-France

Greenstone, M., & Fan, C. Q. (2018). *Introducing the air quality life index*. AQLI Annual Report.

Greenstone. M. (2020). Air quality life index. Energy Policy Institute at the University of Chicago (EPIC) carry forward. David and Christine Large-Loh Eric and Valonia Oei. The Becker Friedman Institute. Donald R. Wilson Jr., the University of Chicago, AB' 88.

Greenstone, M., He, G., Jia, R., & Liu, T. (2022). Can technology solve the principal-agent problem? Evidence from China's war on air pollution. *American Economic Review: Insights, 4*(1), 54–70.

Greenstone, M., Nilekani, J., Pande, R., Ryan, N., Sudarshan, A., & Sugathan, A. (2015). Lower pollution, longer lives: life expectancy gains if India reduced particulate matter pollution. Economic and Political Weekly, 40-46.

Greenstone M., & Hanna R. (2014). Environmental regulations, air and water pollution, and infant mortality in India. *American Economic Review, 104*(10), 3038–3072. https://doi.org/10.1257/aer.104.10.3038

Günter, O., Oberdörster, E., & Oberdörster, J. (2005). A review: Nano toxicology: An emerging discipline evolving from studies of ultrafine particles. *Environmental Health Perspectives, 113*, 823—839. https://doi.org/10.1289/ehp.7339.

Gustavsson, E. K., Follett, J., Trinh, J., Barodia, S. K., Real, R., Liu, Z., Grant-Peters, M., Fox, J. D., Appel-Cresswell, S., Stoessl, A. J., Rajput, A., Rajput, A. H., Auer, R., Tilney, R., Sturm, M., Haack, T. B., Lesage, S., Tesson, C., Brice, A., Vilariño-Güell, C., Ryten, M., Goldberg, M. S., West, A. B., Hu, M. T., Morris, H. R., Sharma, M., Gan-Or, Z., Samanci, B., Lis, P., Periñan, M. T., Amouri, R., Ben Sassi, S., Hentati, F., Tonelli, F., Alessi, D. R., & Farrer, M. J. (2024). RAB32 Ser71Arg in autosomal dominant Parkinson's disease: Linkage, association, and functional analyses. *The Lancet Neurology, 23*(6), 603–614.

HANDLES, National Agency of Public Health, Food, Environment, Work. (2018). *Emergent pollutants "in the ambient air: identification, categorization, and hierarchization of pollutants are currently not regulated for the monitoring of the quality of the air*. Opinion of the Handles Collective Report Expert, Scientific Edition.

Harrison, R. M. Airborne particulate matter. *Philosophical Transactions of the Royal Society A: Mathematical, Physical and Engineering Sciences, 378*(2183), 20190319. https://doi.org/10.1098/rsta.2019.0319. Epub 2020 Sep 28. PMID: 32981435; PMCID: PMC7536032

Hassen, W., Cherif, H., Werheni Ammeri, R., Raddadi, N., Neifar M., Hassen A., & Cherif A. (2021). Exhaustion of pentachlorophenol in soil microcosms with three Pseudomonas species as detoxification agents, *Archives of Microbiology, 203*(7), 4641–4651. https://doi.org/10.1007/s00203-021-02451-y.

Hinds, W. C. (1982). *Aerosol technology. Properties, behavior, and measurement of airborne particles* (424 p). Wiley, Toronto.

International Agency for Research against Cancer. (1994). Monographs on the evaluation of carcinogenic risks to humans. In *Some industrial chemicals* (Vol. 60, pp. 389–433). IARC Publications.

Jaramillo, P., Griffin Michael, W., & Scott Matthews H. (2007). Comparative life-cycle air emissions off coal, domestic natural gas, LNG, and SNG for electricity generation. *Environmental Science & Technology, 41*(17), 6290–6296. PMID 17937317, https://doi.org/10.1021/es063031o

Ko, M., Kurylo, M., Lawrence, M., Logan, J. A., Manning, M., Mauzerall, D., McConnell, J., Mickley, L., Montzka, S., Müller, J. F., Olivier, J., Pickering, K., Pitari, G., Roelofs, G. J., Rogers, H., Rognerud, B., Smith, S., Solomon, S., Staehelin, J., …, Yantosca, R. (2015). In F. Joos, & M. McFarland (Review Eds.), Co-ordinating Lead Authors: Ehhalt, D., Prather, M. *Chapter book: Atmospheric chemistry and greenhouse gases* (50 p).

Kopferschmitt-Kubler C., & Pauli G. (1999). Pollen, «Pollens and pollution». In Re-examined French of Study of allergies and Clinical Immunology (Vol. 39, No. January 4th, pp. 283–288). https://doi.org/10.1016/S0335-7457 [99] 80054-9

Lacour, S. A., de Monte, M., Diot, P., et al. (2006). Relationship between ozone and temperature during the 2003 heat wave in France: Consequences for health data analysis. *BMC Public Health, 6*, 261. https://doi.org/10.1186/1471-2458-6-261

Lidwien, A.M., Smit, M.Sc., Spaan, M.Sc., & Heederik, D. (2005). Endotoxin exposure and symptoms in wastewater treatment workers. *American Newspaper of Industrial Medicine, 48*, 30–39

Lighthartl, B. (1997). Mini review. The ecology of bacteria in the alfresco atmosphere. *FEMS Microbiology Ecology, 23, 263–274.* https://journal.hep.com.cn/fese/EN/10.1007/s11783-023-1627-y

Lighthart, B., & Kim, J. (1989). Simulation of airborne microbial droplet transport. *Applied and Microbiology, 55*(9), 2349–2355. https://doi.org/10.1128/aem.55.9.2349-2355.1989

Lighthart, B., & Shaffer, B. T. (1994). Bacterial flux from chaparral into the atmosphere in mid-summer at a high desert location. *Atmospheric Environment, 28*(7), 1267–1274. https://doi.org/10.1016/1352-2310(94)90273-9

Lorenz, W., et al. (2006). A hypothesis for the origin and pathogenesis of rheumatoid diseases. *Rheumatology International, 26*, 641–654.

Marinussen, M. P., van der Zee, J. C., Sjoerd, E. A., Mr, T., de Haan, A. M., & Frans. (1997). (in) Heavy metal (copper, lead, and zinc) accumulation and excretion by the earthworm, *Dendrobaena veneta. Newspaper of Environmental Quality, 26*, 278–284.

Mayol, E., Jiménez1, M. A., Herndl, G. J., Duarte, C. M., & Arrieta, J. M. (2014). Resolving the abundance and air-sea fluxes of airborne microorganisms in the North Atlantic Ocean. *Frontiers in Microbiology, Sec. aquatic microbiology* (Vol. 5). https://doi.org/10.3389/fmicb.2014.00557

Mohr, M., Krustrup, P., & Bangsbo, J. (2003). Match performance of high-standard soccer players with special reference to development of fatigue. *Journal of Sports Sciences, 21*(7), 519–528. https://doi.org/10.1080/0264041031000071182

Mohr, C., Lopez-Hilfiker, F. D., Yli-Juuti, T., Heitto, A., Lutz, A., Hallquist, M., D'Ambro, E. L., Rissanen, M. P., Hao, L., Schobesberger, S., Kulmala, M., Mauldin, R. L., Makkonen, U., Sipilä, M., & Thornton, J. A. (2017). Ambient observations of dimers from terpene oxidation in the gas phase: Implications for new particle formation and growth. *Geophysical Research Letters, 44*, 2958–2966. https://doi.org/10.1002/2017GL072718

Pedgley D. E. (1991) *Aerobiology: The atmosphere as a source and sink for microbes. Part of the Brock/Springer series in contemporary bioscience book series*(Brock/Springer). Press.

Shang, Mr., Kong, Y., Yang, Z., Cheng, R., Zheng, X., Liu, Y., & Chen Face, T. (2023). Removal of virus aerosols by the combination of filtration and UV-C irradiation. Approximately. *Science and Engineering, 17*(3): 27. https://doi.org/10.1007/s11783-023-1627-y.

Sivaramanan, S. (2015). *Review: Acid rain, causes, effects, and control.* Strategies Environmental Officer, Environmental Impact Assessment Unit, Environmental Management and Access Division, Central Environmental Authority. https://doi.org/10.13140/RG.2.1.1321.4240/1

Straus, D. C. (2009). Molds, mycotoxins, and sick building syndrome. *Toxicology and Industrial Health, 25*(9–10), 617–635. https://doi.org/10.1177/0748233709348287

Strong, J. F., Jeandel, C., Michard, G., Nicolas, E., & Fox, D. (1986). (in) Interstitial toilets chemistry of Villefranche Bay sediments: Trace metal diagenesis. *Marine Chemistry, 18*, 233–247.

Verma, A., Tewari, A., & Azami, A. (2010). An impact of stimulated acid rain levels on different pH-levels on some major vegetable plants in India. *Reports and Opinions, 2*(4), 38–40.

Werheni Ammeri, R., Di Rauso Simeone, G., Hassen, W., Sadfi, N., Hidri, Y., & Hassen, A. (2022a). *Aspergillus sydowii* and *Typha angustifolia* as useful tools for combined bioprocesses of PCP removal in wastewater. *International Journal of Environmental Science and Technology*, 1–14. Published online on January 22, 2022. https://doi.org/10.1007/s13762-021-03853-7

Werheni Ammeri, R., Di Rauso Simeone, G., Hidri, Y., Abassi, M. S., Mehri, I., Costa, S., Hassen, A., & Rao, M. A. (2022b). Combined bioaugmentation and biostimulation techniques in bioremediation of pentachlorophenol contaminated forest soil. *Chemosphere, 290*, 133, 359 PMID: 35149916 https://doi.org/10.1007/s11274-021-03219-8

Werheni Ammeri, R., Di Rauso Simeone, G., Hassen, W., Ibrahim, C., Ammar, R. B., & Hassen, A. (2021a). Bacterial consortium biotransformation of pentachlorophenol contaminated wastewater. *Archives of Microbiology, 203*(04), 6231–6243. https://doi.org/10.1007/s00203-021-02589-9

Werheni Ammeri, R., Kraiem, K., Riahi, K., Eturki, S., Hassen, W., Mehri, I., & Hassen A. (2021b). Removal of pentachlorophenol from contaminated wastewater using phytoremediation and bioaugmentation processes. *Water Science and Technology, 84*(10–11), 3091–3103. https://doi.org/10.2166/wst.2021.328.

Werheni Ammeri, R., Ben Ghorbel, S. K., Mehri, I., Hassen, W., & Hassen, A. (2022c). Bioaugmentation of pentachlorophenol by *Pseudomonas mosseli* HM627603 and P. putida HM627611 in secondary treated wastewater—UVC254 inactivation. *Journal of New Sciences, Agriculture and Biotechnology, 87*(4), 4931–4941. https://doi.org/10.55416/sunb.jns01.2201.08704

World Health Organization. (2018). *Global health observatory off-date repository: Joint effects of air pollution: Deaths by area.*

World Health Organization. (2009). *WHO guidelines for indoor air quality: dampness and mold.* WHO Regional Office of Europe. ISBN 978-92-890-4168-3.

Zhu, T. (2005). *Urban and regional atmospheric compound pollution.* Chemical Industry.

Chapter 2
Comparative Spatial and Temporal Analysis of Atmospheric Pollutants Levels and Energy Consumption Emissions in Brazil: Pre and During COVID-19 Pandemic Periods

Willian Cézar Nadaleti, Anderson Corrêa, Eduarda Gomes De Souza, Emanuélle Soares Cardozo, Maele Costa Dos Santos, Jeferson GOMES, and Diuliana Leandro

Abstract In March 2020, the World Health Organization (WHO) classified COVID-19 as a global pandemic, with the recommendation of social distancing to mitigate the spread of the virus. The reduction in vehicular traffic and industrial activities has resulted in a decline in the emission of certain atmospheric pollutants and greenhouse gases. Atmospheric pollution represents one of the significant challenges of present-day society, and in South America, it is one of the most persistent environmental issues. The COVID-19 measures implemented across the globe, such as imposing lockdowns, restricting movement, and adopting other containment strategies, resulted in a substantial reduction in energy consumption within the transportation sector. Brazil has limited access to electricity and heavily relies on traditional energy sources to meet its domestic energy demand. Biomass combustion, electricity consumption, and the transportation and industrial sectors play significant roles as major contributors to atmospheric pollution. To control and mitigate anthropogenic emissions it is essential to combine technical-scientific knowledge with legal measures. The present study aims to provide a comparative analysis of the concentration of nitrogen dioxide (NO_2), black carbon (BC), carbon monoxide (CO), sulfates (SO_4), and ozone (O_3) during the pre-pandemic and pandemic periods in Brazil,

W. C. Nadaleti (✉) · E. G. De Souza · E. S. Cardozo · M. C. Dos Santos · J. GOMES · D. Leandro
Environmental and Energy Engineering Lab, LEAE. Engineering Center, CEng. Post-Graduation Program in Environmental Sciences, PPGCAmb. Federal University of Pelotas, UFPel. Pelotas, Rio Grande Do Sul, Brazil
e-mail: willancezarnadaleti@gmail.com

A. Corrêa
Environmental Microbiology Laboratory, LAMIC. Institute of Agricultural Sciences, ICIAG and Polymer Recycling Laboratory, LABREPOL. Institute of Chemistry, UFU. Post-Graduation Program in Biofuels, PPGBiocomb, The Federal University of Uberlândia, UFU. Uberlândia, Minas Gerais, Brazil

A. Anouzla et al. (eds.), *Handbook of the Effects of Landfill Emissions on Sustainable Development*, https://doi.org/10.1007/978-3-032-01951-6_2

associated with changes in energy use. The series of compiled data provided by the National Aeronautics and Space Administration (NASA) through the GIOVANNI platform enabled the analysis of the impact of pollutant levels, considering a time interval spanning from January to June of 2018, 2019, and 2020.

Keywords Remote sensing · Energy transition · Renewable energy · Greenhouse gases · Fossil fuels

2.1 Introduction

Air pollution stands as a paramount environmental predicament confronting diverse global regions. It comprises a complex mixture of particulate matter and gaseous pollutants, primarily emitted into the atmosphere by anthropogenic sources, with a minor contribution originating from natural sources (Li et al., 2020). In several developing regions, air pollution exhibits a pronounced correlation with the levels of Nitrogen Dioxide (NO_2), Particulate Material 2.5 ($PM_{2.5}$) and Particulate Material 10 (PM_{10}). These pollutants are generated by industrial activities and transport usage of fossil fuels (He et al., 2020). Besides, landfills could be an additional factor of concern since air pollutants generated as result of landfill sites operations cause damage to the environment and human life (Salami & Popoola, 2023).

A key parameter pollutant for monitoring air quality is the fine particulate material 2.5, along with gaseous emissions concentrations of sulfur dioxide (SO_2) and NO_2. These metrics are commonly utilized to estimate and investigate pollution outbreaks of fossil fuel combustion and vehicle exhaust. Furthermore, the presence of primary pollutants engenders the formation of secondary pollutants, notably O_3. The gas originates from reactions between nitrogen oxides (NO_x) and volatile organic compounds (VOCs) at high temperatures and in the presence of solar rays (Abou El-Magde & Zanati, 2020).

Since the beginning of the COVID-19 epidemic and the blocking measures (Lockdown) to contain the spread of the virus, studies have been conducted on the effect of these restrictions on air quality, with comparisons between pre, during, and after quarantine, as demonstrated by several studies (Dutheil et al., 2020; Li et al., 2020;; Wang et al., 2020a, 2020b; Zambrano-Monserrate et al. 2020; Zhang et al., 2020a).

However, with the beginning of 2020 the closure of various anthropic activities around the world had direct consequences in the emissions of air pollutants, which started to decrease drastically, improving air quality in several countries (Rutz et al., 2020). Although, contrary to the global trend, emissions, especially greenhouse gases, increased in Brazil by 9.5% in 2020 compared to the previous year, totaling approximately 2.16 billion metric tons of CO_2eq. The Land Use Change and Forest sector saw a significant rise of approximately 23.7% in total emissions in 2020, followed by the Agriculture sector at 2.5%. Despite a decrease in emissions from the energy sector in 2020 compared to 2019, methane and carbon dioxide (CO_2)

emissions from fossil fuel combustion increased, as well as fugitive CO_2 emissions, all of which escalated in 2020.

***Further, those improvements depend on when those restrictions were imposed, how public policies were implemented, weather factors and the characteristics of cities, as the number of populations, traffic intensity, among other factors that influence the improvement or worsening of urban air quality (Cheng et al., 2008). Emphasizing the importance to highlight that these improvements in air quality observed in several countries are temporary and short-lived, as soon as industrial activities initiate again and vehicle traffic restrictions end, concentrations of air pollutants tend to return to the previous values or increase, as presented by Beijing, Mumbai and Rome, as soon as the blockages on account of the COVID-19 pandemic were removed, the level of pollutant concentration, as NO_2 and PM $_{2.5}$ and 10 increased considerably (Kumari & Toshniwal, 2020).

Recent studies shed light on this matter. In a study conducted by Nakada and Urban (2020) the influence of quarantine measures on air quality in São Paulo is analyzed. The researchers collected ground-based data on various pollutants, including NO, NO_x, and O_3, among others, from February to April 2020. Similarly, another study by Krecl et al. (2020) investigates the decline in NO_x concentrations across thirteen monitoring stations in São Paulo. Dantas et al., (2020) reported improvements in air quality in some of the districts studied in the city of Rio de Janeiro, with data from monitoring stations. And in a study by Siciliano et al. (2020), also on the city of Rio de Janeiro, the authors identified a small increase in O_3 concentrations during the pandemic blocking period.

Therefore, presenting a study on the entire Brazilian territory, in a broader period, from January to June 2020, is extremely valuable, using satellite data, seeking to analyze data that encompassed all the states of Brazil. Furthermore, several cities do not have air quality monitoring stations. Furthermore, pollutants that are currently the most worrying concerning urban pollution, since they are poorly-studied pollutants in Brazil.

Hence, the presentation of a comprehensive study covering the entire Brazilian territory over an extended period, spanning from January to June 2020, holds immense value. This study leverages satellite data to scrutinize information that encompasses all the states of Brazil. Furthermore, it addresses the absence of air quality monitoring stations in several cities. Moreover, it delves into pollutants that are currently of significant concern in the context of urban pollution, such as NO_2, O_3, CO, Black Carbon (BC), and SO_4, given their limited research attention in Brazil. Additionally, this study provides data on methane (CH_4) and CO_2 emissions, associated with energy consumption and fossil fuel burning.

2.2 Brief Theoretical Framework

The pandemic by SARS-Cov-2 impacted the year of 2020, having as immediate response of the scientific community the implementation of measures to control the global spread of the virus (Copat et al., 2020), nominated by the World Health Organization (WHO) as COVID-19 (Muhammad et al., 2020).

In Brazil, according to the Ministry of Health the first case of contamination by the Coronavirus was announced in the state of São Paulo on February 26, from that moment health authorities and state began to trace measures to contain the virus circulation in the country, as seen since January in various parts of the world, especially in China and Europe. In a short time, the state of São Paulo was quarantined due to the fast growth of the COVID-19 cases in the state, initiating the closing of shopping centers, bars and restaurants, work, classes and meetings remotely. In addition to restrictions on the circulation of domestic and international vehicles and travel, with release for operation only the activities of essential services such as food, health and some industries (Krecl et al., 2020).

In Rio de Janeiro, the first actions to contain the spread of the virus across the state were implemented on March 16, when schools and universities were closed, public events were canceled and work at home recommended. On 19 March, a decree ordered a partial blockade from 21 to 23 March being more stringent, with the closing of bars, restaurants, beaches, shopping malls and general non-essential trade, and public transportation within the city and between states was limited. Industrial and construction activities were not suspended, as were those related to health and basic services (Dantas et al., 2020).

The COVID-19 is an acute respiratory disease that attacks the respiratory system, resulting in severe respiratory distress syndrome in people who contract the virus (Bherwani et al., 2020). In China, the COVID-19 pandemic has been considered the largest outbreak of atypical pneumonia since the outbreak of severe acute respiratory syndrome (SARS) that occurred in 2003 (Wang et al., 2020b).

With the onset of the disease in China at the end of December 2019, on January 23, 2020, a blockade was implemented in Wuhan and neighboring cities with travel restrictions and within a few days these restriction measures were extended to the entire country more rigorously. These quarantine measures have led to a reduction in the spread of the disease, in addition to reducing the number of vehicles in circulation and the operation of industrial activities, contributing to the improvement of air quality in several Chinese cities (Wang et al., 2020b).

The restrictions imposed to contain the spread of the new Coronavirus had severe negative and positive effects in which the population witnessed, therefore, a strong economic, social and cultural impact was experienced by everyone. Nevertheless, there was a positive environmental outcome, evidenced by improved air quality in various cities across different countries (Dutheil et al., 2020).

Following the declaration of COVID-19 as a global pandemic, countries worldwide initiated the development of mitigation plans and measures to curb the disease's spread. These actions varied across nations but predominantly involved recommendations for social distancing, school closures, border restrictions, and the mandatory use of masks by the public in public spaces. These measures, accompanied by the adoption of new daily habits, transformed the lives of most populations. Furthermore, they serve as a valuable tool for studying the short-term effects that reduced vehicular traffic, decreased mobility, and reduced industrial activity have on air quality and the concentrations of air pollutants in major urban centers (Burns et al., 2021).

The reductions observed mainly with NO_2 and other primary pollutants such as CO and $MP_{2.5}$ and 10 were beneficial to the environment and public health, even if these changes are characterized as local and short-term consequences, reflecting the way that decrease in traffic and economic activities directly influence air quality. Moreover, the attention to the formation of secondary pollutants is limited in studies on air pollution during the pandemic, which is the case with O_3, which is a major health hazard pollutant for both short- and long-term exposures, causing severe damage to health (Siciliano et al., 2020). Another pollutant that causes serious health problems is NO_2, which can cause risks of various diseases such as asthma, respiratory and cardiovascular diseases and even increase the mortality rate from lung diseases (Zhao et al., 2020).

Regarding public health, several studies have shown that air pollution, the mixture of pollutants, including carbon oxides, nitrogen oxides, ozone, sulfur oxides and particulate materials may increase the likelihood of respiratory infections, extending the aerosol transmission of several viruses (Chen et al., 2020b). Despite this, with a significant reduction in emissions of air pollutants as occurred around the world during the blockades of COVID-19, possibly resulting in a decrease in air pollution-related morbidity and mortality in several major urban centers (Wyche et al., 2021).

One of the major problems faced by developed countries is air pollution, which is characterized by regions with large industrialized centers, heavy vehicle traffic and urban agglomerations. It usually has several areas that concentrate numerous industries, which end up negatively interfering with air quality (Lovarelli et al., 2020).

The main air pollutants indicators of industrial activities, heavy vehicle traffic and high population indices are NO_2, CO, O_3 and $PM_{2.5}$ and PM_{10}, which have different emission sources, characteristics and air dissemination behaviors (Liu et al., 2020), with SO_2, are the most used air pollutants in the control and analysis of air quality (Briz-Redón et al., 2020), since these pollutants cause problems to the environment, in addition to turns harmful to human health (Manisalidis et al., 2020).

Regarding air pollutants, the CO is a colorless, odorless, tasteless air toxin produced by the incomplete combustion of hydrocarbons. Its source of disposal is mainly by the use of vehicles in urban areas, especially powered by diesel and less gasoline and also by maritime transport, characterized by diffuse sources of difficult control and identification (Chen et al., 2020a).

Another pollutant under focus in current-day air pollution studies is BC, which is characterised by particles originating from the combustion of fossil fuels and biomass. Moreover, is a term used to describe refractory carbonaceous particles that strongly absorb light at all solar wavelengths (Petzold et al., 2013). Further, other pollutants generated through the use of fossil fuels and also during the management of solid waste in landfills include CH_4 and CO_2, gases emitted during these operations with a high global warming potential.

And among the main air pollutants are NO_x, which are emitted mainly as NO from combustion sources, by vehicle exhausts, in industrial activities and in power plants, which is converted to NO_2 after oxidation processes. Furthermore, it influences the formation of photochemical oxidants, such as O_3, and is characterized as a pollutant that causes great damage to health (Zhao et al., 2020).

Studies on air pollution are complex when performed, requiring equipment and effective monitoring to generate meaningful and real data on concentrations of pollutants in a given region, where they also depend on weather conditions, movement of pollutants, deposition and atmospheric chemistry, and where satellite images are used for such a study, their analyses shall be carried out within a time frame, to present adequate study results (Dantas et al., 2020).

In China, Europe and the United States, the blockages and interruption of human and industrial activities have led to a significant reduction of air pollution in several regions, especially in NO_2 concentrations, which is an urban pollutant emitted by anthropic actions (Shrestha et al., 2020; Tobías et al., 2020; Zhang et al., 2020a, 2020b, 2020c). In India significant reductions were found for NO_2 during the blockade (Shehzad et al., 2020). Aman et al. (2020), emphasize that the blockade due to the Coronavirus pandemic in the country was unique and different of any event occurred throughout recorded history, where all industry and trade activities were reduced in addition to restricted vehicle circulation, which has led to a decline in emissions of air pollutants in the most populous regions of India.

In a study conducted by Venter et al. (2020), which found that on May 15, 2020, China determined 113 days of blockade to curb the spread of COVID-19, Italy imposed 84 days and Sweden, a shorter blockade, 40 days, is possible to observe that during the blocking dates, the global concentrations of tropospheric NO_2 were on average 60% lower than expected, as for $PM_{2.5}$, which decreased by 31%, in contrast to O_3 that increased 4% in concentrations. Moreover, the authors emphasize that these changes offer an opportunity to estimate the short-term effects of blocking anthropic activities on air quality.

In South America, according to a study by Mendez-Espinosa et al. (2020), more precisely in the cities of Bogotá and Medellin, the authors analyzed the changes in air quality levels in previous periods and during the blockade, where were found long-term reductions for NO_2, PM_{10}, $PM_{2.5}$ of 60, 44 and 40%, respectively, during the strict lock, already in the short term the decreases were more significant, of 62, 58 and 69%, respectively, and during the relaxed lock the reductions of were 50%, 32% and 9% respectively. Besides, in the work developed by Bolaño-Ortiz et al. (2020), the contagion of new cases and deaths by COVID-19 in the region of Latin America and the Caribbean was evaluated, relating to the indicators of temperature, precipitation, humidity, among others, and with indicators of air pollution PM_{10}, $PM_{2.5}$ and NO_2.

Recent research in Brazil also reported improvements in air quality associated with measures of social distancing, restrictions on vehicle circulation and blockages of industrial activities (Nakada & Urban, 2020). As stated by Krecl et al. (2020), which focused on the air quality of the city of São Paulo, that with the beginning of the blockade imposed in the city, the number of vehicles circulating on roads and flights were reduced, There was an immediate improvement in air quality throughout the metropolitan region of São Paulo, with significant reductions in NO_x concentrations in all monitoring stations evaluated, ranging from 34 to 68%.

In another Brazilian state, in the city of Rio de Janeiro, one of the first to implement the restrictions and blockages due to COVID-19, Dantas et al. (2020) evaluated the impact of the pandemic on air quality. Comparing concentrations of particulate matter, carbon monoxide, nitrogen dioxide and ozone, determined during partial blocking with values obtained in the same period of 2019 and also in the weeks prior to the outbreak of the virus, resulting in CO levels demonstrating the most significant reductions of 30,3 - 48.5% and NO_2 also demonstrated decreases of 16.8–53.8%, while PM_{10} levels were reduced only in the first week of the blockade. Considering tropospheric ozone, the opposite happened, the concentrations of the pollutant increased by up to 67% during the study period.

The main sources of air pollution in urban areas are the use of fossil fuels and industrial activities as mentioned above, therefore, assessing air quality at times when such activities and sources of emissions are not in operation would provide an understanding and possible identification of air pollution sources, and also have an answer to how anthropic actions negatively influence the environment, served as a starting point of what actions should take place to improve the quality of the air (Mor et al., 2020).

2.3 Methodology

2.3.1 *Study Area*

The area of study is the Brazilian territory, which is located in South America, and has a territorial extension of 8,514,876 km^2, being the fifth largest country on the planet, data made available by the Brazilian Institute of Geography and Statistics (IBGE), (IBGE, 2019). Figure 2.1 illustrates a map of location of Brazil, South America.

2.3.2 *Data Survey*

For conducting air pollution studies, in addition to soil-based point observations, satellite-based remote sensing provides data on the spatial density of the vertical tropospheric column (CDV), which is the vertical integral of the measured tropospheric NO_2 concentrations in molecule cm 2, and these are extremely important for this type of study (Krotkov et al., 2017).

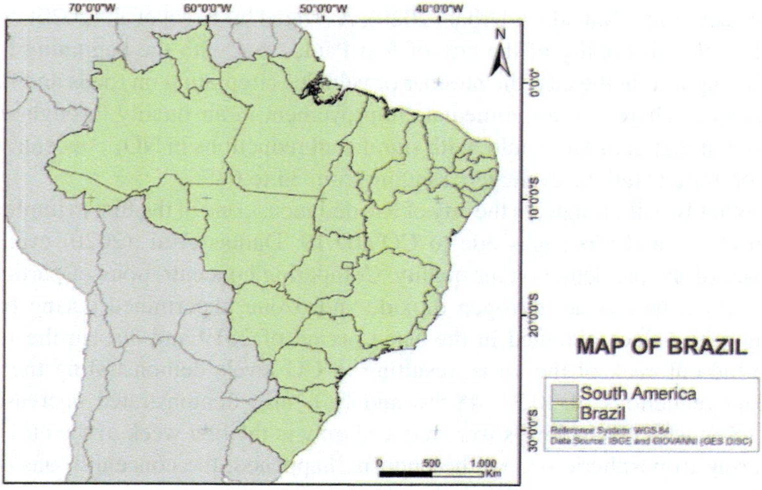

Fig. 2.1 Map of Brazil

The GIOVANNI web application by the NASA Centre for Information and Data Services (DISC) of Goddard Earth Sciences (GES) (GESDISC) (http://disc.sci.gsfc.nasa.Gov/Giovanni) enabled to produce the maps from data obtained in this site, which provide simple and objective visualization and analysis with remote sensing data accesses of earth sciences, especially satellites.

The Giovanni platform includes aerosol data, atmospheric chemistry, temperature, atmospheric humidity and precipitation. Moreover, it includes results from assimilation models that cover a wide variety of atmospheric, terrestrial and oceanographic parameters.

Data from the MERRA-2 (Retrospective Analysis of the Modern Era for Research and Applications, Version 2) and OMI/Aura (Ozone Monitoring Instrument) satellites were analyzed, with characteristics described in Table 2.1, using the measurements of the total ozone column, total NO_2 column, CO concentration on the surface, SO_4 mass concentration on the surface and black carbon mass concentration (CB) on the surface for the periods 2018, 2019 and 2020 respectively for the period from 1 January to 29 June.

The Omno2d data is a Level 3 Grid Product, obtaining a good quality pixel level data, that is grouped and calculated on average in global grids of 0.25×0.25 degrees.

Table 2.1 Satellite database

Data type	Data	Spatial resolution	Temporary resolution	Source
Satellite	IMO/Aura	$0.25° \times 0.25°$	Daily	GIOVANNI
	MERRA-2	$0.5° \times 0.625°$	Per month	GIOVANNI
			Per hour	

Omno2d files are stored in version 5 EOS Hierarchical Data Format (HDF-EOS). Each file contains data from part of the illuminated orbit per day (~ 14 orbits).

Subsequently, the QGIS software version 3.18 allowed us to classify the difference in emission levels of total ozone column, total NO_2 column, surface CO concentration, surface SO_4 mass concentration, and surface black carbon mass concentration. This multiplatform of the Geographical Information System (GIS), enabled to obtain the corresponding maps of the concentrations of air pollutants under study to a spatial and temporal comparative analysis, of the years 2019, 2018 and 2020.

The analysis of CH_4 and CO_2 emissions during the period of 2018–2020 in Brazil was conducted exclusively for the energy sector, encompassing subsectors related to fugitive emissions and emissions from fossil fuel combustion. These data were sourced from the Greenhouse Gas Emissions and Removals Estimation System (SEEG V8.0), an online platform that provides emissions data for all sectors and economic activities within the country. Subsequently, the obtained data were processed using the QGIS program version 3.14.15 to visually depict these emissions before and during the pandemic.

2.4 Results and Discussion

2.4.1 Nitrogen Dioxide (NO_2)

The results shown in Fig. 2.2 demonstrate the concentration values of the NO_2 pollutant, shown in the total and tropospheric column, in 2018, 2019 and 2020, respectively, in the Brazilian territory. The values are classified.

The total and tropospheric column fraction of NO_2 measured remotely by the IMO satellite from January 1 to June 29 in the years 2018, 2019 and 2020 is presented in Table 2.2.

In order to present the difference of the total column fraction and the tropospheric NO_2 values found in Table 2.2 represent the measure remotely calculated by the IMO satellite, where it demonstrates that in 2018 the NO_2 values vary from 1,978 and + 15 to 8,92,803 and + 15 1/cm^2, in 2019 the values were between 1.85911 e + 15 a 7.95864 e + 15 1/cm^2 and during the pandemic period due to blocking measures, with restrictions on vehicle circulation and industrial activities the value decreased to 1.71849 e + 15 a 6.76935 e + 15 1/cm^2, leading to a significant reduction in nitrogen dioxide concentrations in the year 2020 compared to previous years.

In Brazil, analyzing the NO_2 concentration maps allowed to note a 21.8% reduction in NO_2 emissions in 2020, when compared to the years 2018 and 2019, a reduction related to the decrease of human activities due to COVID-19. Demonstrating that industrial activities and the intense circulation of vehicles significantly impacts the air quality, since these activities and mobility were in most Brazilian states with restrictions, imposed by governments and health entities of the country, to contain the spread of the new Coronavirus. However, with the flexibilization of the restriction

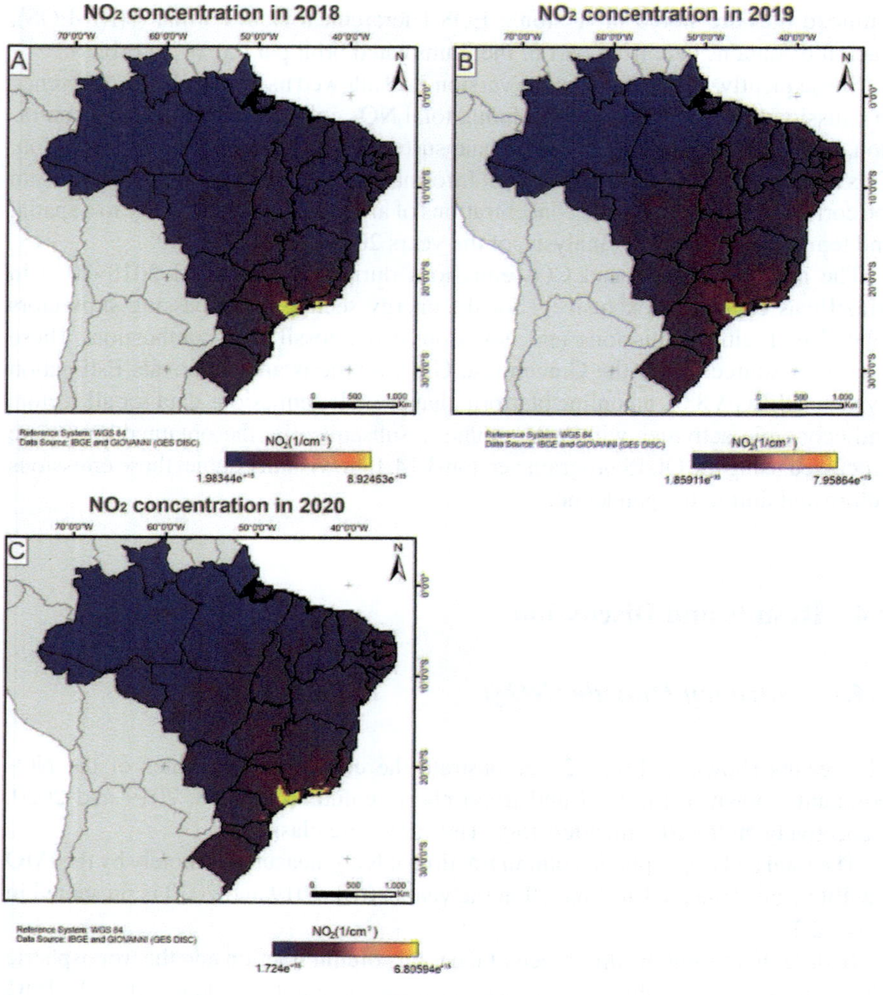

Fig. 2.2 NO_2 concentration: **a** in 2018; **b** in 2019; **c** in 2020

Table 2.2 Total column and tropospheric fraction of NO_2 measured remotely by the IMO satellite

2018 (1/cm^2)	2019 (1/cm^2)	2020 (1/cm^2)
$1.98344e^{+15}$	$1.85911e^{+15}$	$1.724e^{+15}$
$3.7187375e^{+15}$	$3.3839925e^{+15}$	$2.994485e^{+15}$
$5.454035e^{+15}$	$4.908875e^{+15}$	$4.26497e^{+15}$
$7.1893325e^{+15}$	$6.4337575e^{+15}$	$5.535455e^{+15}$
$8.92463e^{+15}$	$7.95864e^{+15}$	$6.80594e^{+15}$

measures in 2021, the values increased again equivalently to the previous years (De Souza et al., 2022).

These results are in line with the literature. Nakada and Urban, (2020) reported variations in concentrations of air pollutants during partial blockade in the state of São Paulo, in the period from February 25 to April 20, comparing with the monthly average of the previous five years and the four weeks prior to partial blocking. The most significant reductions were identified of up to 77.3% NO pollutant and NO_2 with up to 54.3% decrease and CO concentrations showed up to 64,8%. Despite the increase of approximately 30% in ozone concentrations in urban areas with high vehicle traffic.

The IMO measures have been shown to be able to explain how the air quality of a given region is, air and pollution characteristics, and to make variations comparable to surface observations, demonstrating that the use of this type of tool to assess local air quality is extremely important, considering that in several Brazilian cities there are no soil air pollution analysis equipment, making it difficult to carry out studies and to implement environmental policies concerning this pollution, turning harmful to the environment and public health.

The pollutants in question, NO_x, nitric oxide (NO) and NO_2, act strongly in the chemistry of tropospheric pollution and climate change, where the primary source of NO_x is the combustion process, that contributes significantly to anthropogenic NO_x emissions, in addition to emissions from the biosphere, rays and biomass burning that contribute in smaller amounts (Solomon et al., 2011).

Besides, in a study conducted in Ecuador, the authors reported that NO_2 concentrations in 12 cities studied ranged from 37.2 to 66.5 $\mu mol/m^2$ in 2019 and in 2020 from 31.3 to 51.0 $\mu mol/m^2$, analyzing in Guayaquil and Quito, Ecuador's capital and the country's most industrialized cities had the highest levels of NO_2 in 2019, equal to 66.5 and 50.5 $\mu mol/m^2$, respectively and 2020 the levels were slightly lower than 51.0 and 39.2 $\mu mol/m^2$, respectively. However, in an additional analysis conducted in the period from 1 to 15 April 2020, the values of NO_2 presented a reduction equal to 23.4 in Guayaquil and 22.4% in Quito, after the blockade caused by the COVID-19 outbreak. Demonstrating that differences in concentrations of air pollutants could be observed both during blockages and after some time the measures of restrictions were imposed (Pacheco et al., 2020).

In Spain, a study conducetd by Baldasano (2020), air quality measurements were carried out in March 2020, in two of the largest cities in the country, Madrid and Barcelona, which presented reductions in NO_2 levels of 62 and 50%, respectively. With this study on NO_2 air pollution, the authors highlighted that the values found represent the amount of concentrations that are required met and the actions for the two cities to have cleaner and healthier air. Furthermore, the study brought a comparison with NO_2 levels in 2018 and 2019, in the same month in focus (March), with results indicating that due to the blockade, the average NO_2 reduction in Madrid was 56 and 46% compared to 2019 and 2018, while in Barcelona the reduction was slightly higher, 59 and 55 percent.

The pollutant in question, NO_2 is a by-product emitted mainly by vehicles, power plants and industrial installations, consequently the closure of the industrial and commercial sector will result in the reduction of NO_2 emissions in large urban centers, and to a lesser extent in medium-sized cities. In a study similar to those already mentioned above, in Southeast Asia, the authors reported that NO_2 concentrations decreased by 33.7% during the first weeks of blockade in April due to the COVID- 19 pandemic, comparing the mean annual value of atmospheric NO_2 with the previous years from 2010 to 2018, demonstrating that the monthly averages of NO_2 in the month of April present in 2020 the lowest values of the last decade (Stratoulias & Nuthammachot et al., 2020).

In a study carried out in Austria, more precisely in Graz, based on the traffic data available in the city, the authors of the study observed a 45.6% reduction in traffic during blockades compared to the same period in 2017–2019, in areas located near roads. In addition, it enabled to relate this traffic reduction to the high decrease in NO_2 concentrations of 36.9 to 41.6% at the measurement sites, since vehicle traffic is the main source of nitrogen dioxide (Lovrić et al., 2020).

Nakada and Urban (2020), in their study, the authors observed that in the industrial area of São Paulo, even with the measures adopted in the quarantine, the concentrations of air pollutants showed low in relation to their variation levels when compared to the monthly average of five years (2015–2019), and also observed an increase in the concentration of all analyzed pollutants concerning the period before partial blocking. It is important to mention that industrial activities were not affected in the first government-imposed blockades, although some industrial sectors were affected by falling demand and also by labour shortages in the first months of the pandemic, this was not noticed.

NO_2 is a pollutant present in various regions of the world in high concentrations and demonstrated to be released into the atmosphere mainly by the intense traffic of vehicles and industrial activities, whereas reducing their foci of elimination and reduction in concentrations is of utmost importance for both air quality and public health, since according to Achakulwisut et al. (2019), exposure to NO_2 results in 4 million new pediatric asthma cases annually, without taking into account cases of pulmonary diseases that are not diagnosed or reported.

2.4.2 Black Carbon (BC)

Figure 2.3 demonstrates the concentrations of BC mass on the surface, shown in the total and tropospheric column, in 2018, 2019 and 2020, respectively, in the Brazilian territory.

The total column and tropospheric fraction of BC measured remotely by the MERRA-2 Model satellite from January 1 to June 29 in the years 2018, 2019 and 2020 is shown in Table 2.3 the values of the BC mass concentrations on the surface.

The values of the BC mass concentrations on the surface in the Brazilian territory described in Table 2.2 represent the measure remotely made by the MERRA-2 Model satellite, where it shows that in the year 2018 the BC values vary between 1.26675153 and 8.64612 E-12 1/cm^2, in 2019 for the same period the values were between

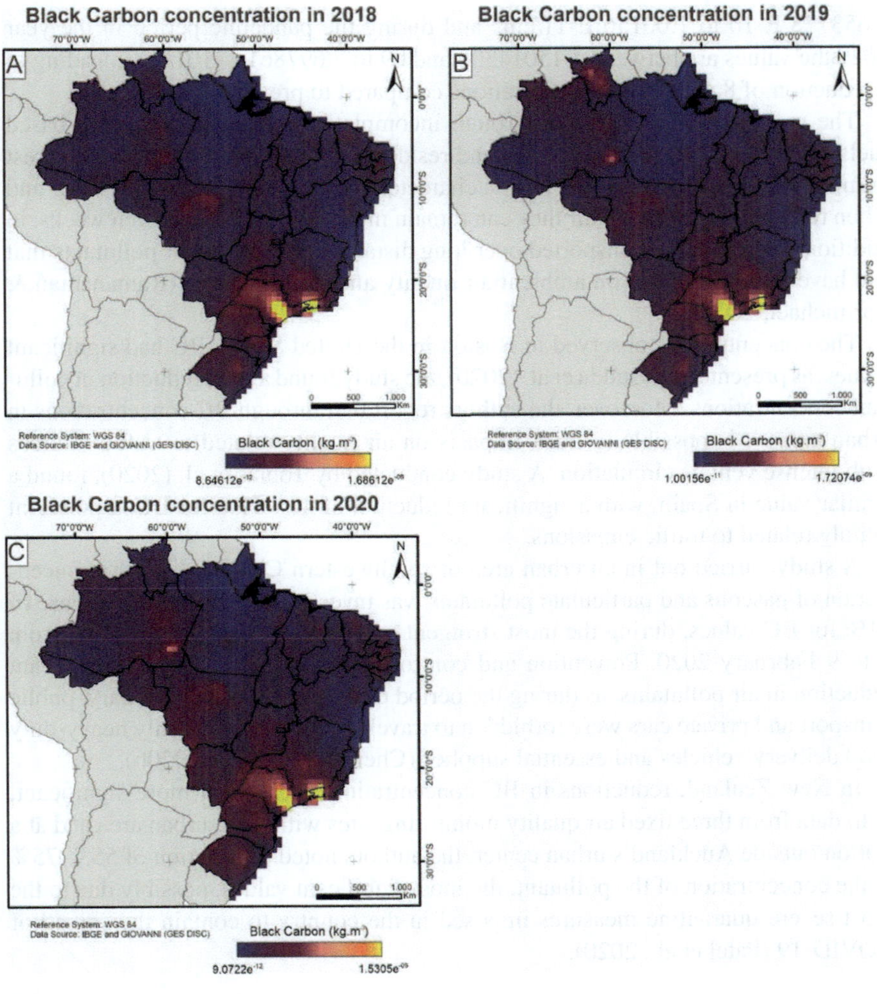

Fig. 2.3 BC concentration: **a** in 2018; **b** in 2019; **c** in 2020

Table 2.3 BC mass concentrations on the surface measured remotely by the MERRA-2 Model satellite

2018 ($kg\ m^{-3}$)	2019 ($kg\ m^{-3}$)	2020 ($kg\ m^{-3}$)
$8.64612e^{-12}$	$1.00156e^{-11}$	$9.0722e^{-12}$
$4.2801459e^{-10}$	$4.376967e^{-10}$	$3.8942915e^{-10}$
$8.4738306e^{-10}$	$8.653778e^{-10}$	$7.697861e^{-10}$
$1.26675153e^{-09}$	$1.2930589e^{-09}$	$1.15014305e^{-09}$
$1.68612e^{-09}$	$1.72074e^{-09}$	$1.5305e^{-09}$

8.653778 E-10 to 1.00156 E-11/cm^2 and during the pandemic period in the year 2020 the values are between 1.15014305 and 09 to 7.697861e−101/cm^2, leading to a reduction of 8.4% in BC concentrations compared to previous years.

The main sources of BC are through incomplete combustion of carbon-based fuels (diesel, gasoline and kerosene) and residential solid fuel, in addition to forest burning practices. BC particles have a characteristic of being chemically inert, and when released into ambient air they can remain in the air for hours or even weeks, in addition they could be transported over long distances, and act as air pollutants that can have harmful effects on ambient air quality and human health (Ramanathan & Carmichael, 2008).

The concentrations observed in Boston in the United States, BC had significant values, as presented by Hudda et al. (2020), the study found a 51% reduction in pollutant concentrations. Moreover, the authors report that through BC concentrations in urban areas it is possible to track impacts on air quality related to traffic in areas with intense vehicle circulation. A study conducted by Tobías et al. (2020), found a similar value in Spain, with a significant reduction of 45 - 51% for BC, a pollutant mainly related to traffic emissions.

A study carried out in an urban area of southwestern China, where the concentration of gaseous and particulate pollutants was investigated, showed a decrease of 61% for BC values, during the most stringent control measures, in the period from 1 to 9 February 2020. Prevention and control measures have led to a significant reduction in air pollutants, as during the period of the first week of February, public transport and private cars were forbidden to travel, with permission only heavy-duty food delivery vehicles and essential supplies (Chen et al., 2020a, 2020b).

In New Zealand, reductions in BC concentrations were even more significant, with data from three fixed air quality monitoring sites within the urban area and at a station outside Auckland's urban center, the authors noted a reduction of 55.7–75% in the concentration of the pollutant, the most significant values, possibly due to the most severe quarantine measures imposed in the country to contain the spread of COVID-19 (Patel et al., 2020).

2.4.3 Carbon Monoxide (CO)

Figure 2.4 shows the CO concentration values on the surface, in the total and tropospheric column, in the years 2018, 2019 and 2020, respectively.

The total column and tropospheric fraction of CO (surface concentration) measured remotely by the MERRA-2 Model satellite from 1 January to 29 June in the years 2018, 2019 and 2020 is presented in Table 2.4.

For CO in the total column and tropospheric fraction, the data are shown in Table 2.4 and demonstrate that in 2018 the values vary from 47,0554 and 8508,706 ppbv, in 2019 the values were between 49.1822 and 524.011 ppbv and during the quarantine period the values reached the range of 47.468 to 468.808 ppbv, leading to a reduction of 7.0% in CO concentrations in the year 2020, related to

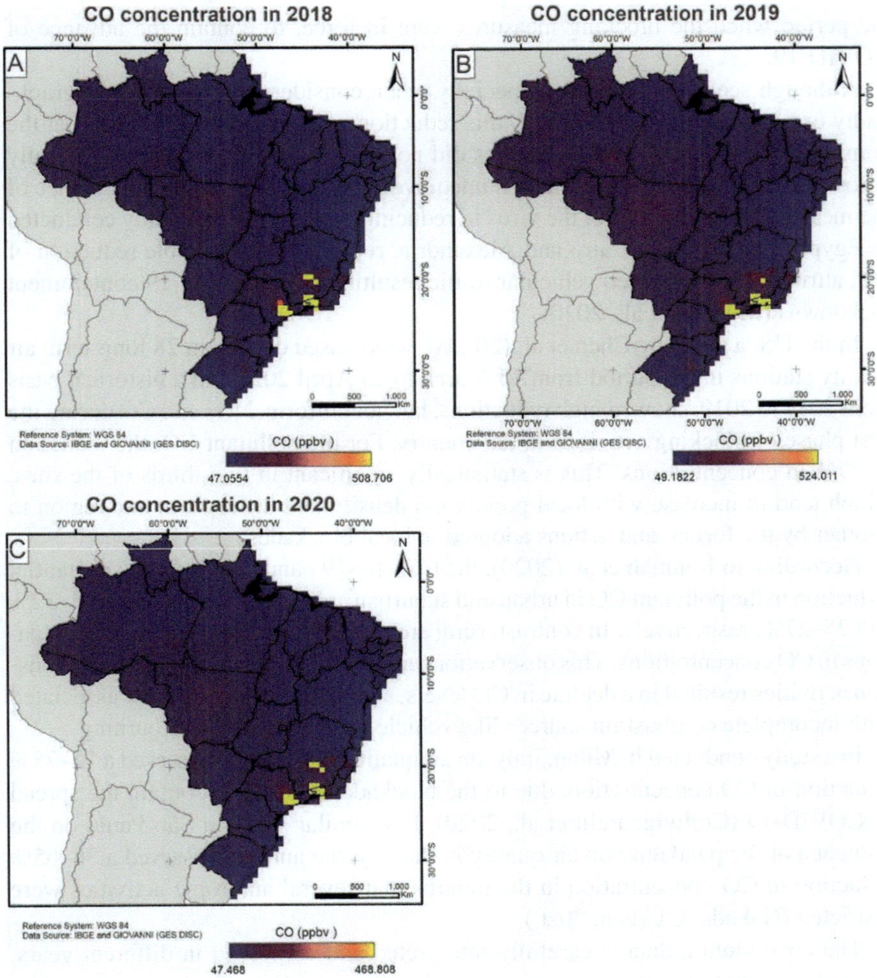

Fig. 2.4 CO concentration: **a** in 2018; **b** in 2019; **c** in 2020

Table 2.4 Total column and tropospheric fraction of CO measured remotely by the MERRA-2 Model satellite

2018 (ppbv)	2019 (ppbv)	2020 (ppbv)
47.0554	49.1822	47.468
160.96805	167.8894	152.803
276.8807	286.5966	258.138
392.79335	405.3038	363.473
508.706	524.011	468.808

the period when the blocking measures were in force, to contain the advance of COVID-19.

Although seemingly modest, especially when considering a vast and geographically extensive country like Brazil, this reduction in air pollution signifies that the commencement of lockdown measures did not occur abruptly but rather gradually across different states. Nonetheless, it unequivocally underscores the significance of the measures taken to control the virus in reducing air pollutants. A study conducted in Egypt, specifically in Cairo and Alexandria, reported a comparable reduction of 5%, attributed to decreased vehicular traffic resulting from COVID-19 containment lockdowns (Mostafa et al., 2020).

In the US, a study by Chen et al. (2020a), which used data from 28 long-term air quality stations in the period from 15 March to 25 April 2020 and a historical basis from 2017 to 2019, showed high reductions, but not uniform, NO_2 and CO during the first phase of blocking activities in the country. For the pollutant CO, the reduction of 37% in concentrations. This is statistically significant in two-thirds of the sites, which tend to increase with local population density and differ from one region to another by the format and actions adopted in local blockades.

According to Kanniah et al. (2020), the COVID-19 pandemic led to a substantial reduction in the pollutant CO in urban and suburban areas, with decreases of 25–32% and 25–27%, respectively. In contrast, rural areas exhibited relatively minor fluctuations in CO concentrations. This observation underscores that restrictions on combustion activities resulted in a decline in CO levels, a pollutant predominantly associated with incomplete combustion sources like vehicle traffic and biomass burning.

In a study conducted in Milan, Italy, on air quality, the authors observed a 55–75% reduction in CO concentration, due to the blockade imposed to contain the spread of COVID-19 (Collivignarelli et al., 2020). In a similar study in São Paulo on the influence of the pandemic on air quality in the city, the authors observed a 36–65% reduction in CO concentration in the months that several anthropic activities were restricted (Nakada & Urban, 2020),

The air pollution data is carefully interpreted and compared in different years, due to the air pollution levels, particularly PM, SO_2 and CO, which are affected by meteorological factors. Moreover, the figures illustrating reductions in air pollution in several countries should not be interpreted, or generalized to other countries, because the form of measures of social remoteness vary depending on the countries, and there are numerous factors that change the air quality of a region (Ju et al., 2021).

2.4.4 Sulphates (SO_4)

The mass concentrations of SO_4 on the surface is demonstrated in Fig. 2.5, presented in the total and tropospheric column, in 2018, 2019 and 2020, respectively, in the Brazilian territory.

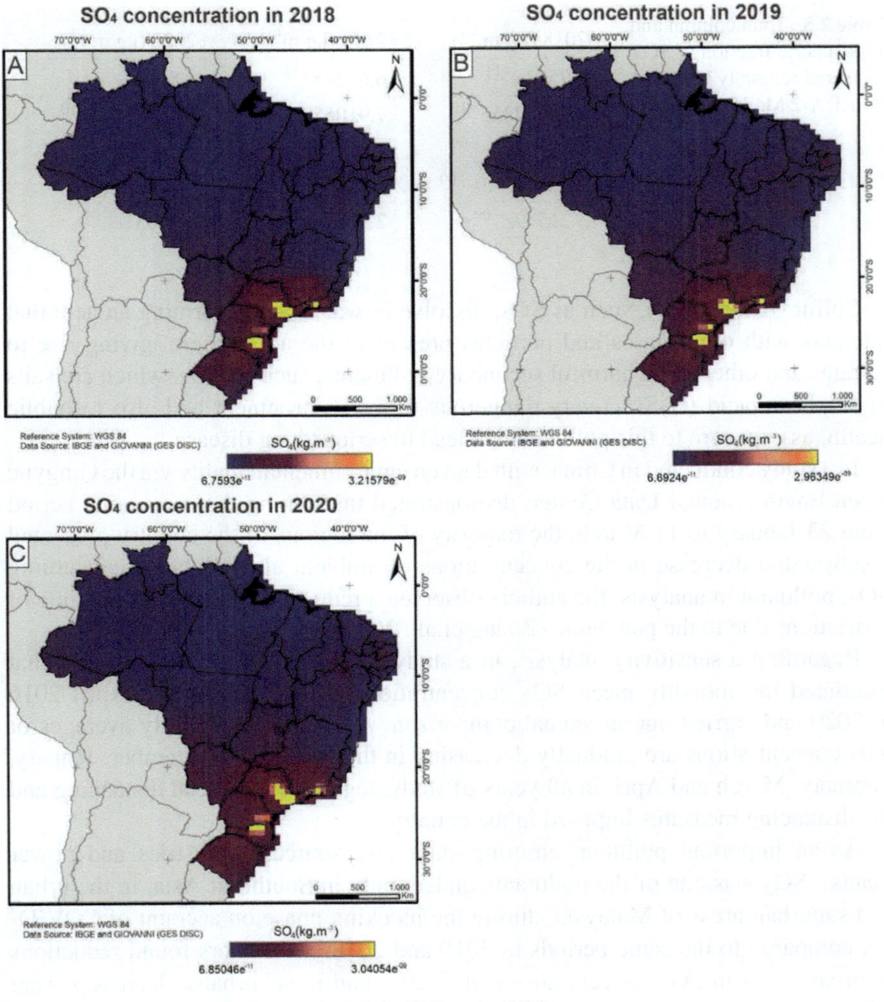

Fig. 2.5 SO_4 concentration: **a** in 2018; **b** in 2019; **c** in 2020

The total and tropospheric column fraction of SO_4 (surface mass concentration) measured remotely by the MERRA-2 Model satellite from 1 January to 29 June in the years 2018, 2019 and 2020, demonstrated in Table 2.5.

Table 2.5 demonstrates the difference of the fraction of total column and tropospheric of SO_4 that represent the measure remotely made by the MERRA-2 Model satellite, where it shows that in 2018 vary from 6.7593e−11 and 1.6416915e−09 kg/m^{-3}, in 2019 the values were between 6.6924e−11e 1.515207e−09 kg/m^{-3} and during the pandemic period in the first half of 2020 the values are between 6.85046e−11e 1.5545223e−09 kg/m^{-3}, leading to a small decrease of 4.76% in pollutant concentrations in the year 2020.

Table 2.5 Total column and tropospheric fraction of SO_4 measured remotely by the MERRA-2 Model satellite

2018 (kg m^{-3})	2019 (kg m^{-3})	2020 (kg m^{-3})
$6.7593e^{-11}$	$6.6924e^{-11}$	$6.85046e^{-11}$
$8.5464225e^{-10}$	$7.910655e^{-10}$	$8.1151345e^{-10}$
$1.6416915e^{-09}$	$1.515207e^{-09}$	$1.5545223e^{-09}$
$2.42874075e^{-09}$	$2.2393485e^{-09}$	$2.29753115e^{-09}$
$3.21579e^{-09}$	$2.96349e^{-09}$	$3.04054e^{-09}$

Sulfur oxides (SO_x), such as SO_2, dissolve in water vapor, forming an acid that interacts with other gases and particles present in the atmosphere, giving rise to sulfates and other more harmful secondary pollutants, such as SO_4, which are salts of sulphuric acid (H_2SO_4) very dangerous to the environment and also to public health, as exposure to this pollutant can lead to serious lung disease.

In a study conducted in China, with data on environmental quality via the Qingyue Open Environmental Data Center, demonstrated that during the quarantine period from 23 January to 11 March, the majority of the regions of the country presented a substantial decrease in the concentrations of ambient air pollutants, concerning SO_2, pollutant in analysis, the authors observed a reduction of 15.9% at the time of restrictions due to the pandemic (Zhang et al., 2020a).

Regarding a sensitivity analysis, in a study conducted by Ju et al. (2021), that examined the monthly mean SO_2 concentrations from December to April 2016 to 2020 and carried out an annual comparison. Accordingly, monthly averages of SO_2 concentrations are gradually decreasing in the months of December, January, February, March and April in all years of study, regardless of social distancing and the distancing measures imposed in the country.

As an important pollutant emitting stationary sources (industries and power plants), SO_2 was one of the pollutants under study in Southeast Asia, in the urban and suburban areas of Malaysia, during the blocking phase on account of COVID-19, compared to the same periods in 2019 and 2018, the authors found reductions in urban areas in SO_2 concentrations of 9–20% and in suburbans, decreases were 17–19% in 2020, compared to 2019 and 2018. Moreover, the authors explain that perceived significant reductions in industrial and rural locations in several regions of the country (Kanniah et al., 2020).

2.4.5 *Tropospheric Ozone (O_3)*

Moreover, Fig. 2.6 illustrates the concentration values of pollutant O_3, shown in the total and tropospheric column, in 2018, 2019 and 2020, respectively.

The total column fraction of O_3 measured remotely by the MERRA-2 Model satellite from 1 January to 29 June in the years 2018, 2019 and 2020 is presented in Table 2.6.

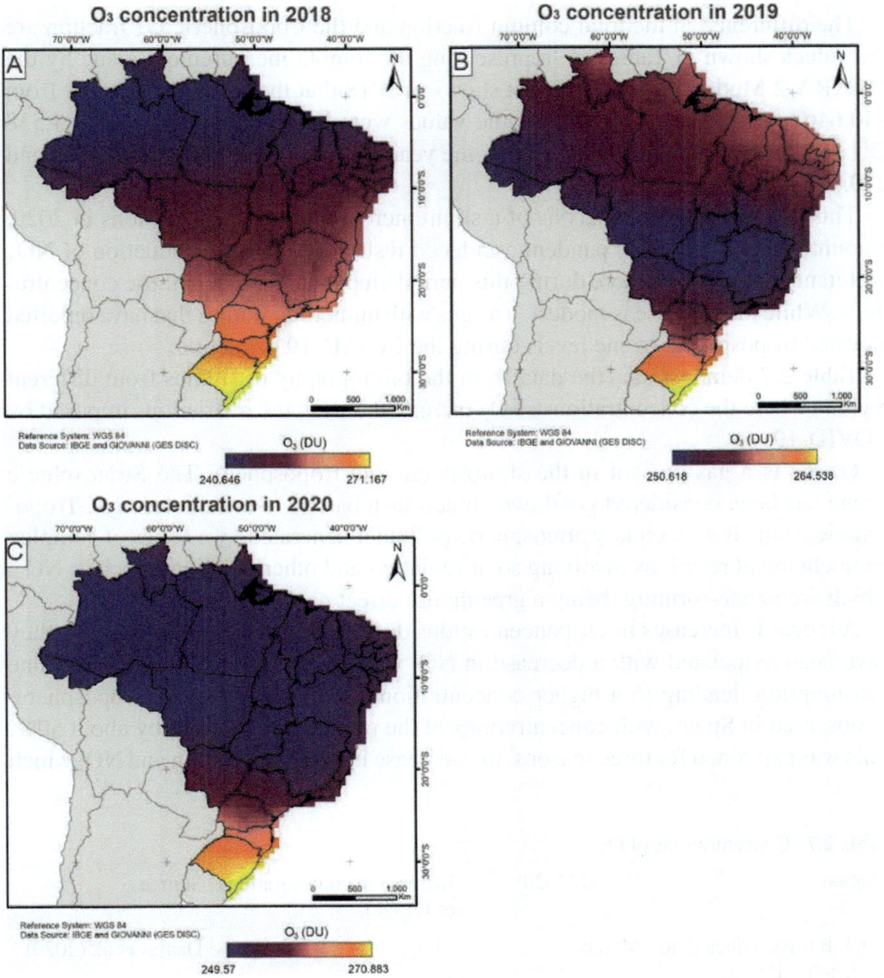

Fig. 2.6 O_3 concentration: **a** in 2018; **b** in 2019; **c** in 2020

Table 2.6 Total column fraction of O_3 measured remotely by the IMO satellite

2018 (DU)	2019 (DU)	2020 (DU)
240.646	250.618	249.57
248.27625	254.098	254.899825
255.9065	257.578	260.2265
263.53675	261.058	265.55475
271.167	264.538	270.883

The difference in the total column fraction and the tropospheric O_3 fraction are the values shown in Table 2.6, representing the remote measurement made by the MERRA-2 Model satellite, where it shows in 2018 that the values of O_3 vary from 240,646 and 271,167 DU, in 2019 the values were between 250,618 and 264,538 DU and during the pandemic period in the year 2020 the values reached 249.57 and 270,883 DU.

This data enabled the analysis of a slight increase in O_3 concentrations in 2020, amounting to 1.68%. The pandemic-induced restrictions and the reduction of NO_x pollutants in the atmosphere during this period elucidate the rise in these concentrations. While the increase is modest, it aligns with numerous studies that have reported elevated tropospheric ozone levels during the COVID-19 pandemic.

Table 2.7 demonstrates the data from the bibliography of studies from different regions where the concentrations of O_3 increased during the restrictions imposed by COVID-19.

Ozone is a gas present in the stratosphere and troposphere. The Stratospheric ozone has been considered good, which acts as a barrier to ultraviolet rays, Tropospheric ozone is a secondary atmospheric pollutant generated by a series of complex photochemical reactions involving solar radiation and other pollutants such as NO_x, which are ozone-forming, being a greenhouse effect gas (Zoran et al., 2020).

As regards increases in O_3 concentrations during the COVID-19 pandemic, they have been associated with a decrease in NO, which can cause a decrease in ozone consumption, leading to a higher concentration of COVID-19 in the troposphere, as observed in Spain, with concentrations of the pollutant increasing by about 50%. This was explained for three reasons, the decrease in O_3 consumption and NO, which

Table 2.7 Concentrations of O_3

Region	Period (2020)	Increase in concentration of O_3 (%)	Source
Irajá; Bangu; Tijuca (Rio de Janeiro, Brazil)	March	31.1 22.5 63.0	Dantas et al. (2020)
São Paulo (Brazil)	February–April	30	Nakada and Urban (2020)
India	March–April	17	Sharma et al. (2020)
Barcelona (Spain)	February–March	57	Tobías et al. (2020)
Nice (France)	January–April	24	Sicard et al. (2020)
Roma (Italy)	January–April	13.6	Sicard et al. (2020)
Wuhan (China)	January–April	36.4	Sicard et al. (2020)
China	January–March	36.5	Zhang et al. (2020b)
Baghdad (Iraq)	March–May	75	Hashim et al. (2020)
United Kingdom	March–May	7.6	Jephcote et al. (2020)
Auckland (New Zealand)	March–April	16.7	Patel et al. (2020)

reduces O_3 consumption (titration, $NO + O_3 = NO_2 + O_2$), causing an increase in O_3 concentrations in the atmosphere, and finally a climate issue, which is the increase in temperature and insolation in the months of February to April (Mahato et al., 2020; Tobías et al., 2020).

In a study conducted in Baghdad, Iraq, the authors also found a significant increase of O_3 during the COVID-19 blockade, linking this increase to the higher levels of solar activity in that region during June and July, months in which the city was in a situation of partial blockade and total blockade (Hashim et al., 2020).

In several studies on air pollution during the pandemic, the authors did not discovered high increases for O_3, as in the case of the one with an increase of 4.52% (Sicard et al., 2020), and in Malaysia the increase in O_3 concentrations was between 3 and 7% in urban locations during 2020, due to the reduction in NO levels, similarly to other urban environments (Kanniah et al., 2020). In Austria, an increase of 11.6–33.8% was observed for O_3, in the blocking period caused by the new coronavirus (Lovrić et al., 2020).

Brazil is a tropical climate country, and the concentrations of the pollutant ozone are a major concern, as in the case of the city of Rio de Janeiro, where high temperatures and solar irradiation favor the atmospheric processes that lead to the formation of O_3. In addition to being a poorly controlled pollutant due to its non-linear dependence on emission sources, even these restrictions have been beneficial for the decrease of several air pollutants. Ozone occurred differently, which suffered contrary action (Siciliano et al., 2020).

2.4.6 Emissions in the Energy Sector: Methane (CH_4)

Due to the unavailability of complete methane concentration data in the atmosphere for the period under analysis, this study exclusively focused on methane emissions within the country's energy sector. This sector stands as one of the major contributors to GHG emissions, and any variations in its emissions have ramifications across all aspects of society. The energy sector is categorized into subsectors or emission categories, specifically fugitive emissions and emissions from fossil fuel combustion.

Methane emissions from fossil fuel combustion results from the combustion of fuels that have the potential to generate methane. On the other hand, fugitive emissions encompass methane emissions during coal mining in the petroleum and natural gas industries, with similar considerations applying to carbon dioxide emissions. Emissions from fossil fuel combustion account for the utilization of fuels across various sectors, including residential, public, transportation, commercial, industrial, agricultural, and public electricity generation.

Table 2.8 presents the total fugitive methane emissions, as well as the total methane emissions associated with fuel combustion for the period from 2018 to 2020, along with the overall emissions from the country's energy sector.

In 2020, a period coinciding with the virus's spread, fugitive emissions decreased by approximately 12%, while emissions from the combustion and use of fuels in the

Table 2.8 Methane emissions within Brazil's energy subsectors

Emissions CH_4(t)				
Category	2018	2019	2020	Total
Fugitive emissions	174.379	177.546	156.207	508.132
Emissions from burning fossil fuels	414.497	409.679	414.944	1.239.120
Total	588.876	587.225	571.151	

energy sector increased by approximately 1.28%. According to the analytical report from SEEG published in 2021, greenhouse gas emissions in the transportation sector (− 5.6%) decreased during the pandemic, as did emissions in the industrial sector, where fuel usage decreased by approximately (− 6.9%), and emissions related to electricity generation (− 10.8%), as the country progressively incorporates renewable energy sources into its electricity grid. However, in contrast, emissions associated with the fuel combustion subsector were driven by fuel production (+1.9%), which increased even during the pandemic and lockdown period. Similarly, residential emissions increased due to the consumption of cooking gas, and in agriculture, emissions rose with the use of fuels in agricultural machinery (SEEG, 2021). Figure 2.7 illustrates the total methane emissions in tons from the entire energy sector for the years 2018 (Fig. 2.7a), 2019 (Fig. 2.7b), and 2020 (Fig. 2.7c).

In general, methane emissions from the energy sector in the country experienced a decrease of 2.73% when compared to the previous year. Emissions from this sector in 2018 amounted to approximately 588,876 tCH_4, after which the values gradually declined over the years, reaching 587,225 tCH_4 in 2019 and 571,151 tCH_4 in 2020, driven primarily by the reduction in fugitive emissions. States, on the whole, showed decreases in methane emissions, notably Rio de Janeiro and São Paulo, with reductions of − 12.03% and − 17.27%, respectively, in 2020. Despite the overall decrease in total emissions in the state of Pernambuco during the period under analysis, it experienced an increase in fugitive emissions in 2020, by approximately + 26.4%. The replacement of fossil fuels with biofuels, reduced exploitation of coal, decreased fugitive emissions in oil and natural gas exploration, and the development of energy-efficient technologies should be considered to mitigate emissions in the energy sector. Methane emissions also occur in significant quantities in the waste sector, particularly in activities related to the final disposal of solid waste in landfills, the incineration of medical waste, open burning of solid waste, and the treatment and disposal of domestic and industrial wastewater. The anaerobic degradation of municipal solid waste, depending on the type of final disposal method such as sanitary landfills, controlled landfills, and open dumps, can produce substantial amounts of methane (SEEG, 2022). In 2020, emissions from the waste sector increased as a result of the pandemic, with solid waste generation rising by approximately 10% in 2020, leading to an increase in methane emissions (+1.8%) amounting to 3,152,593 tCH_4 in the same year. However, some biogas and methane recovery projects in landfills have been implemented in Brazil (SEEG, 2021).

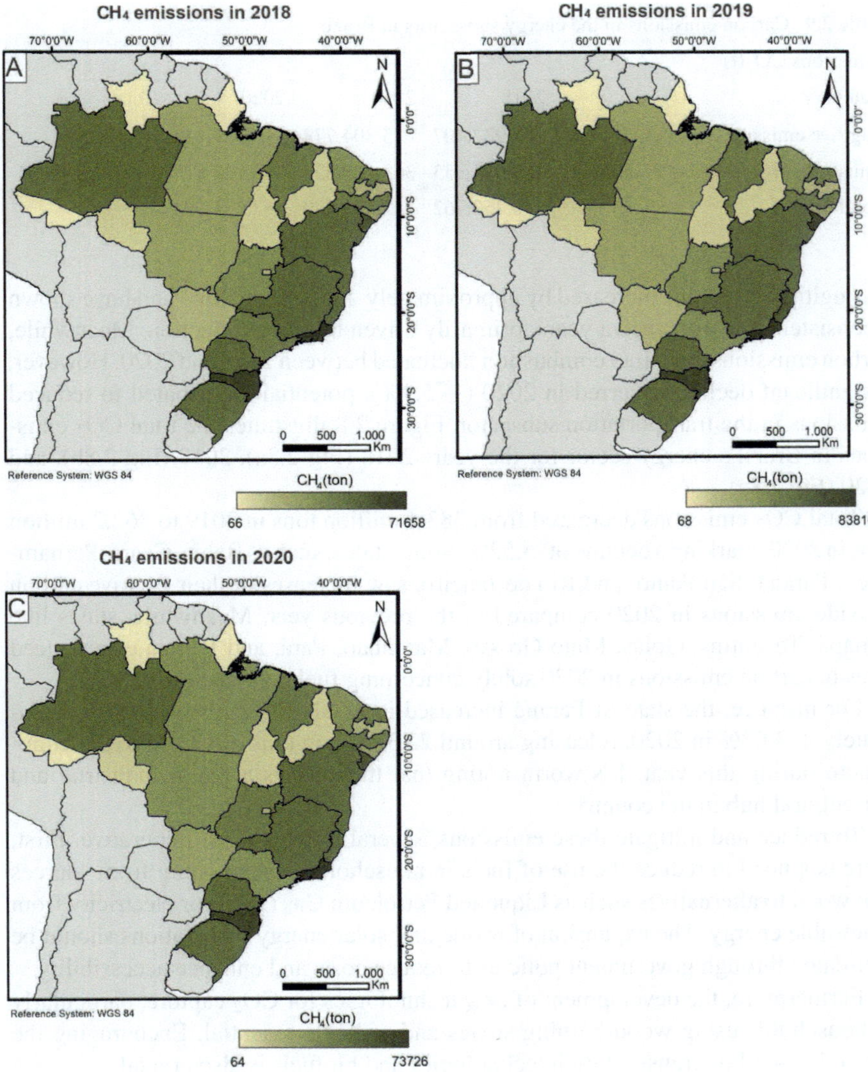

Fig. 2.7 CH_4 emissions in the energy sector in Brazil: **a** in 2018; **b** in 2019; **c** in 2020

2.4.7 *Emissions in the Energy Sector: Carbon Dioxide (CO_2)*

Carbon emissions in the energy sector encompass CO_2 emissions generated during the combustion of fossil fuels and emissions from coal mines that can undergo combustion. Additionally, fugitive emissions account for CO_2 removals during the extraction and production processes of oil and natural gas. Table 2.9 presents a comparison of emissions from 2018 to 2020 in energy subsectors, considering both fugitive emissions and emissions from fossil fuel combustion.

Table 2.9 Carbon emissions in the energy subsectors in Brazil

Emissions CO_2(t)				
Category	2018	2019	2020	Total
Fugitive emissions	14.932.407	15.304.778	16.087.413	46.324.598
Emissions from burning fossil fuels	365.324.355	368.117.821	346.144.575	1.079.586.751
Total	380.256.762	383.422.599	362.231.988	

Fugitive emissions increased by approximately + 5.11% in 2020 and have shown a consistent rise over recent years, primarily driven by fuel production. Meanwhile, carbon emissions from fuel combustion fluctuated between 2018 and 2020. However, a significant decline occurred in 2020 (− 5.9%), potentially attributed to reduced emissions in the transportation subsector. Figure 2.8 illustrates the total CO_2 emissions in Brazil's energy sector for the years 2018 (Fig 2.8a), 2019 (Fig 2.8b), and 2020 (Fig 2.8c).

Total CO_2 emissions decreased from 383.4 million tons in 2019 to 362.2 million tons in 2020, marking a decline of -5.52%. Some states, such as Bahia, Ceará, Pernambuco, Paraná, São Paulo, and Rio de Janeiro, saw increases in their fugitive carbon dioxide emissions in 2020 compared to the previous year. Meanwhile, states like Amapá, Tocantins, Goiás, Mato Grosso, Maranhão, Pará, and Paraná experienced rises in carbon emissions in 2020 solely concerning fuel combustion.

For instance, the state of Paraná increased its total CO_2 emissions by approximately + 3.03% in 2020, releasing around 22.3 million tons of CO_2 into the atmosphere during this year. It's worth noting that the state is a major industrial and agricultural hub in the country.

To reduce and mitigate these emissions, several measures are imperative. First, there is a need to reduce the use of fuels in households, transitioning from sources like wood to alternatives such as Liquefied Petroleum Gas (LPG) or electricity from renewable energy. The expansion of residential solar energy installations should be facilitated through government policies to reduce costs and enhance accessibility.

Furthermore, the development of new technologies for CO_2 capture, particularly in households using wood-burning stoves and grills, is essential. Encouraging the use of low-carbon transportation technologies and biofuels is also crucial.

Industries must strive to minimize their reliance on fossil fuels and explore the generation of their energy from waste products. Additionally, cities and small municipalities should develop urban projects for harnessing energy from municipal solid waste and domestic and public sewage, converting them into economic and environmental benefits. With the implementation of these mitigation measures, it becomes feasible to reduce greenhouse gas emissions and combat global warming effectively.

The COVID-19 pandemic constituted a global event that gave rise to various health and economic challenges across several countries. The mitigation measures implemented universally shed light on the impact of human activities on air quality. This period effectively served as an experimental opportunity to evaluate air quality concerning industrial operations, vehicular traffic, and human mobility.

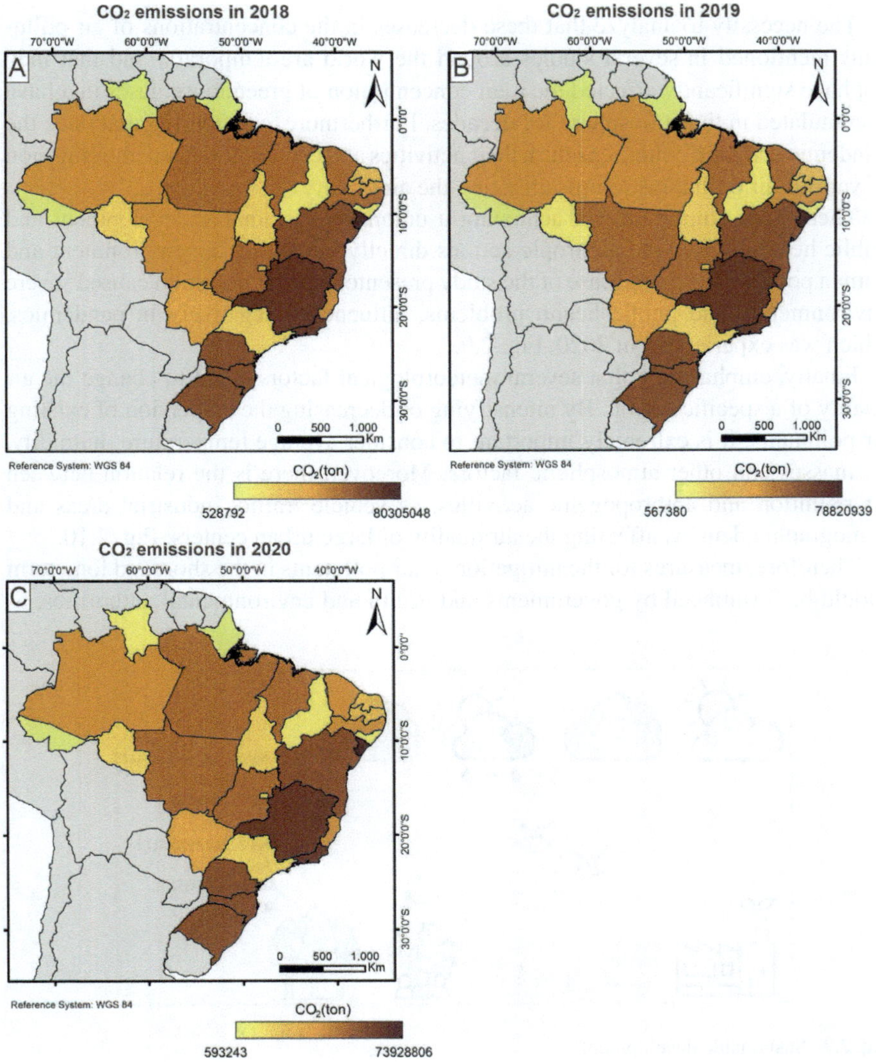

Fig. 2.8 CO_2 emissions in the energy sector in Brazil: **a** in 2018; **b** in 2019; **c** in 2020

The quarantine benefited air quality of large urban centers, and had an extreme importance for the flattening of the new Coronavirus epidemic curve. Despite of, it had a negative impact on lives of populations, with millions of deaths, making daily life full of insecurities and keeping people away from their family and friends, with negative effects on the economy and health of several countries, besides having affected the education of several young people and children around the world, being considered the greatest tragedy of the twenty-first century, which left and is leaving deep marks on humanity.

The necessity to analyze that these decreases in the concentrations of air pollutants mentioned in several studies around the world are temporary, and that may not have significantly affected the total concentration of greenhouse gases that have accumulated in the atmosphere for decades. Furthermore, considering that once the pandemic ends and countries return their activities and economies again, the emission of various air pollutants returns affecting the air quality.

There is an importance of achieving a union for sustainable development and public health, due to the anthropic actions directly impacting the environment and human populations. In the case of the study presented, the air pollution caused severe environmental and public health problems, influencing negatively in pandemics, which was experienced in 2020, Fig. 2.9.

Finally, emphasizing that several meteorological factors assist to change the air quality of a specific region. By intensifying or decreasing the dispersion of existing air pollutants, it is extremely important to consider average temperature, humidity, air masses and other atmospheric factors. Moreover, there is the relation between air pollution and anthropogenic activities, as vehicle traffic, industrial areas and demographic density, affecting the air quality of large urban centers, Fig. 2.10.

Therefore, measures for the mitigation of air pollutants in the short and long term should be formulated by governments and health and environmental authorities, as

Fig. 2.9 Sustainable development

Fig. 2.10 Current air pollution situation in different regions of the world

this is the only form with actions aimed at reducing emissions from industry, traffic and energy generation. Subsequently being able to return to the levels experienced in reducing air pollutants by 2020, otherwise worsening the air quality situation causing several environmental and public health problems.

2.5 Conclusion

The world in the year 2020 has been surprised by the epidemic of the new Coronavirus, where much of the year the population spent in social isolation, having to adopt new habits, along with state and health authorities tracing ways to contain the spread of the new Coronavirus.

FInally, the conclusion from this study indicates that:

- Several studies have reported significant decreases in pollutant concentrations across different regions of the world, which depended only on the time and format The restrictions occurred to indicate the degree of decrease;
- In Brazil, the pandemic was declared a public health emergency in February, with the first case on 26 February, with the state of São Paulo and Rio de Janeiro being the first to implement blockades due to COVID-19, however from March, other states have been implementing their quarantine actions to mitigate the spread of the COVID-19 virus that only spreads throughout the country;
- Regarding the total emissions of CH_4 and CO_2 originating from the Brazilian energy sector, a decreasing trend was also observed during the pandemic period. CH_4 emissions exhibited a progressive reduction of 0.28% between the years 2018, 2019, and 2020. Concerning CO_2 emissions, there was an increase of 0.83% between the years 2018 and 2019, followed by a decline of 5.52% between 2019 and 2020.
- In the Brazilian territory, significant decreases in concentrations of several air pollutants were observed during quarantine, such as for NO_2, BC, CO and SO_4, 21.8, 8.4, 7.0, 4.7%, respectively, and an increase in O_3 concentrations of 1,6% in the period of restrictions to reduce contamination by COVID-19 in the year 2020;
- There is a relation between air pollution and anthropogenic activities, especially vehicle traffic, industrial activities and demographic density, which negatively affect the air quality of large urban centers;
- Certain measures, such as the expansion of energy utilization from municipal solid waste and domestic and public sewage, along with the advancement of carbon capture technologies in residential settings, possess the potential to exert a significant impact on greenhouse gas emissions at the local and regional levels.

Acknowledgements The authors would like to thank the National Council for Scientific and Technological Development - CNPq, the Federal University of Pelotas and the Laboratory for Environmental and Energy Engineering - LEAE for their support in carrying out this research.

The authors have no relevant financial or non-financial interests to disclose. The authors have no conflicts of interest to declare that are relevant to the content of this article. There was no funding support for this work. Non-digital data available.

References

Abou El-Magd, I., & Zanaty, N. (2020). Impacts of short-term lockdown during COVID-19 on air quality in Egypt. *The Egyptian Journal of Remote Sensing and Space Science*. https://doi.org/10.1016/j.ejrs.2020.10.003

Achakulwisut, P., Brauer, M., Hystad, P., & Anenberg, S. C. (2019). Global, national, and urban burdens of paediatric asthma incidence attributable to ambient NO2 pollution: Estimates from global datasets. *The Lancet Planetary Health, 3*(4), e166–e178. https://doi.org/10.1016/S2542-5196(19)30046-4

Aman, M. A., Salman, M. S., & Yunus, A. P. (2020). COVID-19 and its impact on environment: Improved pollution levels during the lockdown period–A case from Ahmedabad, India. *Remote Sensing Applications: Society and Environment, 20*, Article 100382. https://doi.org/10.1016/j.rsase.2020.100382

Baldasano, J. M. (2020). COVID-19 lockdown effects on air quality by NO_2 in the cities of Barcelona and Madrid (Spain). *Science of the Total Environment, 741*, Article 140353. https://doi.org/10.1016/j.scitotenv.2020.140353

Beig, G., Bano, S., Sahu, S. K., Anand, V., Korhale, N., Rathod, A., Yadav, R., Mangaraj, P., Murthy, B.S., Singh, S., Latha, R., & Shinde, R. (2020). COVID-19 and environmental-weather markers: Unfolding baseline levels and veracity of linkages in tropical India. *Environmental Research,* 110121. https://doi.org/10.1016/j.envres.2020.110121.

Bherwani, H., Anjum, S., Kumar, S., Gautam, S., Gupta, A., Kumbhare, H., Anshul, A., & Kumar, R. (2020). Understanding COVID-19 transmission through Bayesian probabilistic modeling and GIS-based Voronoi approach: a policy perspective. *Environment, Development and Sustainability,* 1–19. https://doi.org/10.1007/s10668-020-00849-0

Bolaño-Ortiz, T. R., Camargo-Caicedo, Y., Puliafito, S. E., Ruggeri, M. F., Bolaño-Diaz, S., Pascual-Flores, R., Saturno, J., Ibarra-Espinosa, S., Mayol-Bracero, O.L., Torres-Delgado, E., & Cereceda-Balic, F (2020) Spread of SARS-CoV-2 through Latin America and the Caribbean region: A look from its economic conditions, climate and air pollution indicators. *Environmental Research, 191,* 109938. https://doi.org/10.1016/j.envres.2020.109938

Briz-Redón, Á., Belenguer-Sapiña, C., & Serrano-Aroca, Á. (2020). Changes in air pollution during COVID-19 lockdown in Spain: A multi-city study. *Journal of Environmental Sciences*. https://doi.org/10.1016/j.jes.2020.07.029

Burns, J., Hoffmann, S., Kurz, C., Laxy, M., Polus, S., & Rehfuess, E. (2021). COVID-19 mitigation measures and nitrogen dioxide–A quasi-experimental study of air quality in Munich, Germany. *Atmospheric Environment, 246*, 118089. https://doi.org/10.1016/j.atmosenv.2020.118089

Chen, K., Wang, M., Huang, C., Kinney, P. L., & Anastas, P. T (2020b) Air pollution reduction and mortality benefit during the COVID-19 outbreak in China. *The Lancet Planetary Health, 4*(6), e210–e212. 10.1016/

Chen, L. W. A., Chien, L. C., Li, Y., & Lin, G. (2020b). Nonuniform impacts of COVID-19 lockdown on air quality over the United States. *Science of the Total Environment, 745*, Article 141105. https://doi.org/10.1016/j.scitotenv.2020.141105

Chen, Z., Hao, X., Zhang, X., & Chen, F. (2021). Have traffic restrictions improved air quality? A shock from COVID-19. *Journal of Cleaner Production, 279*, Article 123622. https://doi.org/10.1016/j.jclepro.2020.123622

Cheng, Y. F., Heintzenberg, J., Wehner, B., Wu, Z. J., Su, H., Hu, M., & Mao, J. T. (2008). Traffic restrictions in Beijing during the Sino-African Summit 2006: Aerosol size distribution and visibility compared to long-term in situ observations. *Atmospheric Chemistry and Physics, 8*(24), 7583–7594. https://doi.org/10.5194/acp-8-7583-2008

Collivignarelli, M. C., Abbà, A., Bertanza, G., Pedrazzani, R., Ricciardi, P., & Miino, M. C. (2020). Lockdown for CoViD-2019 in Milan: What are the effects on air quality? *Science of the Total Environment, 732*, Article 139280. https://doi.org/10.1016/j.scitotenv.2020.139280

Copat, C., Cristaldi, A., Fiore, M., Grasso, A., Zuccarello, P., Santo Signorelli, S., & Ferrante, M. (2020). The role of air pollution (PM and NO_2) in COVID-19 spread and lethality: a systematic review. *Environmental Research,* 110129. https://doi.org/10.1016/j.envres.2020.110129.

Dantas, G., Siciliano, B., França, B. B., da Silva, C. M., & Arbilla, G. (2020). The impact of COVID-19 partial lockdown on the air quality of the city of Rio de Janeiro, Brazil. *Science of the Total Environment, 729*, Article 139085. https://doi.org/10.1016/j.scitotenv.2020.139085

De Souza, E. G., Lazuta, A. M., de Souza, T., Marques, L. C., Leandro, D., & Nadaleti, W. C. (2022). Análise Espacial E Temporal da Concentração de Dióxido de Nitrogênio (NO_2) no Brasil (2017–2021): Impacto das Medidas de Restrição da Pandemia de Covid-19 na Redução da Poluição Atmosférica. *Boletim Geográfico do Rio Grande do Sul, 39*, 59–77.

Dutheil, F., Baker, J, S., & Navel, V. (2020). COVID-19 as a factor influencing air pollution? *Environmental Pollution (Barking, Essex: 1987), 263*, 114466. https://doi.org/10.1016/j.envpol.2020.114466

Hashim, B. M., Al-Naseri, S. K., Al-Maliki, A., & Al-Ansari, N. (2020) Impact of COVID-19 lockdown on NO_2, O_3, $PM_{2.5}$ and PM_{10} concentrations and assessing air quality changes in Baghdad, Iraq. *Science of the Total Environment, 754*, 141978. https://doi.org/10.1016/j.scitotenv.2020.141978.

He, L., Zhang, S., Hu, J., Li, Z., Zheng, X., Cao, Y., ... & Wu, Y. (2020) On-road emission measurements of reactive nitrogen compounds from heavy-duty diesel trucks in China. *Environmental Pollution,* 114280. https://doi.org/10.1016/j.envpol.2020.114280.

Hudda, N., Simon, M. C., Patton, A. P., & Durant, J. L. (2020). Reductions in traffic-related black carbon and ultrafine particle number concentrations in an urban neighborhood during the COVID-19 pandemic. *Science of the Total Environment, 742*, Article 140931. https://doi.org/10.1016/j.scitotenv.2020.140931

IBGE. (2019). Brazilian Institute of Geography and Statistics. Ibge.com.br. Accessed 20 November 2020.

Jephcote, C., Hansell, A. L., Adams, K., & Gulliver, J. (2020). Changes in air quality during COVID-19 'lockdown' in the United Kingdom. *Environmental Pollution, 272*, 116011. https://doi.org/10.1016/j.envpol.2020.116011

Ju, M. J., Oh, J., & Choi, Y. H. (2021). Changes in air pollution levels after COVID-19 outbreak in Korea. *Science of the Total Environment, 750*, Article 141521. https://doi.org/10.1016/j.scitotenv.2020.141521

Kanniah, K. D., Zaman, N. A. F. K., Kaskaoutis, D. G., & Latif, M. T. (2020). COVID-19's impact on the atmospheric environment in the Southeast Asia region. *Science of the Total Environment, 736*, 139658. https://doi.org/10.1016/j.scitotenv.2020.139658

Kerimray, A., Baimatova, N., Ibragimova, O. P., Bukenov, B., Kenessov, B., Plotitsyn, P., & Karaca, F. (2020). Assessing air quality changes in large cities during COVID-19 lockdowns: The impacts of traffic-free urban conditions in Almaty, Kazakhstan. *Science of the Total Environment, 730*, 139179. https://doi.org/10.1016/j.scitotenv.2020.139179

Krecl, P., Targino, A. C., Oukawa, G. Y. C., & Junior, R. P. C. (2020). Drop in urban air pollution from COVID-19 pandemic: Policy implications for the megacity of São Paulo. *Environmental Pollution (Barking, Essex: 1987).* https://doi.org/10.1016/j.envpol.2020.114883.

Krotkov, N. A., Lamsal, L. N., Celarier, E. A., Swartz, W. H., Marchenko, S. V., Bucsela, E. J., Chan, K. L., Wenig, M., & Zara, M. (2017). The version 3 OMI NO_2 standard product. *Atmospheric Measurement Techniques, 9*, 3133–3149. https://doi.org/10.5194/amt-10-3133-2017

Kumari, P., & Toshniwal, D. (2020). Impact of lockdown on air quality over major cities across the globe during COVID-19 pandemic. *Urban Climate, 34*, Article 100719.

Li, L., Li, Q., Huang, L., Wang, Q., Zhu, A., Xu, J., Liu, Z., Li, H., Shi, L., Li, R., & Azari, M. (2020). Air quality changes during the COVID-19 lockdown over the Yangtze River Delta Region: An insight into the impact of human activity pattern changes on air pollution variation. *Science of The Total Environment,* 139282. https://doi.org/10.1016/j.uclim.2020.100719.

Liu, Q., Harris, J. T., Chiu, L. S., Sun, D., Houser, P. R., Yu, M., Duffy, D. Q., Little, M. M., & Yang, C. (2020) Spatiotemporal impacts of COVID-19 on air pollution in California, USA. *Science of The Total Environment,* 141592. https://doi.org/10.1016/j.scitotenv.2020.141592

Lovarelli, D., Conti, C., Finzi, A., Bacenetti, J., & Guarino, M. (2020). Describing the trend of ammonia, particulate matter and nitrogen oxides: the role of livestock activities in Northern Italy during Covid-19 quarantine. *Environmental Research,* 110048. https://doi.org/10.1016/j.envres.2020.110048.

Lovrić, M., Pavlović, K., Vuković, M., Grange, S. K., Haberl, M., & Kern, R. (2020). Understanding the true effects of the COVID-19 lockdown on air pollution by means of machine learning. *Environmental Pollution, 274,* 115900. https://doi.org/10.1016/j.envpol.2020.115900

Mahato, S., Pal, S., & Ghosh, K. G. (2020). Effect of lockdown amid COVID-19 pandemic on air quality of the megacity Delhi, India. *Science of the Total Environment, 730*, 139086. https://doi.org/10.1016/j.scitotenv.2020.139086

Manisalidis, I., Stavropoulou, E., Stavropoulos, A., & Bezirtzoglou, E. (2020). Environmental and health impacts of air pollution: A review. *Frontiers in Public Health,* 8. https://doi.org/10.3389/fpubh.2020.00014

Mendez-Espinosa, J. F., Rojas, N. Y., Vargas, J., Pachón, J. E., Belalcazar, L. C., & Ramírez, O. (2020). Air quality variations in Northern South America during the COVID-19 lockdown. *Science of the Total Environment, 749*, Article 141621. https://doi.org/10.1016/j.scitotenv.2020.141621

Ministério da Saúde - COVID19. (2020). *Painel de Coronavírus.* https://covid.saude.gov.br/. Accessed 20 November 2020.

Mor, S., Kumar, S., Singh, T., Dogra, S., Pandey, V., & Ravindra, K. (2020). Impact of COVID-19 lockdown on air quality in Chandigarh, India: Understanding the emission sources during controlled anthropogenic activities. *Chemosphere,* 127978. https://doi.org/10.1016/j.chemosphere.2020.127978.

Mostafa, M. K., Gamal, G., & Wafiq, A. (2020). The impact of COVID 19 on air pollution levels and other environmental indicators—A case study of Egypt. *Journal of Environmental Management, 277*, Article 111496. https://doi.org/10.1016/j.jenvman.2020.111496

Muhammad, S., Long, X., & Salman, M. (2020). COVID-19 pandemic and environmental pollution: A blessing in disguise? *Science of the Total Environment, 728*, Article 138820. https://doi.org/10.1016/j.scitotenv.2020.138820

Nakada, L. Y. K., & Urban, R. C. (2020) COVID-19 pandemic: Impacts on the air quality during the partial lockdown in São Paulo state, Brazil. *Science of the Total Environment,* 139087. https://doi.org/10.1016/j.scitotenv.2020.139087.

Pacheco, H., Díaz-López, S., Jarre, E., Pacheco, H., Méndez, W., & Zamora-Ledezma, E. (2020). NO2 levels after the COVID-19 lockdown in Ecuador: A trade-off between environment and human health. *Urban Climate, 34*, Article 100674. https://doi.org/10.1016/j.uclim.2020.100674

Patel, H., Talbot, N., Salmond, J., Dirks, K., Xie, S., & Davy, P. (2020). Implications for air quality management of changes in air quality during lockdown in Auckland (New Zealand) in response to the 2020 SARS-CoV-2 epidemic. *Science of the Total Environment, 746*, Article 141129. https://doi.org/10.1016/j.scitotenv.2020.141129

Petzold, A., Ogren, J. A., Fiebig, M., Laj, P., Li, S. M., Baltensperger, U., Holzer-Popp, T., Kinne, S., Pappalardo, G., Sugimoto, N., & Wehrli, C (2013). Recommendations for the interpretation of "black carbon" measurements. *Atmospheric Chemistry & Physics Discussions, 13*(4). https://doi.org/10.5194/acp-13-8365-2013

Ramanathan, V., & Carmichael, G. (2008). Global and regional climate changes due to black carbon. *Nature Geoscience, 1*(4), 221–227. https://doi.org/10.1038/ngeo156

Rutz, C., Loretto, M. C., Bates, A. E., Davidson, S. C., Duarte, C. M., Jetz, W., Johnson, M., Kato, A., Kays, R., Mueller, T., & Primack, R. B. (2020). COVID-19 lockdown allows researchers to quantify the effects of human activity on wildlife. *Nature Ecology & Evolution, 4*(9), 1156–1159. https://doi.org/10.1038/s41559-020-1237-z.S2542-5196(20)30107-8

Salami, L., & Popoola, L. T. (2023). A comprehensive review of atmospheric air pollutants assessment around landfill sites. *Air, Soil and Water Research, 16*, 117862212211453. https://doi.org/10.1177/11786221221145379

SEEG. (2021). *Análise das emissões de Gases de Efeito Estufa e suas implicações para as metas climáticas do Brasil 1970–2020.* https://seeg.eco.br/wp-content/uploads/2023/03/SEEG9_DOC_ANALITICO_2021_FINAL.pdf. Accessed August 26, 2023

SEEG. (2022). *Desafios e Oportunidades para Redução das Emissões de Metano no Brasil.* https://seeg.eco.br/wp-content/uploads/2023/03/SEEG_METANO_2022_FINAL.pdf. Accessed August 26, 2023

Sharma, S., Zhang, M., Gao, J., Zhang, H., & Kota, S. H. (2020). Effect of restricted emissions during COVID-19 on air quality in India. *Science of the Total Environment, 728*, Article 138878. https://doi.org/10.1016/j.scitotenv.2020.138878

Shehzad, K., Sarfraz, M., & Shah, S. G. M. (2020). The impact of COVID-19 as a necessary evil on air pollution in India during the lockdown. *Environmental Pollution, 266*, Article 115080. https://doi.org/10.1016/j.envpol.2020.115080

Shrestha, A. M., Shrestha, U. B., Sharma, R., Bhattarai, S., Tran, H. N. T., & Rupakheti, M. (2020). Lockdown caused by COVID-19 pandemic reduces air pollution in cities worldwide. https://doi.org/10.31223/osf.io/edt4j

Sicard, P., De Marco, A., Agathokleous, E., Feng, Z., Xu, X., Paoletti, E., Rodriguez, J. J. D., Calatayud, V. (2020). Amplified ozone pollution in cities during the COVID-19 lockdown. *Science of the Total Environment,* 139542. https://doi.org/10.1016/j.scitotenv.2020.139542

Siciliano, B., Dantas, G., da Silva, C. M., & Arbilla, G. (2020). Increased ozone levels during the COVID-19 lockdown: Analysis for the city of Rio de Janeiro, Brazil. *Science of the Total Environment, 737*, Article 139765. https://doi.org/10.1016/j.scitotenv.2020.139765

Silva, D. (2011). *Rule three: school modeling practice mathematics* (Dissertation Master in Science and Mathematics Education). Institute of Mathematical and Scientific Education, Federal University of Pará.

Solomon, S., Solomon, D., Qin, M. Manning, M., Marquis, K., Averyt, M. M. B., & Tignor, H. J. LeRoy Miller, Z. (2011). *Chen Climate Change Working Group I: the Physical Science Basis.* Technical Report Intergovernmental Panel on Climate Change, Geneva.

Srivastava, A. (2020). COVID-19 and air pollution and meteorology—An intricate relationship: A review. *Chemosphere, 263*, 128297. https://doi.org/10.1016/j.chemosphere.2020.128297

Stratoulias, D., & Nuthammachot, N. (2020). Air quality development during the COVID-19 pandemic over a medium-sized urban area in Thailand. *Science of the Total Environment, 746*, Article 141320. https://doi.org/10.1016/j.scitotenv.2020.141320

Tobías, A., Carnerero, C., Reche, C., Massagué, J., Via, M., Minguillón, M. C., Alastuey, A., & Querol, X. (2020). Changes in air quality during the lockdown in Barcelona (Spain) one month into the SARS-CoV-2 epidemic. *Science of the Total Environment,* 138540. https://doi.org/10.1016/j.scitotenv.2020.138540.

Van der Geer, J., Hanraads, J. A. J., & Lupton, R. A. (2018). A arte de escrever um artigo científico. *Heliyon, 19*, Article e00205. https://doi.org/10.1016/j.heliyon.2018.e00205

Venter, Z. S., Aunan, K., Chowdhury, S., & Lelieveld, J. (2020). Air pollution declines during COVID-19 lockdowns mitigate the global health burden. *Environmental Research, 192*, 110403. https://doi.org/10.1016/j.envres.2020.110403

Wang, P., Chen, K., Zhu, S., Wang, P., & Zhang, H. (2020a). Severe air pollution events not avoided by reduced anthropogenic activities during COVID-19 outbreak. *Resources, Conservation and Recycling, 158*, Article 104814. https://doi.org/10.1016/j.resconrec.2020.104814

Wang, Q., & Su, M. (2020). A preliminary assessment of the impact of COVID-19 on environment—A case study of China. *Science of the Total Environment, 728*, 138915. https://doi.org/10.1016/j.scitotenv.2020.138915

Wang, Y., Yuan, Y., Wang, Q., Liu, C., Zhi, Q., & Cao, J. (2020b). Changes in air quality related to the control of coronavirus in China: Implications for traffic and industrial emissions. *Science of the Total Environment, 731*, 139133. https://doi.org/10.1016/j.scitotenv.2020.139133

Wyche, K. P., Nichols, M., Parfitt, H., Beckett, P., Gregg, D. J., Smallbone, K. L., & Monks, P. S. (2021). Changes in ambient air quality and atmospheric composition and reactivity in the South East of the UK as a result of the COVID-19 lockdown. *Science of the Total Environment, 755*, Article 142526. https://doi.org/10.1016/j.scitotenv.2020.142526

Zambrano-Monserrate, M. A.; Ruano, M. A., & Sanchez-Alcalde, L. (2020). Indirect effects of COVID-19 on the environment. *Science of the Total Environment,* 138813. https://doi.org/10.1016/j.scitotenv.2020.138813.

Zhang, Q., Pan, Y., He, Y., Walters, W. W., Ni, Q., Liu, X., Xu, G., Shao, J., & Jiang, C. (2020a). Substantial nitrogen oxides emission reduction from China due to COVID-19 and its impact on surface ozone and aerosol pollution. *Science of the Total Environment, 753*, Article 142238. https://doi.org/10.1016/j.scitotenv.2020.142238

Zhang, R., Zhang, Y., Lin, H., Feng, X., Fu, T. M., & Wang, Y. (2020b). NO_x emission reduction and recovery during COVID-19 in East China. *Atmosphere, 11*(4), 433. https://doi.org/10.3390/atmos11040433

Zhang, X., Tang, M., Guo, F., Wei, F., Yu, Z., Gao, K., Jin, M., Wang, J., & Chen, K. (2020c). Associations between air pollution and COVID-19 epidemic during quarantine period in China. *Environmental Pollution, 268*, Article 115897. https://doi.org/10.1016/j.envpol.2020.115897

Zhao, S., Liu, S., Sun, Y., Liu, Y., Beazley, R., & Hou, X. (2020). Assessing NO_2-related health effects by non-linear and linear methods on a national level. *Science of the Total Environment, 744*, Article 140909. https://doi.org/10.1016/j.scitotenv.2020.140909

Zoran, M. A., Savastru, R. S., Savastru, D. M., & Tautan, M. N. (2020). Assessing the relationship between ground levels of ozone (O_3) and nitrogen dioxide (NO_2) with coronavirus (COVID-19) in Milan, Italy. *Science of the Total Environment,* 140005. https://doi.org/10.1016/j.scitotenv.2020.140005.

Chapter 3
A Review of Odor Problems from Landfills

Aysen Muezzinoglu

Abstract Environmental sanitation is a very crucial problem in a crowding world. Special attention to waste management is necessary in areas where intense production of any kind takes place: in agriculture, industry, mining, construction, too. Odor can be thought as the signal of the nature to warn us if microbial or chemical contamination risk is nearby. Public health and welfare can be sustained only by proper handling and management of wastes. In many cases, waste problems arise due to gas emissions, liquid discharges or solid disposal. Besides, they can co-exist and have the potential of converting into one another with or without human interference. They are accompanied by noxious and offensive odor releases in many cases. Odor, among several others, is a serious problem in solid waste disposal. Odor recognition by the population is likely in collecting, transferring, pre-screening and classification of solid wastes. Ultimate removal of the wastes at solid waste disposal units is an odorous operation, too. Leachates coming from different types of solid wastes may emit unacceptable odors. Odor detection is a subjective experience as it is based on human smelling ability. Its perception, therefore the level of complaint varies from one individual to another. It depends on the hedonic tone of the odor so it is subject to human preferences, too. If the intensity of the stimulant increases above a certain ceiling, it is possible that smelling ability is paralyzed temporarily or sometimes permanently. But the noxious odor problems around waste facilities are to be resolved as if the odor stimulation affects everybody alike. Normally an average smelling capacity is taken as the reference recognition level in odor studies. In practice there is no escaping from odorous gas releases at one or the other stage of the collection, handling and disposal of solid wastes. Odor characteristics and intensity differ from one step or method to another. Presently, odor control in waste management may be accomplished by state-of-the-art technologies. In the landfills these technologies may be applied on collected gas streams arising from anaerobic decomposition and continues with transfer, cleaning/separation, treatment, handling/storage at the landfill degasification systems and finally converted into heat and power. Odorous components are simply adsorbed by suitable cover materials or masked by deodorants. In older landfills fugitive landfill gas emissions may continue for a long

A. Muezzinoglu (✉)
Dokuz Eylul University, Izmir, Turkey
e-mail: aysen.muezzin@deu.edu.tr

A. Anouzla et al. (eds.), *Handbook of the Effects of Landfill Emissions on Sustainable Development*, https://doi.org/10.1007/978-3-032-01951-6_3

time unless they are collected. In any case the wastes should not be disposed-off in wild storage areas and open or uncontrolled dumps without due care. When this happens, solid wastes have the worst impacts on the environment and public well-being. Unfortunately, in several cases acceptable methods of management of solid wastes are difficult to apply. So, public disputes and annoyance are seen frequently. To solve them it is recommended that more public participation such as through citizen science applications be applied to challenges of landfill.

3.1 Introduction

Solid waste management is a challenge that becomes increasingly critical as the urban land-use, pollution, waste transport, urbanization difficulties in overcrowding cities are increasing. In a world threatened by rapid climate change, it is getting even more vital.

Solid wastes originate from various sources and activities such as households, agriculture, industry, mining, workplaces, water treatment plants, hospitals, construction and demolition wastes, coal combustion, commercial transportation residues, etc. They may contain vast quantities of ashes from households or large coal-based energy conversion facilities, debris from mining, or sludges from wastewater handling and treatment. Depending on its composition, environmental implications of the waste disposal facilities may differ. Waste composition is a very important factor in choosing the method of disposal. Of course, there are other factors in determining the method of disposal such as the topography, groundwater table, proximity to water resources, land-use plan details, urbanization patterns, transportation means, land cost, waste collection network, climate, meteorology, etc. Yet, in practice one must take into account perhaps the most important determining factor in selecting the method of disposal: the finance and the funding capacity. Next to them the technical competence of the responsible private and public organizations are of significance. Local laws and regulations trying to protect people from negative impacts of waste management are among the basic instruments in selection of methods of disposal. Any method of disposal of wastes including the landfill practice almost invariably creates public interest. These may include land use disputes, public arousal of strong objections mostly ending in political provocations, etc.

A landfill, also called a "dump", sometimes rubbish or garbage dump, is a site for the disposal of wastes with technical provisions. Although it is the oldest and most common form of waste disposal, historical https://doi.org/10.1007/s12665-022-10295-yones were rather "wild" dumps with not much technical and constructional precautions in the past. Systematic burial of the waste with daily, intermediate and final covers only began in the second half of the twentieth century.

Although solid waste management is an essential service to the community, it must always be kept in mind that it produces an unpleasant side effect of objectionable odors. Landfill gas is formed after some time when burried solid waste begins to decompose. Models showing the decomposition process has been mentioned in

Sect. 3.3.1 of this article Anaerobic process begins after all oxygen is used up and this process creates gaseous products, the most common of which are methane (CH_4) and carbon dioxide (CO_2). These two greenhouse gases typically combine to make up more than 95% of the landfill gas. Greenhouse gases emitted from the landfill contribute to the global climate change and although they are odorless, the unpleasant smells are caused by the small percentage of other gases next to them. The most common odors from a landfill are due to hydrogen sulfide (H_2S), ammonia (NH_3) and volatile organic compounds (VOCs). Hydrogen sulfide is recognized as a strong "rotten egg" smell. The volatile organic compounds in composition can vary in composition so different odors may be generated, but typically they have a characteristic pungent smell.

Each year increasing volumes of landfill gases are being emitted to the surroundings from landfill areas. A landfill site is known to be a significant source of greenhouse gases such as CH_4, CO_2, and N_2O and several odorous compounds. Kaza et al. (2018) estimated 1.6 billion tons of CO_2-equivalent greenhouse gases were emitted worldwide from landfills in 2016. Landfills and solid wastes were calculated to constitute nearly 18% of total anthropogenic CH_4 emissions in 2017 (Jackson et al., 2020). As the number of landfills increase due to population growth and development all over the world and unless the off gases are flared or evaluated as biogas or burned for heat and power generation, impacts of methane as a greenhouse gas in climate change challenge will build up.

Mitigation of methane emissions as well as landfill gas emission reduction require good monitoring guidelines be applied. Scheutz and Kjeldsen (2019) suggested a minimum methane mitigation efficiency of 80%. To achieve this a new monitoring method based on the tracer gas dispersion measuring technique for methane emissions from landfills was used. They have presented a guideline for *best practice* measurement performance.

Landfill gases contain VOCs which are precursors of photochemical transformations in the air. This is measured by the ozone forming potential of the gas emissions.

Barlaz et al. (2002) mentioned that the regulations require landfills be monitored for 30 years after closure unless otherwise required by the official authorities. Different conditions may impose extended periods until the termination of postclosure monitoring. In the monitoring program for landfill gas emissions, ambient air odors, individual odorous gases, greenhouse gases and ozone precursors, subsurface migration of gases and seeping leachate impacts as a potential pollution under the ground, would be necessary.

Landfill gas emissions are mainly composed of methane (CH_4) (55–60%, v/v), carbon dioxide (CO_2) (40–45%, v/v), non-methane volatile organic compounds (NMVOCs), sulfides (e.g., hydrogen sulfide H_2S, dimethyl sulfide DMS, and mercaptans), as well as nitrogen compounds (e.g., nitrous oxide N_2O, and acrylonitrile under anaerobic degradation conditions (Scheutz et al., 2008). Landfills have an appreciable ozone forming potential due to the non-methane volatile organic compound emissions, might have detrimental health impacts on workers and nearby residents (Majumdar et al., 2013).

Although odorous compound percentage in the landfill air are way too low compared to the major components of non-odorous methane and carbon dioxide, their low olfactory thresholds make the air very unpleasant for people living nearby.

Some odorous compounds in the composition of the landfill gas are shown to be carcinogens. For example, Fang et al. (2022) showed that the occupational carcinogenic risk (2.79×10^{-5}–1.12×10^{-3}) from landfill gases, and these probabilities cannot be neglected. Aromatic hydrocarbons and halogenated hydrocarbons were shown to be the contributors to the carcinogenic risk of landfill gases. Among the adverse impacts of municipal solid waste facilities, the carcinogenic risk was the most critical, with studied impact distances of about 1.5 km for waste transfer stations and around 5 km from landfills by these authors. They indicated that these safety distances should be even higher for the regions with low wind speeds and higher atmospheric stability.

In a recent review Pan et al. (2023) submitted analysis of volatile organic carbon compounds (VOCs) and semi-volatile organic compounds (SVOCs) from 72 landfills in 20 countries. They found that the aromatic and halogenated compounds were responsible for carcinogenic, mutagenic and teratogenic health risks on landfill workers and nearby population. They also underlined the risks due to ethyl acetate and acetone in landfill air. The VOCs and SVOCs were found at levels that might affect health of the residents within one kilometer of landfills.

An engineered landfill is required to have a gas collection system to minimize gas emissions. But (Kaza et al., 2018) reported that only 8% of waste is disposed of in sanitary landfills with a gas collection infrastructure. Besides, the efficiency of a gas collection system is shown to be generally insufficient (Duan et al., 2021). The latter group measured the gas collection system efficiencies of 23 landfills in Denmark to range between 13 and 86% with an average of 50% in contrast to similar values lower than Swedish (58%), UK (64%) and US (63%) landfills. From the feasibility point of view old and small landfills where CH_4 generation is too low, the application of a landfill gas collection system is neither technically feasible nor economically productive. So older landfills are not restored with gas collectors although gas emissions might be ongoing. On the other hand, Einhäupl (2021) in a PhD dissertation work reports on the concept of 'landfill mining' described as the excavation and processing of formerly buried waste after closure to obtain energy and materials.

Next to the odor and leachate problems, another design criterion is avoiding the landslides which may become catastrophic. For prevention of landslides sites must have stable ground away from areas prone to landslides, slopes must be worked out for not being too steep and the waste must be compacted to prevent sliding, and adequate drainage must be installed to remove excess water.

Landfills may be designed as mixed waste input sites with separate cells or for toxic/hazardous and municipal waste dumps separately. In any case water seepage must be handled with care in order to preserve the soil and protect the groundwater from pollution. Free flow of leachates above and under the ground must be avoided and such polluted waters must be channeled out into a treatment facility. In the landfill the wastes are dumped into cells with geo-membrane or at least compacted clay bottom layers and walls. Cells are covered with odor adsorbing materials and gas release must be made available. The wastes are kept for a long while in order to keep harmful materials and bio-degradable wastes. After a landfill area is full, it must be properly closed. Gas extraction structures need to be provided. Odorous gas release and leachates usually continue even long after the termination of the active operation of the landfill. After closure the site should not be opened to construction but green recreational areas and forests are formed after closure.

A properly designed and working sanitary landfill is perhaps the most reliable setup among all solid waste management options. Landfilling is not the only option in waste management however, and may follow or backed up by other methods. The solid waste management methods may include incineration, pyrolysis, composting and be better economically secured by a sound management policy of recovery/recycling/remediation starting from the origin of the wastes. All of these may be solo actions or used in combination with each other. Yet, all methods require proper segregation/sorting, separation and transfer steps as prerequisites.

3.2 Materials and Methods

The variants of gas and odor formation and its characteristics originating from dumping sites are discussed in this review. The odor impacts of landfills and precautions against odorous gas releases are summarized with disposal operations and proper design. The collection, transfer and screening prerequisites, which are ending in circular economy objectives are mentioned.

Method used in this review is using the published data and ideas in scientific and engineering project reports on the present status of landfills. To the best of the knowledge a wide geographical coverage has been chosen over the world to reflect different social approaches and climatic, developmental and regulatory circumstances.

Examples were collected across the landfill-gas-release and landfill-odor-based literature from academic databases and internet sources, the ideas and conclusions of the scientists and solid waste service providers were conveyed and evaluated. State-of-the-art techniques and latest scientific findings are worked out for different aspects of the landfill technology and their implications on the urban environment as well as human health and welfare. In occasional cases the odor studies from other odor sources such as industries, wastewater treatment plants, compost plants, etc. were included as long as their findings are relevant in landfill odor problem handling. Examples are given and novel developments and expectations for a better world are covered.

Waste-to-energy technology is not covered in this article as it has less direct relationship with the odor problem.

3.3 Landfill Gases

3.3.1 Landfill Gas Emissions and Their Formation

There are several gas emissions which cause pungent odors in relation to the composition of the solid waste degradation. Gases are produced either by evaporation of volatile material present in the waste surface or are formed during the decomposition of organic matter. There are hundreds of different molecules that can be formed. But major ones are methane (CH_4), carbon dioxide (CO_2), volatile organic compounds (VOCs), ammonia (NH_3), hydrogen sulfide (H_2S) and a variety of other reduced sulfur compounds (RSCs). Also, air pollution with sulfur dioxide (SO_2) is formed via oxidation of reduced sulfur gases, and reduction of sulfates under anoxic conditions.

Several researchers determined specific components of their interest in the landfill gas emissions. Nair et al. (2019) reviewed the VOC emissions and other research groups studied current methods for measuring fugitive CH_4 (Mønster et al., 2019) and other gases as well as the trace-gas emissions (i.e., Duan et al., 2022).

Huang et al. (2022) claimed that odor emissions could be determined indirectly by CH_4 emission quantification thinking that methane emission rate is crucial information for designing field monitoring campaigns, anyway. If the correlations between methane and other gases are dependable this might give an idea on the other gases. Moreover, different emission mitigation strategies are available for controlling different emission gases. This makes it possible to adopt combined strategies for the simultaneous control of mixed gaseous emissions from a landfill, such as operating landfills with aeration or leachate recirculation and designing a landfill bio-cover.

Full coverage of the microbiology during the landfilled waste decompostion is given (Meyer-Dombard et al., 2020). Microbial activity turns anaerobic within days of burial when diverse, indigenous, hydrolytic, fermentative, acidogenic, and acetogenic microorganisms begin to degrade hydrocarbons.

Among the major biodegradable constituents of municipal solid wastes cellulose and hemicellulose constitute 45–60% of the dry weight (Barlaz et al., 2002). These authors summarized the basic microbiology explained this biodegradation mechanism in four phases:

In the first phase of decomposition, the O_2 entrained in the wastes at burial is used up for production of large quantities of CO_2. This is causing elevated temperatures. At the end of this phase the system becomes devoid of oxygen.

In phase two where anaerobic conditions prevail, after the hydrolytic bacteria convert cellulose and hemicellulose into soluble intermediates, acetogenic and methanogenic bacteria work together to convert these intermediates to CH_4 and CO_2. Because of an imbalance in action between the hydrolytic-acetogenic-methanogenic bacterial activity succession, in a first sub-stage short-chain carboxylic acids accumulate thus the pH decreases, and there is little solids decomposition. CH_4 production begins only in the third phase of decomposition. This results in a decrease in the accumulated carboxylic acids with corresponding decreases in the leachate COD and BOD and an increase in pH. During this phase, significant decomposition of cellulose and hemicellulose occurs.

In the fourth phase of decomposition, the accumulation of carboxylic acids is depleted. The rate of CH_4 production is dependent on the rate of cellulose and hemicellulose hydrolysis. In this phase, the BOD/COD ratio is relatively low because any dissolved organic that is degradable is consumed by the indigenous microbial population. Wastes will continue to decompose until little degradable material is left.

3.3.2 Landfill Gas Composition

Methane and carbon dioxide coming from anaerobic degradation of organic matter make up a very high portion of the landfill gas. The remaining part includes nitrogen and oxygen from air as well as ammonia, sulfides and other reduces sulfur gases, hydrogen, etc. The actual amount and percentages of these gases depend on the type of waste present in the landfill, the age of the landfill, oxygen content, the amount of moisture, and temperature. As a general rule gas production increases when the temperature and/or moisture contents of the wastes increase. Though amount of production of these gases generally reaches a peak in five to seven years, a landfill can continue producing gases for more than 50 years (New York State, Department of Health website).

Odorous air samples from a landfill in Beijing were collected seasonally and measured by both gas chromatography and an olfactometer by Chuandong et al. (2017). Evaluations showed that hydrogen sulfide (65.9%), dimethyl sulfide (14.4%) and trimethylamine (8.6%) contributed most to the odor at the landfill. Odour thresholds of 51 common compounds were measured by a dynamic olfactometer. From concentration versus the threshold data it was shown that H_2S (65.9%), dimethylsulfide and trimethylamine were the major contributors of odor in the landfill.

Odor emissions from landfills in different regions in China (Shanghai, Hangzhou and Incheon) had the average NH_3 concentrations ranging from 4.02 to 11.32 mgm^{-3} and average H_2S concentrations ranging from 0.5 to 1.73 mgm^{-3}, ending in serious impacts on the environment (Gao et al., 2021). In this study, the authors monitored municipal solid waste (MSW)-related areas, sludge-related areas and waste areas with the most malodorous compounds.They have also studied the variety and complexity of volatile organic compounds (VOCs) in relation to the landfill odors. In terms of all 120 VOCs, 67 malodorous VOCs with significant variations were categorized into six classes of compounds under all test conditions. Test results were compared. In the MSW landfill, 33 samples were collected, 80 VOCs were identified out of which a total of 100 VOCs using the standard methods. The samples were found to contain eight organic sulfides, five aldehydes-ketones, six organic acids, and one organic amine.

Li et al. (2015) investigated the relative concentrations of ethanol and other odorous gas compounds emitted from the working surface at a landfill in China.

Gas sampling was conducted at the landfill on selected days from March 2012 to March 2014 to allow different period coverage between the two years. A total of 41, 59, 66, 54, 63, 54, 41, and 42 species of odorous compounds were identified at eight different sampling days. These researchers quantified concentrations of the species in the study. They detected a total number of 86 species of odorous compounds throughout the study. These compounds belonged to six different categories such as oxygenated compounds, halogenated compounds, terpenes, sulfur compounds, aromatics, and hydrocarbons. In each category gases added up to 2.450, 0.246, 0.203, 0.319, 0.530, and 0.217 mgm^{-3}, respectively. The relative concentrations of 59 odorous compounds with respect to ethyl alcohol were determined. The dominant contaminants that cause odor pollution around the landfill were found as ethyl sulfide, methyl mercaptan, acetaldehyde, and hydrogen sulfide. Dimethyl disulfide and dimethyl sulfide also contributed to the odor to a certain degree.

In Ph.D. thesis research Dincer (2007), studied the gas emissions and odor releases from Izmir Harmandali landfill. This landfill is in operation since 1990 and still accepts pre-classified solid wastes. It disposes 3000 tons d^{-1} of domestic, medical and industrial wastes as well as wastewater treatment sludges but not the hazardous wastes. The field sampling was conducted on May 6, 2005; September 1, 2005 and May 29, 2006 to represent the beginning of the hot and dry summer season in the area that has a typical Mediterranean climate (with hot, dry summers and wet, warm winters). Six points were fixed in the landfill to represent different stages and ages of operation and one at the headspace of the WWTP leachate entrance. Gas samples were collected using a sampling box from the air adjacent to the area of the landfill cover and headspace using a specially designed hood. Samples in special odor free bags were taken to dynamic olfactometry lab for analysis according to European Standard EN 13725. VOCs predominant in the odorous gases were studied and tested for odor. Adsorbent tubes (1/4 in OD, 7 in length tubes) were used following the ambient air sampling methods recommended by US-EPA (1999) study included testing and analyses of pure compounds in the gases, to find their odor thresholds. Thus, the dimensionless number defined as "odor activity value" (OAV)s showing the odor contribution of a measured component in the mixture were established.

In Dincer's study most odorous RSCs found in the landfill gas were H_2S, CS_2, dimethyl sulfide (DMS), dimethyl disulfide (DMDS). In different days of sampling 53, 48, 46 VOC species were detected, too. They are classified as monoaromatics (e.g. benzene, toluene), halogenated compounds (e.g. chlorobenzene, trichloroethene), aldehydes (e.g. hexanal, propanal), ketones (e.g. acetone), VFAs (volatile fatty acids, such as acetic acid, formic acid), esters (e.g. butyl acetate, butyl formate). In May 2005, the leading group in VOCs was ketones (25% of total VOC) followed by monoaromatics (21%), aldehydes (20%), VFAs (17%), halogenated compounds (14%) and esters (2%). In September 2005, aldehydes had the highest concentration (37%) of total VOC concentrations. Ketones (36%), monoaromatics (13%), halogenated compounds (6%), VFAs (5%), and esters (1%). The VOC concentrations at the landfill were much higher than at the other sampling sites in May 2006 samplings and at sampling point #1 which is the burial site of industrial wastes and the wastewater sludges. At this location concentrations reached up to mg m-3 levels. This may be due to the burial of industrial wastes containing high amounts of hydrocarbon contamination. In May 2006, the most abundant group was monoaromatics (41% of total VOC concentrations) followed by ketones (19%), halogenated compounds (18%), aldehydes (16%) and RSCs (4%). Toluene was the most abundant aromatic compound Izmir landfill. It was noted that this might be related with the heavy truck traffic during sampling. Chlorinated organics in the landfill might be related to the residues of cleaning and disinfection agents widely used in urban and industrial sources. The decrease in concentrations of VOCs with respect to time was thought due to loss by evaporation during the long hot period before the wastes arrived. BTX compounds were also determined in the landfill; propanal, acetone, hexanal, butanal, *p*-xylene, styrene, *o,m*-xylene, tetrachloroethene, butyl acetate, ethylbenzene. They have also been found to be interrelated. Correlations between the olfactometric odor levels and groups of gas components were evaluated using Pearson coefficients which indicated that the most effective group on odor were the reduced sulfur gases (0.82). Total VOCs, esters, aldehydes/ketones, halogenated compounds and monoaromatics were impacting the odor by 0.52, 0.53, 0.54, 0.46, 0.51 and 0.54 respectively.

In an article by Dincer and Muezzinoglu (2006) the data and analysis in this thesis for different odor complaint areas in the city including the Harmandali landfill of the city of Izmir was given.

3.3.3 *Landfill Gas Emission Rates*

Gas emission rates from landfills are quite variable depending on operational details, disposal cell structure, climate, meteorology, characteristics of wastes, season, age of the wastes in the landfill cells. However, percentage of putrefiable organic materials is the main issue when odor emission rates are concerned.

Emission factors (EFs) can be defined as the amount (volume or mass) of the total gas or of the component in the mixture in cubic meters/kilograms per ton of waste dumped. Specific EFs from area sources are defined as the amount per unit surface area per unit time such as (cubic meters/square meter/hour. If the EFs are available for a landfill it becomes easier to estimate the emission rates to be used in dispersion calculations of the gases. This is an important input into the models that can be used to find the maximum ground level concentration of odors and to determine the buffer distance for urban plans so that the built environment will be suitable for healthy life. These models are computer programs to simulate the emission conditions in the landfill and estimate the air quality around. Models are commercially available or are developed by scientists for use. They require representative meteorological, climatic and emission data for realistic results. It is important to consider the temporal and spatial variability of the air quality using multiple runs of the model to understand the variations around the landfill.

To be able to find the gas emission factors of an existing landfill, the amount of waste already disposed and the emission rates of produced gases must be known. This requires a sound measurement program for volume and composition of the gases. Measurement data can be used to calculate emission factors for specific gases, such as methane and carbon dioxide that are the main ingredients of the landfill gases.

For assessment of the EFs of odorous gases such as H_2S, NH_3, VOCs, mercaptans and other RSCs, more careful measurement programs are needed as they are present in the gas mixture in minute quantities only. Their emissions are quite variable with time so a monitoring program is necessary to understand their dispersion behavior.

Although the EFs change from one landfill to another and with respect to time, a general guide of South Coast Air Basin of California for methane and carbon dioxide can be mentioned:

Methane: 0.03 kg CH_4/ton of waste
Carbon Dioxide: 0.23 kg CO_2/ton of waste are given (SCS Engineers, 2017).

3.3.4 Methane Gas Elimination and Gas Collection for Power Production

Methane gas emissions begin after the anaerobic degradation begins in the landfill which may be delayed for an appreciable lag time considering the decomposition mechanism described above. When gas releases begin it must be handled with care. Gas can be collected, sold as biogas after cleaning or burned to obtain heat and power at suitable regeneration units. Methane in mixture with air is an explosive gas and can easily start fire if it is released under uncontrollable conditions. Explosion hazard of methane gas occurs at about 5% (lower explosive limit, LEL) up to about 17% (upper explosive limit, UEL) by volume in air. Methane in air may reach to these levels in covered areas unless precautions are taken. So, it must be vented off to avoid explosions and fires.

Technology for power and heat production from landfill gases is pretty well developed. It is good practice for circular economy and also better for the climate change dilemma. This way carbon in the wastes is ultimately converted into CO_2 instead of CH_4 which is a comparably less problematic as greenhouse gas. Methane is more than 25 times as potent as carbon dioxide at trapping heat in the atmosphere. CO_2 is the combustion end product and is also responsible for radiative forcing only less so. Therefore, it requires abatement, too. Again when landfill gas is burned some of the other odorous gases are oxidized, too. And the sulfur and nitrogen containing odorous components are oxidized into air pollutants (mainly SOx and NOx) which must be cleaned out from the stack gases. Regeneration technology and its effects are left out of discussion in this chapter which must cover odors from landfills.

3.4 Landfill Odors

Landfill operations caring for full sanitation and odor mitigation plans include certain basic steps. First of all, collected fresh inputs of wastes are pre-screened, classified, transported and buried in cells under layers of soil or other adsorptive material in the landfill. To sustain the sanitary conditions and control the odor releases, liquid and gaseous end products of the disposal cells are to be kept under control. Leachates should be diverted from percolating through the soil and reach the groundwater table. So, they must be collected in a pipeline system and sent to treatment. Collected gases should be deodorized and treated by using biofilters, adsorbers, scrubbers or condensers, and when the emission rates prove feasible either are flared or converted to heat and power. This will add to the circular economy and also reduce climatic impacts of greenhouse gas emissions and diminish odor nuisance. Trees and other green areas could be established around the periphery of the site for more sightly appearance. In some landfills chemicals are sprayed onto the landfill to reduce the odor emissions from wastes.

Post-closure landfill sites are not totally free from odorous gas releases, too. For a long period of time after closure of a landfill, the gas emissions continue and therefore should be under control and venting procedures. Such places are not allowed for buildings but only green areas can be established for a foreseeable future.

3.4.1 *Odor Sensing and Olfactometry*

Odor intensities at receptor sites can be detected by air quality monitoring. There is good measurement datasets presenting the odorous component concentrations in the air. A number of them has been mentioned in this article. But continuous odorous gas monitoring except for H_2S or NH_3 is not common practice for several reasons: it requires expensive and high-tech laboratory equipment such as GC/MS. Depending on the composition of the mixture further even more sophisticated laboratory techniques can be necessary. These analyses can be carried out by highly trained lab personnel.

Mixtures of gases do not necessarily reflect the odor levels of molecular concentrations of the individual gases added up, as usually there is synergy. Besides individual gas components even below odor threshold levels might add up to unpredictable odor intensities. So, these techniques are rather for infrequent checkups of simpler field sensory techniques and not for monitoring requirements.

Bax et al. (2020) discussed the issue of odor measurements at receptor sites. They summarized the state-of-the-art of odor measurement techniques, their applicability, advantages and limits. They have given special attention to experienced- as well as non-experienced user requirements of the methods. These methods cover:

Dynamic olfactometry
Chemical analysis (with speciation or non-specific)

Gas chromatography–olfactometry
Tracer analysis
Instrumental odor monitoring by E-Noses
Field odor inspection
Olfactometry
Citizen science.

Among these methods, dynamic olfactometry and field odor inspection are the only methods specifically developed for odor measurements and used by regulatory systems in many countries. Chemical characterization, like the gas chromatography–olfactometry is based on consolidated analytical techniques, commonly adopted to provide information about the odor properties of gaseous mixtures.

In European countries olfactometry for measuring odor perception of human beings is extensively in use. It is an instrumental technique which uses a dilution device called an olfactometer. In olfactometry human nose must have standard responses, therefore must be standardized. For sniffing panelists only the candidates fulfilling the requirements and meeting the predetermined repeatability and accuracy criteria can work. Panel members are trained to comply with the odor standards, to learn their responsibilities and they are pre-tested by using unknown concentrations of odorous reference gases. Panelists who do not find the given reference gas dilution and mix the zero gas with odorous mixture cannot continue testing. The testing continues until each allowed panelist positively detects (yes or no) the odor in the diluted mixture and at this stage the panelist has reached the detection threshold for that odor. Panelists who are outliers in the group during sample measurement results cannot continue either. Human sensory system may change from one day to another, so all these preliminary standardization tests are repeated prior to actual sample analysis in an odor measurement run. Each panelist can work only a few number of times and between measurements there must be a resting time.

These tests are carried out in odor-free, clean room by trained panelists sniffing the odorous gas + air mixtures at certain dilutions. It is based on a preselected panel of four persons meeting certain conditions working with the instrument which dilutes the odor samples with odor-free air at precise ratios. Samples are diluted several times differing from each other by a factor of two and presented to the panelists three times. Odor-free (neutral) air is supplied by a compressor fitted with carbon filters and an air dryer. The olfactometer is a computer controlled semi-automatic instrument with four panel member places. Perception responses of the panelists are recorded and evaluated. Panel members judge for the presence of odor in the diluted sample. The procedures are run simultaneously for the four panelists each having either neutral air or the diluted sample without knowing what they are sniffing. The concentration sequence is started from low (threshold) and is increased. The result is the dilution factor of the sample upon which all panel members agreed on after three sniffing rounds.

The duration of the measurement sequence is held at a possible minimum. The odor concentration in (OU m^{-3}) is the multiple of the threshold concentration. Olfactometric samples are taken into Tedlar® (polyvinyl fluoride) or Nalophan®

(polyethylene terephthalate) bags by means of an odor free vacuum pump with the 'lung' technique. Odorless containers, sampling lines and lab rooms are used. The bags filled with samples from the field are brought to the testing room for panel study. Samples must be analyzed within a maximum of 24 h.

For area sources, it is necessary to use auxiliary devices to collect odor samples through sampling boxes or chambers one side of which is open. Boxes are placed open side down onto the odor emitting solid or liquid surface. These investigations are sometimes conducted using a hood or a wind tunnel, depending on the field conditions.

Although dynamic olfactometry is the standard method for determination of odor concentration in many countries, it has several disadvantages, too. First of all, it does not allow continuous measurements so that detecting the fluctuations with respect to time is not easy. This is a disadvantage in source emission and field odor measurements. Moreover, dynamic olfactometry considers mixed odors and do not identify single odorous compounds. Olfactometric odor samples are difficult to store, because of their instability, so samples require immediate analysis after sampling. This means the panel and the lab should be ready to work within a couple of hours depending on the technique. And the last but not the least disadvantage of olfactometry is that it is time-consuming and expensive.

Other methods based on human odor detection measurements like the odor patrol monitoring and simulated instrumental techniques like the artificial olfactory instruments (E-Noses) mimicking the human smelling function can be used, too. E-Noses contain an array of electrochemical sensors specific to odorant species and are working with a pattern recognition system for identification of the odorous molecules and measuring their concentration. After a training phase, electronic noses are able to preview the class of an unknown samples and consequently to associate environmental odors to the specific source (Sankaran et al., 2012).

Citizen participation with complaint forms or similar methods can be a good and cheap method of odor evaluations around landfills. Odor measurement by using the systematics of citizen science, seems to be a promising technique. It is based on the active involvement of citizens, have been proposed in recent years to address socio-environmental conflicts within the impacted communities. They rely on the active involvement of citizens in the mapping and management of odor pollution problems around facilities. These projects allow co-creation of local solutions together with the local authorities, as well as odor experts. This method proposes a holistic approach, based on the cooperation of key stakeholders (citizens, regional and local authorities, industries, and odor experts, scientists, etc.) for the mapping and co-design of solutions, to investigating, building, and suggesting an appropriate regulatory framework.

Resident sensing data on odor occurrences versus the air quality measurement data around an industrial site were correlated by Lee et al. (2013). They determined the concentrations of the odorants by instrumental analysis and used resident monitoring records of direct olfactory observations. A strong correlation was found between the odor intensities by instrumental analysis and odor intensity values obtained from residents monitoring records. Latter was found higher than the chemical concentration values obtained from instrumental analysis. They showed results of one of the earliest citizen science methods mentioned above. They concluded that the resident monitoring of odors around odor emitting sources was effective to understand the characteristics of odorants emissions from the sources.

Full chemical characterization of odor samples provides the chemical composition in a mixture and their individual concentrations. Characteristic compounds that are responsible from odors in a landfill are ammonia, hydrogen sulfide and VOCs (volatile organic compounds) like amines, mercaptans, sulfur compounds, saturated and unsaturated fat acids, aldehydes, ketones, hydrocarbons, limonene, chlorinated compounds, alcohols, etc (many researchers e.g. Capelli et al., 2013; Dincer, 2007).

For VOC analysis the air samples are collected into canisters or polymer bags or if micro level pollutants are to be analyzed into adsorbent tubes. In the latter case, the critical point is the choice of adsorbents that depends on the chemical feature of compounds. Active samples by adsorbent tubes are taken from odorous air by sucking the odorous gases through the tubes at a certain flow rate. Passive samples can also be used for monitoring. After opening the ends these tubes are left to the environment and samples are taken by directly exposing the suitable adsorbents to sampled air for a preselected period of time. Time of exposure depends on the expected concentrations and the quality assurance demands. Passive sampling is a well-developed method of sampling at large areas with long enough exposure times.

Malovanyy et al. (2021) studied air pollution and odors arising from Zhytomyr Ukranian landfill by using a human sensor system and by using lichen bioindicators. This study showed an exponential dependence between the distance from the source of gas formation and the intensity of odor. The level of atmospheric air pollution was also determined based on the assessment of the coverage of the tree trunks with lichens.

3.4.2 Predicting Odors with Atmospheric Dispersion Models

Atmospheric dispersion models are computerized programs based on algorithms that simulate pollutant dispersion around emission sources and help estimate the concentrations reaching receptors at ground (or elevated) levels. The use of dispersion models is necessary in studies for authorization of facilities, and impact evaluations at the receptor sites. An appropriate high-quality data sets of emission strengths and meteorology are crucial to get dependable results. Emissions are given on the basis of certain averaging periods determined by the duration of emission sampling(s) or calculations. The same is true for meteorological data that is usually provided by meteorology departments and are given as averaged by hours, days, months, seasons, years, etc.

Mathematical form of the dispersion model can be either Gaussian, Lagrangian or Eulerian. There are many ready-to-use software packages with successful applications in air pollution studies. Some are preferred methods by regulatory authorities.

Some are even compulsory for impact management (such as EIA) studies. Beyond those mentioned above puff models have been developed which assume that pollutants are emitted as large number of puffs in succession. This assumption suits well with the behavior of odor emissions that are intermittent and differing in frequency in a given period of time. CALPUFF models can be preferred on wide domains or areas with complex terrain.

There is a basic difficulty in odor dispersion models compared to other air pollutants. As the meteorological conditions are highly variable over short periods of time, the use of for example 1-h average data is not short enough to reflect the short-term peak odor sensations that are experienced by the sniffs of human beings (Nicell, 2009). Thus, the emission dispersion estimates of the odor intensities may not match with the levels measured by human sensory system as the odor stimulus may persist for durations of a few seconds only (Elbir et al., 2007; Dincer et al., 2004; Schauberger et al., 2002).

Thus, the average time used by dispersion models for the estimation of odor concentration at receptors is a serious problem. Odor produces quicker response in the receptors' nose. Besides not only the odor sensing is very quick and meteorology is highly fluctuating but also odor emissions are mostly discontinuous and frequently changing in intensity in time. Besides, corresponding odor fluctuations at receptors are even more annoying to human beings according to the Weber-Fechner psychophysics law. This law suggests that a greater annoyance is caused by short periods of odor than by longer lasting odor emissions.

Dispersion models are generally used in conjunction with odor monitoring techniques to evaluate the landfill odor impacts on the surrounding areas. Li (2006) studied the CFD (Computational Fluid Dynamics) model to simulate odor dispersion from a 3000-sow farrowing farm and compared the results with outputs obtained with CALPUFF model in shorter distances than 300 m. They found that CFD model gave higher predictions than the CALPUFF model.

Air pollution dispersion models of short term are normally based on a 1-h averaging time data, and therefore predicted concentrations represent air pollution data over a minimum of 1-h period. On the other hand, odor complaints datasets are represent a series of much shorter detectable exposures (Brattoli et al., 2011a, 2011b). Since meteorological conditions are highly variable in an hour, and because the odor is sensed over a couple of sniffs in the order of seconds only, using 1-h average does not represent the short-term peak odor sensations that are experienced by the population. However, 1-h averaging time must normally be used because the frequently available meteorological data are recorded as hourly averages.

In order to overcome this draw back it was thought to use shorter averaging times in order to involve wind fluctuations and for obtaining more sensitive predictions from odor dispersion models. For example, Elbir et al. (2007) carried out olfactometric samplings with 10 s intervals and measured the odor concentrations around a rendering plant. For the same sampling period they ran a CALPUFF model for predicting the odor levels at various locations around the plant. Simultaneous odor measurements and predictions were found to be comparable. Although the olfactometric odor measurements are not normally applied in ambient air studies, it was possible in this study as the concentrations were well above the minimum detectable ambient odor levels of 10 odor units OU (m-3) that can be safely measured by this technique.

This was made possible thanks to the meteorological data which was continuously generated on-site which could be worked out into 10 s interval means. Likewise new generation air dispersion models can run at averaging times of less than 1 h, for example half-hourly means (Schauberger et al., 2002) or 10 min averages (Nicell, 2009). But such very short averaging times are not normally requested by the regulators due to the lack of meteorological data precision.

3.4.3 Landfill Operation for Better Odor Control

Preliminary classification of the wastes is a good operational remedy for controlling the odor and other gas releases in a landfill. This is a beneficial step on top of the economy it involves. Gao et al. (2023) indicated the air pollution improvements with classified wastes received by landfills. In their report they evaluated the changes in air pollution levels before and after waste classification based on a three-year field observation using high-resolution online instruments in a municipal solid waste comprehensive disposal site. They found that thanks to the classification of wastes the concentrations of hydrogen sulfide (H2S) and volatile organic compounds (VOCs) in ambient air were decreased by 48% and 43%, respectively. Also, the spike frequency was controlled by approximately 0.01. The average H2S and VOC concentrations were generally low during the daytime, but relatively high during the nighttime because of meteorology and photochemistry. They also calculated that the overall ozone forming potential of the site was decreased by 56%. However, emissions from transportation and combustion sources might have been increased due to additional operations.

Herath et al. (2023) studied Karadiyana municipal solid waste dumpsite in Colombo, Sri Lanka, which has been in operation for over 30 years and evaluated surface gases for volatile organic compounds (VOCs), ammonia (NH_3), and hydrogen sulfide (H_2S) emissions. Based on the surface conditions and waste characteristics, the dump surface was divided into eight parcels, and multiple samplings were done using static flux chamber methods. The study observed that the average flux rates of VOCs, H_2S, and NH_3 were 137.2 ± 243.8, 6.63 ± 15.9, and 14.2 ± 16.2 mg m^{-2} h^{-1} at the dump.

The maximum average VOCs and H_2S flux rates of 828.6, 24.3 mg m^{-2} h^{-1} were recorded. A considerable fraction (62.5, 35.6%) of the gases were found to come from new organic waste in the daily total gas emissions (61.0, 3.1 kg d^{-1}). The moisture content of the organic waste layers was positively correlated with the trace gas flux rates and the NH^3 flux rates depending on the pH of the surface. Results showed that the age of the waste determines the trace gas emission rate, and leachate surfaces which are active odor sources. They concluded that gas collection and treatment systems are necessary for the initial waste dumping area and leachate surfaces in order to mitigate odor pollution in the landfill. Establishing a proper drainage system at the dump site would reduce trace gas emissions.

Under the surface physicochemical heterogeneity

2023 Waste Management

Arman et al. (2022) studied the characteristics, sources, odor impacts, and health risks of the odorous compounds in the solid waste management facility of Tehran and neighboring areas. In this study, 25 odorous compounds were measured using both active and passive sampling methods at 10 sampling points. Points were chosen as follows: five points within the facility, three points near the facility, and two at nearby livestock sites. The contributions of the odorous groups in the total average air concentration were 33% for halogenated compounds, 21% for ammonia, 14% for reduced sulfur compounds, 13% for aromatics, 8% for oxygenated compounds, 8% for hydrocarbons, and 3% for terpenes. The average sum of odor activity values of the odorous compounds within the facility, near the facility, and at the livestock sites were determined as 3363.48, 720.65, and 308.43 OU (m^{-3}), respectively. Out of the sum of odor intensities, 80% compounds were not perceptible and among the rest only hydrogen sulfide and methyl mercaptan were described as extremely strong. The sum of the average hazard quotients of hydrogen sulfide, trichloroethylene, tetrachloroethylene, and ammonia were 84.35 for occupational exposure in the facility and 70.49 for public which were both much higher than the acceptable values. The total average incremental lifetime cancer risks of the odorous compounds were determined as 2.02×10^{-3} for occupational and 2.36×10^{-3} for public exposure. They concluded that the landfill gas constituents caused significant carcinogenic and non-carcinogenic health risks along with the challenges in welfare of the public and the economic losses. The results indicated that mitigation measures for control of the odorous compounds in the solid waste management facility of Tehran should be reinforced. In addition to the solid waste management facility, the livestock sites emitted strong odors to the neighboring areas.

3.5 Technical Guidelines of Solid Waste Management and Landfill Operations

A good environmental approach for managing municipal solid waste is to reduce the amounts of waste at the place of origin and wherever possible reuse, recycle or reclaim by composting at source to maximum extend. Disposal in landfills must only be for the residual parts.

Every day, the fresh solid wastes are spreaded and compacted in the active area of the landfill. At the end of the day, wastes are covered with a thick layer of soil. This cover helps to control the odor, avoid fine litter around, prevent fires, and ward off scavenging animals.

In large landfills, gases must be vented and captured before they are released into the atmosphere. Collected gas may be routed to a facility that cleans it and pumps the biogas where it is sold or converted to renewable energy.

The migration of landfill gas emissions depends on a variety of atmospheric conditions. These are detailed in Sect. 4.4.3 of this chapter. Barometric pressure drops are often associated with rainfall which is why more pronounced odors are noticeable before and after rainfall events. During colder winter months, temperature inversions may occur where warm air is trapped above cold air closer to the ground. This slows or stops the normal rise of gas and create a stronger smell at ground levels. Horizontal air movements help spread any escaped gas and lessen their concentration, but days with little wind mean that odors may be more noticeable. On a calm day with low barometric pressure, the landfill gas and their associated odors will be the hardest to control.

Landfills may have specialized designs according to disposed waste contents. Most of the landfills are municipal waste collection facilities that are located near urban agglomerations. They are public service facilities for disposal of household wastes and mostly accept nonhazardous wastes. Medical wastes require special attendance and there are special design rules for their disposal. They may take place in municipal landfills at special sections and have special designs. Likewise, urban landfills may or may not accept industrial wastes, demolition or ruined building rubbles, ashes, depending upon its size.

The odorous gases are emitted through the landfill surface as fugitive gases. This means that landfills are area sources and covering surface of the wastes helps reduce odor emissions. This increases the amount of gas collected for further operations such as power conversion.

Fine fractions of the wastes can cause more odor than the normal size wastes. It is because of the large surface area of the fines. So, fines should not be left on the surface but immediately covered by mixed wastes or covers.

Odors are different depending on different ages of disposed wastes. For example, fresh waste produces distinct odors so daily wastes must be covered each night because they generate more nasty odors. Good housekeeping recommends that wastes must be covered in an intermediate term (better immediate) followed by final covering. At the end of the operation all wastes should be under cover.

Another focus of odor emissions is the treatment sludge wastes in the landfill if they are accepted. These wastes should be immediately covered, too. Some of the landfills receiving sludges keep them in wells or trenches prepared for storage until properly handled.

It might be usual practice to install special industrial waste landfills designed and operated next to the industrial agglomerations. In some cases, large industrial facilities build them in their own yards. For example, coal-based power plants are major producers of coal ash and coal combustion residues of very high quantities and they handle these wastes in special dump areas resembling dams. Sometimes, as in the case of mixed industrial districts a common industrial landfill is operated. And the last but not the least significant are hazardous waste landfills which must have specialized designs, construction and operation. Of course, each landfill will have different properties, and the characteristics and rates of odor emissions due to the types of debris and wastes collected are different.

Landfill gas is produced through bacterial decomposition, volatilization and chemical reactions. Most landfill gas is produced by bacterial decomposition that occurs when organic waste solids, food residues (i.e. meat, vegetables), garden waste (i.e. leaf and yard waste), wood and paper products, are broken down by bacteria naturally present in the waste and in soils. Volatilization generates landfill gas when certain wastes change from a liquid or solid into a vapor. Chemical reactions occur when different waste materials are mixed together during disposal operations. Additionally, moisture plays an important role in the speed of decomposition. Generally, the more moisture, the more landfill gas is generated, both during the aerobic and anaerobic conditions.

3.5.1 Treatment of Odorous Landfill Gases

Gas treatment technologies were considered for controlling the unpleasant smells coming from landfills that contain several odorous ingredients Treatment technology research is underway for the synthetic gas/air mixtures simulated to the gases emitted from landfills. For example, for mercaptans non-thermal catalysis; for NH_3 and H_2S simultaneous treatment and for H_2S biofiltration (Rabbani et al., 2016; Vikrant et al., 2018); for benzene photocatalytic oxidation (Dursun & Ayturan, 2022) worked on treatment options. This topic has been considered as discussed by Wei et al. (2022).

There are several plasma oxidation, catalytic oxidation and similar high technology studies to oxidize the odorous gases being released from landfills. (like Huang & Ye, 2009; Maxime et al., 2014). Usually, synthetic air mixtures of NH_3 and H_2S in laboratory models were tested with these techniques. Synergy between the two gases is mentioned so it is better to have these two gases together than alone to obtain higher treatment efficiencies. As end products of the oxidation O_3, SOx and NOx are produced. These gases are typical air pollutants which need to be abated. Efficiencies of treatment of odorous gases showed that it may be a promising technology for removal of these two odorous compounds in the landfill gas with these provisions. Catalyst selection, regeneration of the spent catalyst, energy density and frequencies used are important parameters.

3.5.2 Evaluation of the Odor Reduction Potential of Alternative Cover Materials at Landfills

Ding et al. (2019) tested a novel landfill cover material consisting of biochar-added sludge compost under various operating parameters for odor control. They used 9 laboratory-scale landfill reactors to simulate in-situ odor control by this cover material. They noted that bacterial genera in the cover were responsible for reduction of odorous emissions and explained the relationship between the structure of microbial communities and odor removal performances.

For simultaneous control of landfill odors and methane emissions, Lee et al. (2018) made a pilot-scale test by spreading a bio-cover consisting of a mixture of soil:perlite:earthworm cast:compost (6:2:1:1, v/v) in a landfill in South Korea. They followed its performance and bacterial community dynamics for 240 days. The removal efficiencies of odor and methane were evaluated to compare the odor dilution ratios and methane concentrations from the bio-cover surface. The measured odor dilution ratios ranged from 300 to 3000 at the bio-cover surface, but they were 6694–20801 at the landfill soil cover surface used as control. They found odor removal efficiency was above control by 85% in all seasons. Methane removal efficiency (35–43% in winter) was good, although it was influenced by the ambient temperature. Methylobacter was the dominant methanotroph in all seasons. They stated that the seasonal characteristics of the on-site bio-cover performance and its bacterial community are useful for designing a full-scale bio-cover for simultaneous mitigation of odors and methane at landfills.

Evaluation of the odor reduction potential of alternative cover materials at a commercial landfill was reviewed by Solan et al. (2009). Instead of the virgin soils and traditional landfill covers that are costly and increasingly scarce, and also ending in diminished landfill storage, they proposed using certain suitable waste materials as alternative. They studied the suitability of construction and demolition (C&D) fines, commercial and industrial fines (C&I) and woodchip (WC) as potential landfill cover materials for use in odor control. Before and after application observations at an active landfill followed by background odor analyses, negligible residual odors were notable from all of these three cover materials. For daily wastes the application woodchip was found to be more effective as cover.

3.6 1Application of Deodorants in Odor Control

In order to reduce the gas release from surfaces in the landfill, several chemicals were sprayed or spread over the area. A number of studies exist in literature in this direction, such as Dai et al. (2018) which showed that the deodorant they developed could remove 47% of ammonia, 40% of isobutylene and 20% of other odorous gases within 20 min in a model reactor.

Field work was carried out at the landfill at Harmandali site in Izmir using a calcium hydroxide-based material they named Alkeo-C by a private company (Ege Biyoteknoloji, Izmir, Turkey) with a sufficient odor reduction efficiency. But the economy and some practical field work drawbacks were mentioned (personal communication).

Berihun (2017) reported applicability of BioMist, a Swiss Pharm chemical in removing the odor from leachates at specified environmental conditions in the landfill in Addis Ababa, Ethiopia. Field observations, measurements, experimental details, sampling and analysis results were given. The result showed that BioMist is effective in removing bad odor and improving water quality even at low concentrations.

3.7 Landfill Site Selection

3.7.1 Site Selection Criteria

Landfills are not allowed on environmentally sensitive areas. Sites must be easily reachable with trucks or other means of mass transport for bringing in the wastes. Immediate area surrounding the site must not be densely populated and environmental conditions such as the meteorology, topography, groundwater levels, ground formations and soil types covering it must be well known. Slopes must not easily give way to slides, leachates must not form easily flowing streams of wastewater and flood the downstream areas. Impermeable ground material is preferred and full geochemistry must be known.

A site with natural hydrogeological bottom material for the landfill is an extremely helpful and economic asset. It will minimize the risk of seeping leachates through soil. Sites near sensitive water bodies such as wetlands, or water sources used for drinking, irrigation, industrial uses or watering stocks will be more costly to comply with the environmental requirements.

In many countries based on scientific recommendations and local experiences regulatory guidelines are prepared for landfill site selection. For example, a developed one is in Australia South Wales (Australia South Waste Environment Protection Authority, 2016) that brings in siting restrictions to landfills declaring that the judicious location of a landfill is the single most effective environmental management tool.

In order to protect the soils and water bodies from contamination the Australian authorities foresee proximity distances in which landfills are not allowed:

1. within 250 m (or other calculated protection distance) of an area of significant environmental or conservation significance, in areas such as national parks, historic and cultural heritage areas, conservation areas, wilderness areas, wetlands, rainforests, critical habitats, scenic areas, scientifically valuable areas;
2. within specially reserved drinking water catchments areas;
3. within 250 m of a residential zone or dwellings, schools or hospitals;
4. within permanent or intermittent water bodies or in an area overlying an aquifer that contains drinking water quality;
5. within a karstic region or areas containing substrata that might slip or subside;
6. within a floodway to protect the downstream areas during a major flood (100-year-frequency).

Similarly, some local or national guidelines foresee buffer distances to protect areas nearby. They are devised and monitoring stations are to be installed. Monitoring requirements for the on-site environmental monitoring systems must be able to check for groundwater contamination and for landfill gas emissions, as well as provide additional safeguards if necessary. These are stringent requirements for site selection and seldom ends in successful results in fixing a site between alternatives.

There are scientific methods of site selection which are mentioned in this chapter such as described below. Realizing that the municipal solid waste disposal site selection is one of the most important steps in urban waste management that gives service providers the opportunity to control future odor problems. Bilgilioglu et al. (2022) studied geographic information systems (GIS) and multi-criteria decision-making (MCDM) tools in the planning phase of landfill site selection in Mersin, Turkey. They chose eleven evaluation criteria such as the elevation, slope, permeability, groundwater level, distance from lineaments, from rivers and water surfaces, distance from roads, distance from settlements, distance from protected areas, and land cover properties. Considering the evaluation and exclusion criteria, 19.12% of the proposed areas were deemed suitable, and 80.88% were unsuitable A municipal solid waste site suitability map was created in this study. The outcomes indicate that 80,377 ha and 83,022 ha of the study area were classified as high and very high suitability, respectively. Between the criteria social factors were not included.

Likewise, Tulun et al. (2021) prepared a GIS-based landfill site suitability map for the Aksaray province of Turkey. In this study an alternative method with a modifed Analytical Hierarchy Process was applied with Geographic Information System. Taking into consideration the economic criteria, such as slope and distance to roads, as well as environmental factors, including land use/land cover, geology/hydrogeology, distance to residential areas, distance to surface water and distance to protected areas were tested. According to the results of the study, environmental factors were determined (W: 0.869769) to be more important than the economic factors (W: 0.130230). Eight percent of the studied alternative areas were not suitable, 20% were of low suitability, 14% were moderately suitable, 51% were suitable and 7% were highly suitable for landfill construction. Among the highly suitable areas, two alternative locations were identifed in the southern part of the province. The frst alternative (Alternative-I) has an area of 184 km^2 and is about 35 km away from the urban center, while the second alternative (Alternative-II) has an area of 3133 km^2, and is 65 km away from the urban center. Both alternative areas are suitable within 30 years of waste storage period. The fndings obtained from the study might serve as a guide in the selection of landfill.

Ahire et al. (2022) conducted a site selection study for the upcoming waste management project of Nashik and Environs, Maharashtra, India, by using remote sensing and geographical information system (GIS) techniques with multi-criteria decision analysis. It considered environmental, socio-economical and technical criteria to maintain sustainability and improve the standard of living of people at nearby vicinity. Fifteen parameters were considered while selecting a potential landfill site. To assist city planners in identifying potential landfill sites several potential landfill sites were proposed and their suitability was discussed.

3.8 1Buffer Zones Around the Landfills

Landfill buffer zone might be described as a separation distance between the outer periphery of the waste site and the areas that can be safely accommodated. Buffer zone identifies an area around the plant boundaries after which safe odor levels is acquired. A statistically prescribed limit is to be regulated with its probability of occurrence. Usually odor intensity in odor units (OU) is set by regulators.

Wind speed and direction, temperature, pressure, humidity, frequency of precipitation, temperatures and temperature lapse rates with altitude, etc. are the meteorological parameters affecting the selection of boundaries of the area at which odors disperse down to a safe level. Topography is also important in dispersion phenomena. These parameters are used in dispersion models to find unacceptably high concentrations of pollution and odor at ground level.

Some scientists prefer using chemical monitoring or full characterization of the odor for determination of the safe buffer zone dimensions that must be set. But for practical applications it is better to use reports of human smelling ability measured by olfactometry, public participation, or odor patrol reports to check the presence of odor in already established facilities.

Maximum ground level concentration of the odor, C_{max}, to be found by computations using dispersion models must take place within the buffer zone boundaries. The buffer zone is not necessarily circular in shape but can have irregular shapes around the landfill, possibly extending towards downstream direction of the predominant wind, the location of receptors and the topography. The buffer zone can be defined using dispersion models based on realistic meteorological scenarios, too. If in one wind direction regulated odor limit is fulfilled inside the plant boundaries, the buffer zone boundary overlaps with the plant perimeter.

Impact zone analyses were conducted based on the odor perception thresholds. Tansel and Inanloo (2019) developed an assessment and delineation tool to identify the possible odor impact zone around landfills. They based their study on the land use and atmospheric conditions to find the number of people who would be impacted. They studied the odor emission quantities and found the dispersion estimates in three different landfills. They used Landfill Gas Emissions Model (LandGEM) software for estimations. To delineate the odor impact zones, air dispersion characteristics of the odorous gases were analyzed using the dispersion modeling software ALOHA (Areal Locations of Hazardous Atmospheres) and mapped the results using ArcGIS. Results of using their developed odor impact zone tool indicated that the atmospheric stability can increase the odor impact radius around the three landfills by 340–1100% depending on the land use characteristics of the surrounding areas.

In a Ph.D. thesis work Vidovic (2017, Florida Atlantic University) studied the statistical data file for the complaints and combined the complaint data with meteorology in the area. They needed to warn that if this data will be used it must be made sure that the reception, storage and handling of the complaints must be made by respectable third persons or organizations.

Relationships on the complaint statistics were studied by Muezzinoglu and Dincer (2005) who correlated the measured air pollutant data with complaints on the sampling date. Pearson coefficients between the odorous gas measurements and public complaint data at the measurement day around industries in Louisville, Kentucky, USA showed only a few of these chemicals have weak dependencies.

It might be thought that the difference in time averages of sniffs upon which the complaints are based by definition and intermittent measurements or the dispersion predictions causes the discrepancy between the odor datasets of air quality measurement/monitoring and dispersion estimates. This issue has already been discussed at the odor dispersion modelling section.

3.9 Landfill Impact Management

3.9.1 Health Impact Management of Landfills

Odor of the landfill gas is caused primarily by hydrogen sulfide and ammonia, which are produced during breakdown of waste material. For example, if construction and demolition debris contain large quantities of wallboard (also called gypsum board), large amounts of sulfur containing gases can be formed.

Organic wastes containing sulfur and nitrogen containing compounds decompose to form these gases. Hydrogen sulfide has the foul smell of rotten eggs, while ammonia has a pungent odor. Humans can detect hydrogen sulfide and ammonia odors at very low levels in air because their odor thresholds are very low. The odors generally occur below limits that would cause health effects.

New York State Department of Health website mentions the possible health impacts due to landfill gases. According to this website, carcinogenic and non-carcinogenic health effects can be separately considered. For health impacts odorous gases produced in landfills, such as ammonia, sulfides, methane, and carbon dioxide are mentioned to be of highest concern. Whilst ammonia and hydrogen sulfide are responsible for most of the odors at landfills, flammability and explosiveness of methane is of concern, too.

In the website mentioned above there is message to the public against hazards that might materialize at the buildings nearby as the landfill gas might move through soil and collect in nearby buildings.

Methane and carbon dioxide are the major constituents of landfill gas. Also, they are greenhouse gases that drive climate change to have impacts indirectly on the health of the people. Although these gases are not toxic and are colorless/odorless, they can displace oxygen in enclosed spaces ending in lack of oxygen to breath. Health effects caused by a reduced oxygen level include a faster heartbeat and having to take deeper breaths, similar to the effects felt after vigorous exercise. A greatly reduced oxygen level can cause reduced coordination, fatigue, nausea, vomiting, and unconsciousness.

Durmusoglu et al. (2010) discussed the health risk assessment by inhalation of BTEX compounds at a studied landfill site. They showed that the mean concentrations of benzene, toluene, ethylbenzene, and xylenes were 140.3, 1271.7, 239.9, and 341.3 μgm^{-3}, respectively. The concentrations for benzene were not in the carcinogenic levels, and the others both individually and altogether were in the safe side in the landfill area. The risks on the population in the neighborhood area of the landfill was on the safe side after dilution of BTEX group emissions in the air.

At critical concentrations of methane in air occurring in enclosed spaces with poor ventilation explosive limits may be exceeded to create explosions and wildfires. This issue was discussed above. Methane and carbon dioxide can also seep underground and accumulate in nearby buildings and displace oxygen to create indoor pollution problems especially in the basements and other semi-enclosed places.

Landslides, floods, pest (rats, flies, etc.) invasions, unsightly appearances, open leachate flows and sometimes floods, leakage into the ground and surface waters are among the other health dangers landfills may get involved.

Studies have been conducted in communities near landfills and waste lagoons to evaluate health effects associated with inhalation of landfill gases. Exposures of up to about two weeks to elevated levels of ammonia and hydrogen sulfide under the inhalation toxicity levels in air were found to cause serious effects. Among the symptoms coughing, eye irritation, nose and throat discomfort and breathing difficulty, headache, nausea can be counted. These effects usually go away after the exposure stops. When the studies went on for several months and reported health complaints with elevated levels of hydrogen sulfide and landfill odors, the reported health complaints included eye, throat and lung irritation, nausea, headache, nasal blockage, sleeping difficulties, weight loss, chest pain, and aggravation of asthma may increase (New York State Department of Health website).

When landfills reach the maximum amount of waste they can hold, cover materials are laid over the landfill surface to safely terminate the operation. Gas collection wells are then installed throughout the area. These wells are made of perforated vertical pipes reaching the surface. Gases entering these wells are vented into the open air. In better practice they are flared or incinerated, passed through a filter, or used to obtain heat and power.

Quantification of landfill gas releases and estimation of methane emissions are essential for planning and design of gas collection systems. There are well known models for estimating these quantities. For example, Sarptas and Erdin (2014) studied Harmandali landfill in Izmir by using landfill gas generation models of Tabasaran/Rettenberger, TNO, LandGEM and Multiphase in order to estimate the total and remaining energy potential. Based on the waste composition, deposited gas generation potential was calculated as 109 m^3 CH_4 per ton of municipal solid waste by these authors.

Kanhai et al. (2021) studied the case of landfill at Accra, Ghana from point of view of air pollution, waste management and health perspectives. They proposed a method by way of collecting data, modeling the emissions arising from waste management facility in order to obtain concentration estimates and evaluated the risks of the air pollutants on community health. The results of this health impact assessment study indicated that, based on the emissions of $PM_{2.5}$ from the waste sector in Accra, a change from the business-as-usual technology to more sustainable options would reduce air pollutants emissions and avert 120 premature deaths in 2030.

3.9.2 Environmental Impact Assessment (EIA) Requirements

Impact assessment report is requested to minimize any environmental harm in site selection as well as the establishment and operation of landfills. Applications for licence to construct and operate landfill works (including new cells at existing sites) are included in the list of operations that need environmental impact assessment EIA. Each new landfill proposal must be supported by a comprehensive environmental impact assessment (EIA) report to show that the project would meet the standards depicted in pollution control regulations. This report must evaluate the details of the engineering project of the landfill construction and operation. It must cover the plans, specifications as well as detailed engineering blue prints and land-use maps. Landfill cell construction and operation details, locations of receptors such as nearby housings and interphases with other urban facilities such as drinking water, stormwater and wastewater networks, roads, offices etc. All of the relevant infrastructure must be shown in the blueprints of the proposed project.

3.10 Results and Conclusions

There are a few outputs that can be learned from the present research and engineering research and engineering reports about the landfill odors aiming at odor sensing, effects and mitigation:

Firstly, one should underline the necessity of waste classification, having in mind the possibility of recycling and reuse. It must begin as early as accumulation at the waste bins of the sources to guarantee that only unusable fractions will have to be landfilled at the end. In other words, classification should begin at the households before delivering the wastes to the garbage collectors. They must be delivered in separate parcels stored inside homes until collected. This classification must be guided by authorities perhaps by public incentives to the households. Collected wastes must first be sent to classification facilities that must segregate further the reusable fractions from the non-reusables before they go their way to the final disposal units. Finally, they must be disposed in final facilities as segregated wastes, as their future behavior and threat to the environment varies with their quality.

Circular economy is a must in a world rapidly depleting its material and energy resources being exploited by the steadily overcrowding and over-producing system. So, a reuse and recovery plan should most definitely include wastes. In order to establish the circular economy, reuse and recycling of resources and closing the material cycles as well as obtaining renewable energy and fuels is crucial. As was covered in his dissertation book by Einhäupl (2021) even the closed landfills must be drilled (mined) and the contents must be reevaluated.

Secondly, research showed that resident report data on odor occurrences such as the complaint dataset in their present quality versus the air quality measurement data do not conform nicely as expected. However, if participatory studies are designed more scientifically and data is collected by the contribution of the knowledgeable residents, better correlation could be found between the public data and the odor intensities. Researchers studying this topic found that citizen science projects have shown better conformity between the datasets of sensory evaluations from participatory resident projects and the odor emissions. For example, citizen participatory projects have been mentioned by Bax et al. (2020) as a good way of assessment of odors around landfills. They mentioned citizen science would work for this purpose.

Elias Hernandez Valera (2020) in his MSc thesis research report has analysed the role of citizen participation towards an inclusive and circular solid waste management system. Recommended participatory process on the solid waste management system at Pachuca, Mexico, was studied by this author. The project 'Multi-Level Perspective and the Citizen Participation' was based on the connection between citizen participation and the introduction to the circular economy and indicated opportunities for promotion of better management of solid waste practices and circularity. By connecting the current situation of solid waste in Pachuca with the participatory process of citizens, an opportunity to promote sustainability is mentioned. Cooperation between citizens, government, public and private institutions related to solid waste management system is exemplified.

Regulatory organizations have already announced voluntary projects with predetermined rules such as the US EPA that introduced a vision of "participatory science" (PS). In EPA definitions a PS project engages the public in advancing scientific knowledge. Other terms include citizen science, community science, or public participation in scientific research. EPA supports initiatives through a range of resources including funding, technical support, and tools. In their website US EPA states that if all parts of society are more engaged and empowered to help, advancement of scientific knowledge and better environmental protection actions will be possible. For example, they offer an Equipment Loan Program for those participants ready to collect own participatory science data. US EPA also help participants with a 'EPA's QA Toolkit' containing a Quality Assurance (QA) Handbook together with a video series, and two fact sheets that can help complete a Quality Assurance Project Plan and successfully start a project.

From the perspective of citizen science applications environmental policies including the waste management schemes and obviously the landfills were the first area to demonstrate an awareness of citizen science. The director of the Global Citizens' Observatory for Environmental Change, (an initiative of the European Union) is showing the way to integrate citizens' observations on the environment. She noted that: 'Often the best information comes from those who are closest to it'.

Haklay (2015) reported the intersections of citizen science and policy at local, regional, and national levels and across policy domains, on the basis of the emerging experience in Europe. This covers formulating research questions, recording observations, analyzing data, and using the resulting knowledge. Activities to come from the public on site selection, waste generation and classification, operational issues of waste management and landfills, odor detection and abatement are suitable topics for citizen science projects.

Many public and regulatory disputes on the waste management and landfill site selection can be handled with the methods of citizen science.

Thirdly, the waste management system studies from human health and welfare perspective must be increased all over the world. Waste is not something you dump somewhere and forget about for the rest of your life.

References

Ahire, V., Befhera, D. K., Saxena, M. R., Patil, S., Endait, M., & Poduri, H. (2022). Potential landfill site suitability study for environmental sustainability using GIS-based multi-criteria techniques for nashik and environs. *Environment and Earth Science, 81*, 178. https://doi.org/10.1007/s12665-022-10295-y

Arman, F., Mohseni Bandpey, A., Shahsavani, A., Saadani, M., Saadei, R., & Abtahi, M. (2022). Characterization, source identification, and health risk assessment of odorous compounds in solid waste management facility of Tehran. *Air Quality, Atmosphere & Health., 15*, 1609–1621. https://doi.org/10.1007/s11869-022-01205-z

Australia South Wales Environment Protection Authority. (2016). *Environmental guidelines for solid waste landfills* (2nd ed.). www.epa.nsw.gov.au

Barlaz, M., Rooker, A. P., Kjeldsen, P., Gabr, M. A., & Borden, R. C. (2002). Critical evaluation of factors required to terminate the postclosure monitoring period at solid waste landfills. *Environmental Science & Technology, 36*(16), 3457–3464. https://doi.org/10.1021/es011245u

Bax, C., Sironi, S., & Capelli, L. (2020). How can odors be measured? An overview of methods and their applications. *Atmosphere, 11*, 92. https://doi.org/10.3390/atmos11010092

Brattoli, M., de Gennaro, G., de Pinto, V., Loiotile, A. D., Lovascio, S., & Penza, M. (2011a). Odour detection methods: Olfactometry and chemical sensors. *Sensors (Basel)., 11*(5), 5290–5322. https://doi.org/10.3390/s110505290

Berihun, D. (2017). Control mechanism of bad odor generated from sanitary landfill leachate using Swiss pharm BioMist. In *Proceedings 2017*. https://doi.org/10.4172/2475-7675.1000129

Bilgilioglu, S., Gezgin, C., Orhan, O., et al. (2022). A GIS-based multi-criteria decision-making method for the selection of potential municipal solid waste disposal sites in Mersin, Turkey. *Environmental Science and Pollution Research, 29*, 5313–5329. https://doi.org/10.1007/s11356-021-15859-2

Brattoli, M., de Gennaro, & de Pinto. (2011b). Odour impact monitoring for landfill. In S. Kumar (Ed.), *Integrated Waste Management* (Vol I). https://doi.org/10.5772/17653

Capelli, L., Sironi, S., Del Rosso, R., & Guillot, J.-M. (2013). Measuring odours in the environment versus dispersion modelling: a review. *Atmospheric Environment, 79*, 731–743. https://doi.org/10.1016/j.atmosenv.2013.07.029

Chuandong, W., Liu, J., Zhao, P., Li, W., & Luchun. (2017). Evaluation of the chemical composition and correlation between the calculated and measured odour concentration of odorous gases from a landfill in Beijing, China. *Atmospheric Environment 164*, 337–347. https://doi.org/10.1016/j.atmosenv.2017.06.010

Dai, Z. N., Zeng, F. S., Liu, J., Yang, A. L., & Fu, H. Y. (2018) Application of Deodorant in odor control of municipal solid waste. In *Second International Conference on Power and Energy*

Engineering 3–5 Sept 2018 (ICPEE) (p. 192012063). Xiamen, China: IOP Conference Series: Earth and Environment Science. https://doi.org/10.1088/1755-1315/192/1/012063

Dincer, F. (2007). *Characteristics and chemistry of odors from selected industrial facilities in Izmir* (Ph.D. thesis). Dokuz Eylül University Graduate School of Natural and Applied Sciences. Izmir, Turkey. https://www.apikam.org.tr/YuklenenDosyalar/Dokumanlar/17f4c4db-3ab5-4e83-b0a3-beabe5664834213209.pdf

Dincer, F., Muezzinoglu, A., Elbir, T., & Demircioglu, H. (2004). Comparison of model predictions with the observed odor concentrations from a meat rendering plant. In: *International Conference on Environmental Odour Management*, Cologne, Germany, November 17–19, 2004 (Vol. 1850, pp.197–208).

Dincer, F., & Muezzinoglu, A. (2006). Chemical characterization of odors due to some industrial and urban facilities in Izmir, Turkey. *Atmospheric Environment, 40*(22), 4210–4219. https://doi.org/10.1016/j.atmosenv.2005.12.067

Ding, Y., Xiong, J., Zhou, B., Wei, J., Qian, A., Zhang, H., Zhu, W., & Zhu, J. (2019). Odor removal by and microbial community in the enhanced landfill cover materials containing biochar-added sludge compost under different operating parameters. *Waste Management, 87*, 679–690. https://doi.org/10.1016/j.wasman.2019.03.009

Duan, Z., Kjeldsen, P., & Scheutz, C. (2022). Efficiency of gas collection systems at danish landfills and implications for regulations. *Waste Management, 139*, 269–278. https://doi.org/10.1016/j.wasman.2021.12.023

Durmusoglu, E., Taspinar, F., & Karademir, A. (2010). Health risk assessment of BTEX emissions in the landfill environment. *Journal of Hazardous Materials, 176*(1–3), 870–877. https://doi.org/10.1016/j.jhazmat.2009.11.117

Dursun, S., & Ayturan, Z. C. (2022). Simultaneous removal of gaseous benzene and toluene with photocatalytic oxidation process at high temperatures under UVC irradiation. *Environmental Science and Pollution Research*. https://doi.org/10.1007/s11356-022-18790-2

Ege Biyoteknoloji A. S. Izmir, Turkey (Personal communication) (info@egebiyoteknoloji.com).

Einhäupl, P. (2021). Landfill mining: implementing the missing link to a circular economy: An evaluation of the societal sustainability. Book of dissertation, University of Antwerp, 2021. https://lirias.kuleuven.be/3540230

Elbir, T., Dincer, F., & Muezzinoglu, A. (2007). Evaluation of measured and predicted odor concentrations around a meat packaging and rendering plant. *Environmental Engineering Science, 24*(3), 313–320. https://doi.org/10.1089/ees.2005.0105

Fang, W., Huang, Y., Ding, Y., Qi, X., Liu, Y., & Bi, J. (2022). Health risks of odorous compounds during the whole process of municipal solid waste collection and treatment in China. *Environment International, 158*, Article 106951. https://doi.org/10.1016/J.Envint.2021.106951

Gao, S., Meng, L., Ge, X., Li, Y., Yang, Y., Duan, Y., Fu, Q., Zhang, S., Yang, X., Fei, W., & Jiao, Z. (2023). Role of garbage classification in air pollution improvement of a municipal solid waste disposal base. *Journal of Cleaner Production, 423*, 138737. https://doi.org/10.1016/j.jclepro.2023.138737

Gao, S., Zhao, P., Li, Y., Hu, J., Zhou, S., Huanyan, Z., Wu, J., Jiao, Z., Wang, S. F., & Zhou, B. (2021). Characterization and influence of odorous gases on the working surface of a typical landfill site: A case study in a Chinese megacity. *Atmospheric Environment, 262*, Article 118628. https://doi.org/10.1016/j.atmosenv.2021.118628

Global Citizens' Observatory for Environmental Change (an initiative of the European Union) https://www.eea.europa.eu/media/speeches/global-citizen-observatory-the-role-of-individuals-in-observing-and-understanding-our-changing-world

Haklay, M. E. *Citizen science and policy: A European perspective (Case study series 4)*. Woodrow (2015) Wilson International Center for Scholars: https://www.wilsoncenter.org/publication/citizen-s

Herath, P. L., Jayawardana, D. T., & Bandara, N. J. G. J. (2023). Surface emission determination of selected trace gases from an active municipal solid waste dumpsite under the surface physicochemical heterogeneity. *Waste Management*. https://doi.org/10.1016/j.wasman.2023.02.006

Huang, D., Du, Y., Xu, Q., & Ko, J. H. (2022). Quantification and control of gaseous emissions from solid waste landfill surfaces. *Journal of Environmental Management, 302*(Part A), 114001. https://doi.org/10.1016/j.jenvman.2021.114001

Huang, H., & Ye, D. (2009). Combination of photocatalysis downstream the non-thermal plasma reactor for oxidation of gas-phase toluene. *Journal of Hazardous Materials, 171*(1–3), 535–541. https://doi.org/10.1016/j.jhazmat.2009.06.033

Jackson, R. B., Saunois, M., Bousquet, P., Canadell, J. G., Poulter, B., Stavert, A. R., Bergamaschi, P., Niwa, Y., Segers, A., & Tsuruta, A. (2020). Increasing anthropogenic methane emissions arise equally from agricultural and fossil fuel sources. *Environmental Research Letters, 15*, 71002. https://doi.org/10.1088/1748-9326/ab9ed2

Kanhai, G., Fobil, J. N., Nartey, B. A., Spadaro, J. V., & Mudu, P. (2021). urban municipal solid waste management: Modeling air pollution scenarios and health impacts in the case of Accra, Ghana. *Waste Management (IF 8.1)*. https://doi.org/10.1016/j.wasman.2021.01.005

Kaza, S., Yao, L., Tata, P. B., & Van Woerden, F. (With Ionkova, K., Morton, J., Poveda, R. A., Sarraf, M., Malkawi, F., A. S. Harinath, Banna, An, G., Imoto, H., & Levine, D.) In *What a waste 2.0: a global snapshot of solid waste management to 2050*. 2018. https://openknowledge.worldbank.org/handle/10986/2174.

Lee, H.-D., Jeon, S. B., Choi, W.-J., Lee, S.-S., Lee, M.-H., & Oh, K.-J. (2013). A novel assessment of odor sources using instrumental analysis combined with resident monitoring records for an industrial area in Korea. *Atmospheric Environment, 74*, 277–290. https://doi.org/10.1016/j.atmosenv.2013.04.001

Lee, Y. Y., Jung, H., Ryu, H. W., Oh, K. C., Jeon, J. M., & Cho, K. S. (2018). Seasonal characteristics of odor and methane mitigation and the bacterial community dynamics in an on-site biocover at a sanitary landfill. *Waste Management., 71*, 277–286. https://doi.org/10.1016/j.wasman.2017.10.037

Li, H. G. (2006). Comparison of odor dispersion predictions between CFD and CALPUFF models. *Transactions of the ASABE, 49*(6), 1915–1926. https://doi.org/10.13031/2013.22293 (Published by the American Society of Agricultural and Biological Engineers, St. Joseph, Michigan, www.asabe.org)

Li, D., Lu, W., Liu, Y., Guo, H., Xu, S., Ming, Z., & Wang, H. (2015). Analysis of relative concentration of ethanol and other odorous compounds (OCs) emitted from the working surface at a landfill in China. *PLoS ONE*. https://doi.org/10.1371/journal.pone.0119305

Malovanyy, M., Korbut, M., Davydova, I., & Tymchuk, I. Monitoring of the influence of landfills on the atmospheric air using bioindication methods on the example of the Zhytomyr landfill, Ukraine. *Journal of Ecological Engineering. 22*(6), 36–49. https://doi.org/10.12911/22998993/137446

Majumdar, D., Ray, S., Chakraborty, S., Rao, P. S., Akolkar, A. B., Chowdhury, M., & Srivastava, A. (2013). Emission, speciation, and evaluation of impacts of non-methane volatile organic compounds from open dump site. *Journal of the Air & Waste Management Association, 64*(7), 834–845. https://doi.org/10.1080/10962247.2013.873747

Maxime, G., Aymen Amine, A., Abdelkrim, B., et al. (2014). Removal of gas-phase ammonia and hydrogen sulfide using photocatalysis, nonthermal plasma, and combined plasma and photocatalysis at pilot scale. *Environmental Science and Pollution Research, 21*, 13127–13137. https://doi.org/10.1007/s11356-014-3244-6

Meyer-Dombard, D'.A. R., Bogner, J. E., & Malas, J. (2020). A review of landfill microbiology and ecology: a call for modernization with 'next generation' technology. *Frontiers in Microbiology, 11*, 1127. https://doi.org/10.3389/fmicb.2020.01127

Mønster, J., Kjeldsen, P., & Scheutz, P. (2019). Methodologies for measuring fugitive methane emissions from landfills—A review. *Waste Management, 87*, 835–859. https://doi.org/10.1016/j.wasman.2018.12.047

Muezzinoglu, A., & Dincer, F. Correlation between the public odor complaints and toxic air pollutants. In *Proceedings of the Third International Symposium on Air Quality Management at Urban, Regional and Global Scales* (pp. 970–975). September, 26–30 2005, Istanbul-Turkey.

Nair, A. T., Senthilnathan, J., & Shiva, N. S. M. (2019). Emerging perspectives on VOC emissions from landfill sites: Impact on tropospheric chemistry and local air quality. *Process Safety and Environmental Protection, 121*, 143–154. https://doi.org/10.1016/j.psep.2018.10.026

New York State Health Department Website. *Important things to know about landfill gas*. https://www.health.ny.gov/environmental/outdoors/air/landfill_gas.htm

Nicell, J. A. (2009). Assessment and regulation of odour impacts. *Atmospheric Environment, 43*(1), 196–206. https://doi.org/10.1016/j.atmosenv.2008.09.033

Pan, Q., Liu, Q.-Y., Zheng, J., Li, Y.-H., Song, X., Sun, X.-J., & He, X.-S. Volatile and semi-volatile organic compounds in landfill gas: Composition characteristics and health risks. *Environment International,* 174. https://doi.org/10.1016/j.envint.2023.107886

Rabbani, K. A., Charles, W., Kayaalp, A., Cord-Ruwisch, R., & Ho, G. (2016). Pilot-scale biofilter for the simultaneous removal of hydrogen sulphide and ammonia at a wastewater treatment plant. *Biochemical Engineering Journal, 107*, 1–10. https://doi.org/10.1016/j.bej.2015.11.018

Sankaran, S., Lav, R., Khot, L. R., & Panigrahi, S. (2012). Biology and applications of olfactory sensing system: A review. *Sensors and Actuators B: Chemical, 171–172*, 1–17. https://doi.org/10.1016/j.snb.2012.03.029

Sarptas, H., & Erdin, E. (2014). Estimation of landfill gas energy potential for Izmir Harmandali Landfill Site. In *Conference Paper*. DepoTech. https://doi.org/10.13140/2.1.2316.4489

SCS Engineers. (2017). *Review of odor management at western regional sanitary landfill*, November 9, 2017. File No. 01217279.00.

Schauberger, G., Piringer, M., & Petz, E. (2002). Calculating direction-dependent separation distance by a dispersion model to avoid livestock odour annoyance. *Biosystems Engineering, 82*(1), 25–37. https://doi.org/10.1006/bioe.2001.0043

Schauberger, G., Piringer, M., Wu, C., & Koziel, J. A. (2002). Environmental odour. *Atmosphere, 12*, 1293. https://doi.org/10.3390/atmos12101293

Scheutz, C., & Kjeldsen, P. (2019). Guidelines for landfill gas emission monitoring using the tracer gas dispersion method. *Waste Management, 85*(15), 351–360. https://doi.org/10.1016/j.wasman.2018.12.048

Scheutz, J. C., Bogner, J., Chanton, J. P., Blake, D., Morcet, M., Aran, C., & Kjeldsen, P. (2008). Atmospheric emissions and attenuation of non-methane organic compounds in cover soils at a French landfill. *Waste Management, 28*(10), 1892–1908. https://doi.org/10.1016/j.wasman.2007.09.010

Solan, P. J., Dodd, V. A., & Curran, T. P. Evaluation of the odour reduction potential of alternative cover. Materials at a commercial landfill. *Bioresource Technology, 101*(4), 1115–1119. https://doi.org/10.1016/j.biortech.2009.09.030

Tansel, B., & Inanloo, B. (2019). Odor impact zones around landfills: Delineation based on atmospheric conditions and land use characteristics. *Waste Management, 88*, 39–47. https://doi.org/10.1016/j.wasman.2019.03.028

Tulun, S., Gurbuz, E., & Arsu, T. (2021). Developing a GIS-based landfll site suitability map for the Aksaray province, Turkey. *Environmental Earth Sciences, 80*, 310. https://doi.org/10.1007/s12665-021-09598-3

US-EPA. Compendium Method TO-17. (1999). *Determination of volatile organic compounds in ambient air using active sampling onto sorbent tubes*. USEPA ORD.

US-EPA. *Participatory science for environmental protection*. https://www.epa.gov/participatory-science

Valera, E. H. (2020). *Assessing the role of citizen participation in solid waste management (practices) towards a circular economy model: A case study of Pachuca City in Mexico* (M.sc thesis Land Use Planning, LUP Code: 80436). Wageningen University and Research.

Vidovic, M. (2017). *Meteorological conditions affecting the dispersion of landfill odor complaints* (Ph.D. thesis). Florida Atlantic University.

Vikrant, K., Kailasa, S. K., Tsang, D. C. W., Lee, S. S., Kumar, P., Giri, B. S., Singh, R. S., & Kim, K. H. (2018). Biofiltration of hydrogen sulfide: Trends and challenges. *Journal of Cleaner Production, 187*, 131–147. https://doi.org/10.1016/j.jclepro.2018.03.188

Wei, X., Tao, Z., Xinrui, J., Huan, X. (2022). Degradation of mixed typical odour gases via non-thermal plasma catalysis. *Journal of Hazardous Materials, 440*, 129751. https://doi.org/10.1016/j.jhazmat.2022.129751

Chapter 4
Emissions from Engineered Landfills and Their Impacts on Climate Change and Sustainable Development

Doaa A. El-Emam

Abstract Landfills continue to be a significant part of waste management strategies even though more waste is being reused, recycled, or energetically valorized. Due to its easy management and capacity for mass manipulation, the landfill is particularly important in many developing countries, where it serves as the primary method for the disposal of biodegradable waste. Recently, researches have demonstrated that landfills are a significant source of greenhouse gas (GHG) emissions. Leachate and gases are produced as a result of waste biodegradation in landfills. Methane and carbon dioxide, both significant greenhouse gases, are the main components of landfill gas. 20% of the world's anthropogenic methane emissions come from landfills. Additionally, it frequently contains a large number of additional gases, some of which are toxic, at low concentrations. These emissions may jeopardise both the quality of the environment and the public's health. However, there aren't many reviews in the related fields, so there isn't a complete understanding of the engineered landfills' emissions accumulation and their impacts on climate. This study provides comprehensive information on GHG emissions from landfills over the past 20 years and highlights the significance of creating and disseminating up-to-date knowledge frameworks in light of the growing attention being paid to GHG emissions and their mitigation methods.

Keywords Landfills · Sustainable development · Solid waste · Climate change · Greenhouse gas

D. A. El-Emam (✉)
Environmental Sciences Department, Faculty of Science, Damietta University, New Damietta City, Egypt
e-mail: doaa_alemam@du.edu.eg

A. Anouzla et al. (eds.), *Handbook of the Effects of Landfill Emissions on Sustainable Development*, https://doi.org/10.1007/978-3-032-01951-6_4

4.1 Introduction

The rise in population and economic well-being are two significant factors contributing to the generation of waste, thus rendering waste management a prominent challenge in the twenty-first century. Among the regions, East Asia and the Pacific region accounts for the highest proportion at 23%, followed by Europe and Central Asia at 20%, and South Asian countries at 17% in terms of waste generation (Kaur et al., 2021). The average daily per capita waste generation on a global scale is 0.74 kg, with a significant range spanning from 0.11 to 4.54 kg. The generation of waste is influenced by various factors, including but not limited to population size, level of development, and national income. There exists a generally positive correlation between the generation of waste and the level of income (Gautam & Agrawal, 2021). High-income nations produce approximately 683 million metric tonnes (Mt) of municipal solid waste (MSW) only. Given the shared objective of countries to enhance the quality of life for their citizens, there has been a concurrent rise in industrialization, urbanization, commercialization, and population growth. According to the cited source, it is projected that the generation of municipal solid waste (MSW) will only increase to 3.40 billion metric tonnes by the end of the year 2050 (Malav et al., 2020). The regions experiencing the most rapid growth in terms of waste generation are projected to be the Middle East, North Africa, South Asia, and Sub-Saharan Africa. It is anticipated that by the year 2050, the total amount of waste generated in these regions will increase by a factor of 2 in the Middle East, 2 in North Africa, and 3 in both South Asia and Sub-Saharan Africa (Kaza et al., 2018; Malav et al., 2020).

Prior to 1979, landfills were utilized as locations for the disposal of diverse types of waste, including those that were categorized as hazardous. They were frequently described as open dumps that lacked sufficient measures to prevent the migration of contaminants into the surrounding environment. Open landfills pose significant environmental and public health hazards. Previous studies documented the presence of flies and vermin, as well as frequent and prolonged occurrences of fires. The landfill sites under consideration exhibited a lack of gas collection infrastructure and were also deficient in protective liners to mitigate potential contamination of the underlying groundwater (Tirkolaee & Torkayesh, 2022).

Several waste disposal sites have been officially designated as "Superfund" sites and are presently undergoing or awaiting remediation measures. Open dumps, once deemed illegal, have undergone closure or conversion into municipal solid waste (MSW) landfills. The primary areas of concern in terms of potential gas emissions are landfill sites that were previously devoid of gas control systems (Shareefdeen & Bhojwani, 2022).

Regulation and enforcement of landfills by global and local governments have been established since the mid-twentieth century. Nevertheless, prior to the year 1979, there existed notable disparities in the degree and uniformity of regulation and enforcement among various landfill locations. The federal government implemented the regulation of siting, construction, operation, and closure requirements for landfills in accordance with the Resource Conservation and Recovery Act (RCRA) in 1979. Subtitle D of the Resource Conservation and Recovery Act (RCRA) focuses on the management of municipal solid waste (MSW) and nonhazardous landfills. This section of the act includes provisions that address the monitoring of methane gas emissions at the boundaries of these landfills (Ross & Law, 2023). The Resource Conservation and Recovery Act (RCRA) is a legislative framework designed to effectively address and mitigate the diverse concerns and issues associated with hazardous waste landfills. Regulations were established by the Environmental Protection Agency (EPA) in 1996, in alignment with the provisions of the Clean Air Act (CAA). The regulations in question, referred to as the New Source Performance Standards and Emissions Guidelines (NSPS/EG), were specifically developed to address the issue of methane and non-methane organic compound (NMOC) emissions that arise from landfills handling municipal solid waste (MSW) (Firestone, 2020).

The issue of greenhouse gas (GHG) emissions from various anthropogenic sources is of significant global importance, given its implications for environmental health. The per capita annual emission of carbon dioxide (CO_2) experienced a significant increase from 2.2 tonnes (t) in 1990 to 7.5 t in 2014, surpassing the global average per capita level by a substantial margin. China, followed by the United States and India, are the largest contributors of greenhouse gas emissions. According to the study referenced as (He et al., 2019), waste management emerges as the primary research domain in the context of emissions, with energy reduction being the subsequent area of focus on a global scale. The waste sector, which encompasses solid and wastewater treatment, is projected to account for approximately 3–4% of global anthropogenic greenhouse gas (GHG) emissions. The municipal solid waste landfill is recognized as the fourth most significant source of global greenhouse gas (GHG) emissions, contributing approximately 5% to the overall global greenhouse gas budget. The 5% mentioned comprises of various gases, namely methane (CH_4), carbon dioxide (CO_2), nitric oxide (N_2O), and fluorinated gases, including perfluorocarbons (PFCs) and hydrofluorocarbons (Gautam & Agrawal, 2021; Sovacool et al., 2021).

Landfills have been widely recognized as significant sources of greenhouse gas (GHG) emissions worldwide. It can be attributed to the significant emission of gases, such as carbon dioxide (CO_2) and methane, from landfills into the atmosphere that contributes to global warming. The degradation process is responsible for the generation of these gases, which subsequently contribute to environmental pollution. Furthermore, the activities conducted at landfills have been linked to the pollution of subterranean water sources due to the generation of landfill leachate. This phenomenon is particularly observed when the liners within landfills do not meet the necessary adequacy standards. Moreover, emanations originating from landfills contribute to air pollution, particularly affecting individuals residing in close proximity. Additional pollutants commonly found in landfills include particulate matter, solid waste debris, and vermin posing risks to human health and ecosystem integrity. Therefore, it is crucial to implement effective waste management strategies and promote sustainable alternatives to minimize the negative impacts of landfills on climate change and ensure long-term sustainable development (Guo et al., 2022).

The objective of this chapter is to present a comprehensive overview of the landfill, including an analysis of the emissions of greenhouse gases that arise from the diverse processes associated with landfill operations and their impacts on climate change and sustainable development. The assessment methodology employed in this study involves an investigation of previous research and analysis of globally recognized data. Additionally, the study will consider the subsequent effects on socioeconomic and ecological systems. The study also examines the strategies implemented to address sustainable management practices with the aim of reducing greenhouse gas emissions.

4.2 Landfill

The World Bank has identified landfilling as a widely adopted approach for the management of solid waste at a global level. A contemporary landfill is a purposefully engineered and protected system used for the disposal of waste, achieved through the placement of waste in specially constructed cells on the ground or through excavation into the ground. The landfill environment supports a range of biological, chemical, and physical processes that contribute to the decomposition of waste materials, resulting in the production of leachate and gases (Ghasemzade & Pazoki, 2017).

Landfills in the pre-1950s era were primarily used as simple dumpsites, and it is worth noting that numerous developing nations continue to maintain waste dumpsites. The lack of serious consideration given to contamination issues can be attributed to the prevailing belief that the leachate generated by waste undergoes complete purification through the processes of soil and groundwater (Malav et al., 2020). However, subsequent studies conducted after the mid-1950s have provided evidence to the contrary. Consequently, this resulted in the implementation of more stringent regulations. In light of the fact that municipal solid waste (MSW) is comparatively less detrimental than leachate generated by industrial waste containing chemicals, waste in its entirety has been classified into two categories: hazardous and nonhazardous (Siddiqua et al., 2022). The implementation of new regulations ultimately led to the conversion of nonhazardous waste dumpsites into what is now commonly referred to as "sanitary landfills". Historically, landfills that implemented a daily cover were regarded as sanitary landfills. In contemporary usage, the term "sanitary landfill" refers to a purposefully constructed and managed solid waste disposal facility that aims to mitigate adverse effects on the environment and public health. The practice of sanitary landfilling remains a cost-effective and ecologically sound approach for the management of solid waste (Mazzei & Specchia, 2023).

A contemporary engineered landfill refers to a methodical approach for disposing of waste by placing it in specifically designed and safeguarded compartments on the surface of the land or in excavated areas within the land surface. Several processes take place within it, facilitating the decomposition of waste materials and leading to the generation of leachate and gases. The landfill ecosystem exhibits significant diversity as a result of the heterogeneous composition of waste materials and the diverse operational characteristics of landfills. The promotion of stability within the ecosystem is facilitated by its diversity. Nevertheless, the system is significantly impacted by various environmental factors, including temperature, pH levels, the presence of toxins, moisture content, and oxidation reduction potential (Mondal et al., 2023).

4.2.1 Landfill Types

Landfills are classified into four categories:

4.2.1.1 Cleanfill

A cleanfill, also referred to as an inert waste landfill, is a type of landfill facility that exclusively accepts cleanfill materials and inert wastes, which may include clean excavated natural materials. In a broad sense, the primary means of regulating discharges to land and water from cleanfills are through the implementation of waste acceptance criteria, which have proven to be the sole effective environmental controls. Cleanfill material, also known as inert waste, does not undergo any substantial environmental alterations in terms of its physical, chemical, or biological properties, and does not contain any potentially dangerous substances once it is disposed of in a landfill. The substance must remain free from contamination or any form of adulteration (Townsend & Anshassi, 2023).

4.2.1.2 Municipal Solid Waste Landfill (MSWL)

A municipal solid waste landfill (MSWL), commonly known as a sanitary landfill, is a designated facility that receives and manages municipal solid waste. In addition to municipal solid waste, a landfill site may also accept inert waste. Approximately 70% of the global solid waste produced consists of municipal solid waste (MSW) (Gautam & Agrawal, 2021). Incineration is the prevailing choice for waste disposal in numerous regions across the globe. The primary factor contributing to the prevalence of MSW landfills, particularly in developing economies, is the substantial expenses associated with treatment and disposal options (Batista et al., 2021). However, it is worth noting that several highly industrialized nations, including the United States, Australia, the United Kingdom, and Finland, continue to rely significantly on the practice of landfilling. As an illustration, it can be observed that in the United States, approximately 54% of the 250 Tg (1Tg = 106 mt) of municipal solid waste (MSW) produced in 2008 was disposed of in landfills (Maalouf et al., 2020). Recycling and composting practices, on the other hand, constituted approximately 33% of the overall management of MSW during that period (Lino et al., 2023). According to the Productivity Commission (2006), approximately 70% of municipal solid waste (MSW) in

Australia was disposed of in landfills without undergoing any pre-treatment measures in the year 2002 (Khan et al., 2022). In the context of Japan, it was observed that the proportion of municipal solid waste (MSW) directly disposed of was less than 30% of the total MSW generated in the year 2000 (Singh et al., 2011). This trend can be attributed to the significant reliance on incineration methods for waste management, which has been prevalent over the past few decades. The high incineration rates can be attributed to the historical scarcity of land in Japan. Within the European Union, Greece, the United Kingdom, and Finland are identified as countries that exhibit a notable reliance on direct landfilling as a waste management practice (Tam & Lu, 2016). According to Banias et al. (2020), the proportion of municipal solid waste (MSW) that was disposed of in landfills in Greece was 77% in 2008. Similarly, the UK and Finland had landfill rates of 55% and 51% respectively during the same period (Banias et al., 2020). By way of comparison, the management of municipal solid waste (MSW) in Germany, the Netherlands, Sweden, Denmark, and Austria in 2008 saw landfilling comprising a minor fraction, specifically less than 5% (Banias et al., 2020).

4.2.1.3 Industrial Waste Landfill (IWL)

An industrial waste landfill (IWL) refers to a specific type of landfill that is specifically engineered to accommodate primarily industrial waste materials. Industrial waste landfills are often referred to as monofills, as they are typically associated with a particular industry or industrial site, such as mining, forestry, or smelting. These landfills are specifically designed and operated to accommodate the specific types of waste generated by the respective industries. The stringency of design, operation, and monitoring requirements for non-municipal solid waste landfills may vary, either being more or less stringent in comparison to those imposed on municipal solid waste landfills (Palaniandy et al., 2022). Depending on its design, an industrial waste landfill has the potential to accept municipal solid waste and inert waste in addition to industrial waste. C&D landfills primarily serve as disposal sites for construction and demolition (C&D) waste, although they are capable of accommodating various forms of solid industrial waste. The category of waste encompasses various building components such as concrete, gypsum, asphalt, bricks, and others (USEPA, 2013; Hassan et al., 2022).

4.2.1.4 Hazardous Waste Landfill or Hazardous Waste Containment Facility (HWL)

A hazardous waste landfill (HWL), also known as a hazardous waste containment facility (HWCF), refers to a designated landfill that receives waste materials that are officially classified as "hazardous waste" according to statutory instruments or specific criteria established by the Environmental Risk Management Authority (ERMA). The regulatory requirements for landfills that accept hazardous waste are significantly more stringent compared to those for landfills that solely handle municipal solid waste (Gautam et al., 2019). These requirements encompass aspects such as site selection, design, operations, and monitoring. Due to their inherent characteristics, these particular landfills are subject to rigorous regulation and exhibit a well-defined structure. These containers are purposefully engineered to securely contain hazardous wastes, greatly minimizing the risk of leaching or accidental release into the surrounding environment. Several design requirements for hazardous waste landfills include the implementation of double liners, double leachate collection and removal systems, leak detection systems, dispersal controls, and construction quality assurance measures, among others. In addition to the aforementioned design specifications, it is imperative to note that hazardous waste landfills are subject to regular inspections throughout the year in order to ascertain compliance with the most up-to-date and rigorous standards (Epa, 2022).

4.2.1.5 Open Dumps

It is referring to waste disposal sites that were established prior to 1979, lacking any engineering design or siting criteria, and possessing minimal, if any, regulatory oversight. The open dumps fail to comply with the regulations outlined in the Resource Conservation and Recovery Act (RCRA) Subtitle D (Sharma & Kumar, 2021). Open dumps have been known to accept various types of waste, including household waste, which is similar to the waste found in municipal solid waste (MSW) landfills, as well as commercial and industrial waste. The landfill sites in question were devoid of liners and exhibited infrequent utilization of daily cover for the disposal of sanitary wastes.

Insufficient measures were implemented to mitigate the potential migration of contaminants into the surrounding environment. The majority of open dumps were discontinued and subsequently covered during the 1960s. Regrettably, the precise locations of numerous antiquated waste disposal sites are not indicated on regional planning maps. Several of the existing municipal solid waste (MSW) landfills that are currently in operation have their origins dating back to the 1960s when they were initially established as open dumps or were situated in close proximity to closed dumps (Propp et al., 2021).

4.2.1.6 Green Waste Landfills

Although these landfills are not formally authorized by the Environmental Protection Agency (EPA), numerous local governments are increasingly embracing their use for the disposal of organic materials, allowing for natural decomposition. The prevalence of composting sites has increased due to the limited acceptance of organic waste, such as fruits and vegetables, in conventional landfills and transfer stations. Typical forms of green waste encompass various organic materials such as mulch, weeds, leaves, tree branches, flowers, biodegradable food waste, and grass trimmings, among others (Ramli et al., 2021). According to the United States Environmental Protection Agency (EPA) in 2017, green waste landfills were found to have a discernible impact, as they received over 24,000 tonnes of garden trimmings (US EPA, 2017). The primary objective of green waste landfills is to mitigate the strain on municipal solid waste (MSW) landfills by diverting a specific material that is inherently capable of undergoing natural decomposition.

4.2.2 Landfills Design and Operation

The investigation of construction and operation techniques to improve biological activities within landfills has been a subject of sustained research attention over the last two decades. Numerous experiments have been conducted to investigate methods for enhancing landfill decomposition. Landfills can also be classified according to the design and operation areas (Nanda & Berruti, 2021) to.

4.2.2.1 Dry Landfill

The present system can be characterized as a conventional approach to the design and management of modem landfills, featuring a restricted inflow of water. The phenomenon is distinguished by prolonged periods of low rates of leachate and methane production. The process of waste stabilization is characterized by a significant duration, which can be further impeded by the presence of limited moisture. Dry landfills typically incur significant expenses for the installation of caps and liners, as well as for the management and mitigation of emissions. Consequently, they cannot be regarded as suitable locations for the long-term storage of waste (Madon et al., 2019).

4.2.2.2 Bioreactor Landfill

In this particular landfill, the rates of decomposition and methane production are augmented. The utilization of methane is not the sole advantage associated with this particular landfill, as the rapid rates of stabilization can result in decreased costs for post-closure monitoring. There is potential for cost reduction in landfill cap construction by reconsidering the necessity of impeding water flow into the landfill. The reduction of the acidogenic phase, which is a crucial requirement for enhancing methane production, leads to a decrease in the release of leachable organic compounds. Consequently, this reduction results in lowered expenses associated with the treatment of leachate. Ultimately, expeditious and regulated decomposition processes have the potential to facilitate subsequent reclamation of the landfill within a reduced timeframe.

Bioreactor landfills can be classified into three distinct categories, namely anaerobic, aerobic, and hybrid. Anaerobic bioreactor landfills expedite the process of biodegradation through the utilisation of anaerobic microorganisms that rely on cellular respiration without the need for oxygen (Mondal et al., 2023; Nanda & Berruti, 2021). According to Wang and Wang (2022), these microorganisms facilitate the conversion of organic wastes into organic acids, which subsequently undergo a transformation into methane and carbon dioxide. Aerobic bioreactor landfills employ aerobic microorganisms that rely on oxygen for cellular respiration and generate carbon dioxide as a byproduct (Mondal et al., 2023). Hybrid bioreactor landfills employ a synergistic integration of the aforementioned two methodologies (Wang & Wang, 2022).

4.2.2.3 Final Storage Quality Landfill

In this type of landfill, the material to be landfilled is in such a state that minimal environmental emissions occur after landfilling. Waste selection or pretreatment techniques are required to produce this type of waste. The goal is typically to reduce the biodegradable organic load prior to land-filling, so that organic landfill emissions are minimal, however, reductions in inorganic emissions may also be involved (Ozbay et al., 2021). Landfill monitoring is not necessary over an extended period of time, and leachate and biogas treatment and collection costs are expected to be reduced. Cost, emissions and other savings for the landfill must be weighed against the cost and emissions associated with waste separation and pretreatment (Pivato et al., 2022).

In recent times, there has been a notable transition in the field of landfill design, specifically in the conceptual framework employed. The traditional dry storage concept has been gradually supplanted by the bioreactor approach, as highlighted by Barati et al. (2022). The bioreactor approach involves the augmentation of moisture content in municipal solid waste (MSW) through the recirculation of leachate, thereby promoting enhanced biodegradation. Recirculating leachate presents a financially advantageous alternative due to the high costs associated with securely disposing of collected leachate. By implementing this approach, one can effectively eliminate the costly treatment expenses typically incurred when dealing with leachate. Furthermore, the presence of elevated moisture levels resulting from the recirculation of leachate contributes to the accelerated decomposition of waste materials. Therefore, the implementation of bioreactor landfills presents a notable decrease in the duration of post-closure management and the associated operational expenses (Jagaba et al., 2021).

4.2.3 Landfill Wastes Stabilization

The stabilization of wastes proceeds in five sequential and distinct phases (Simon & Ogunlowo, 2022):

4.2.3.1 Initial Adjustment Phase

This phase is characterized by the initial deposition of solid waste and the subsequent accumulation of moisture within landfill sites. A period of acclimatization is observed until an adequate amount of moisture is present to sustain a thriving microbial community. This phase was only achieved for a limited duration following the landfilling of municipal solid waste. The duration required for microorganisms to acclimatize to their new surroundings, including the availability of food and adequate moisture levels, within a landfill ecosystem. During the initial stages of adjustment, the easily biodegradable organic fraction of municipal solid waste undergoes aerobic decomposition by aerobic microbes. These microbes consume the free available oxygen that is trapped in the cover soil and the porosity of the solid waste. The aerobic decomposition process in landfill ecology results in the generation of carbon dioxide and an increase in temperature. During this phase, the generation of landfill leachate occurs solely as a result of the compaction of municipal solid waste during landfilling, as well as the occurrence of short-circuiting, where external or recirculated leachate flows through the water pipe network within the municipal solid waste matrices. The primary source of microbial community in landfill ecology may be derived from anaerobically digested sludge or anaerobically decomposed municipal solid waste (Al Raisi, 2022; Mor & Ravindra, 2023).

4.2.3.2 Transition Phase

During the transition phase, a shift from an aerobic to an anaerobic environment takes place. The current phase is characterized by a transition from aerobic to anaerobic degradation processes, which is a result of the depletion of oxygen that was initially present in the waste and the subsequent increase in moisture content beyond the field capacity of municipal solid waste (MSW). As the ecology of landfills underwent changes, it was observed that nitrate and sulphate could function as electron acceptors in the reduction reaction of oxygen, resulting in the reduction of nitrogen gas and hydrogen sulphate. The anticipated reduction/oxidation potentials for the reduction of nitrate and sulphate are typically within the range of − 50 to 100 MV (Milyutina et al., 2020). However, for the generation of methane gas, the reduction/oxidation potentials should fall within the range of − 150 to − 300 MV. The gradual decline in redox potential within landfill ecosystems can enhance the activity of anaerobic microorganisms, facilitating the conversion of organic materials into methane and carbon dioxide gases during subsequent stages. The transition phase of anaerobic degradation processes is characterized by distinct leachate compositions, including high levels of chemical oxygen demand (COD), volatile organic acids (VOA), biochemical oxygen demand (BOD), and low pH values. These specific conditions facilitate the mobilization of heavy metals within the leachate. Furthermore, during the transition period, there is a significant decrease in the concentration of nitrogen gas, while the concentration of carbon dioxide exhibits a gradual increase (Milyutina et al., 2020; Vavilin & Lokshina, 2023; Zhao et al., 2022).

4.2.3.3 Acid Formation Phase

During the ongoing process of hydrolyzing solid waste, the subsequent microbial conversion of biodegradable organic matter leads to the generation of volatile organic acids at elevated concentrations throughout this stage. During the subsequence phase of the transition period, anaerobic microbes persist and intensify the production of substantial amounts of organic acids and hydrogen gas. The initial phase of acid formation involves enzyme-mediated transformation, also known as hydrolysis. This process entails the conversion of complex organic compounds such as lipids, polysaccharides, proteins, and nucleic acids into their constituent elements (Abdel-Shafy et al., 2023a). During the subsequent stage of acidogenesis, the hydrolysis of molecular compounds leads to the conversion into more basic molecular compounds, which eventually transform into acetic acid, fluvic acid, and other intricate organic acids through the activity of acidogenic microorganisms. The leachate generated during these phases exhibits notable characteristics, including elevated levels of BOD_5 and COD, which can lead to a decrease in pH and an increase in heavy metal concentrations. In addition, a significant amount of hydrogen gas and a small amount of nitrogen gas were produced (Aziz et al., 2020).

4.2.3.4 Methane Fermentation Phase

Methanogenic bacteria metabolize intermediate acids, resulting in the production of methane and carbon dioxide. This stage involves the utilization of intermediate acids by methanogenic microorganisms, resulting in the production of methane and carbon dioxide gases. The elevation of organic acid concentration in this phase resulted in an increase in pH values within the range of 6.8–8.0, while simultaneously leading to a decrease in the levels of BOD_5, COD, and conductivity. Another manifestation observed during this phase is the decrease in concentration and mobilization of heavy metals in the leachate. This can be attributed to the complexation and precipitation processes that occur as a result of the higher pH value (Aziz et al., 2020; Dimiskovska & Berisha, 2021).

4.2.3.5 Maturation Phase

During the final state of landfill stabilization, the availability of nutrients and substrate becomes constrained, resulting in a significant decrease in gas production. Additionally, the concentration of leachate remains relatively constant, albeit at significantly reduced levels. This stage involves the utilization of intermediate acids by methanogenic microorganisms, resulting in the production of methane and carbon dioxide gases. The elevation of organic acid concentration at this particular stage resulted in an increase in pH values within the range of 6.8–8.0, while simultaneously causing a decrease in the levels of BOD_5, COD, and conductivity. Another

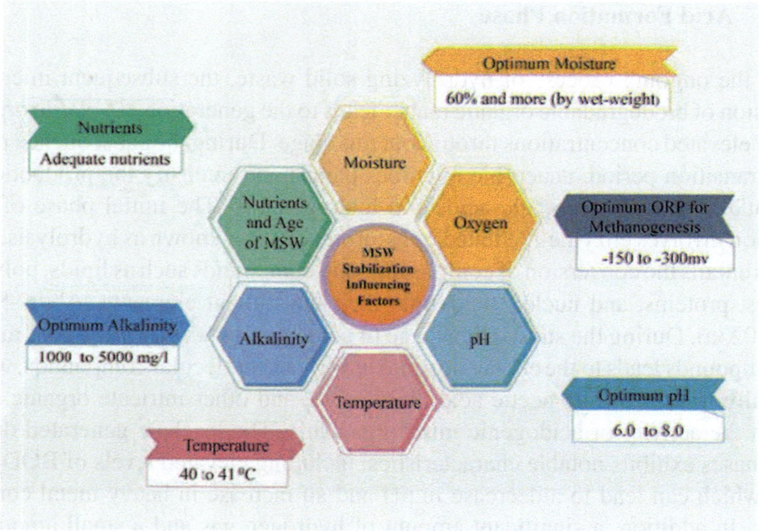

Fig. 4.1 Environmental factors and their optimal values for landfill stabilization (Tesseme & Chakma, 2020)

manifestation observed during this phase is the decrease in concentration and mobilization of heavy metals in the leachate. This can be attributed to the complexation and precipitation processes resulting from the higher pH value (Aziz et al., 2020; Jagaba et al., 2021).

The process of landfill stabilization is influenced by various variables and fundamental environmental factors. The factors that influence waste management processes include moisture content, waste composition, compaction, shredding/size reduction, seeding of microorganisms, buffering capacity, pretreatment of wastes, oxygen levels, and temperature (Chaturvedi et al., 2023). Figure 4.1 presents a concise overview of the optimal values.

4.3 Implications of Landfill Byproducts and Emissions

In addition to the emission of landfill gas and leachate, the presence of wind-blown litter, vermin, and insects has also been recognized as contributing to minor emissions from landfills. However, landfill gas and leachate are the primary contributors to a wide range of negative environmental and socioeconomic outcomes and so they receive the bulk of attention (Danthurebandara et al., 2012).

4.3.1 Landfill Leachate

Landfill leachate refers to a solution consisting of both inorganic and organic components. It is generated when water infiltrates through the buried municipal solid waste (MSW), thereby transporting both suspended and dissolved solids that are formed when the waste's absorbent properties are surpassed. The phenomenon under consideration is a result of a mass transfer mechanism, encompassing the hydrolysis of macromolecules and the biodegradation of their organic components. Additionally, it involves the solubilization of salts present in the waste material, as well as the transport of particulate matter. The organic strength of landfill leachate is typically ten times greater than that of municipal sewage. leachate comprises a multifaceted blend encompassing various components, including dissolved organic matter. This organic matter can be measured in terms of Chemical Oxygen Demand (COD) or Total Organic Carbon (TOC). Additionally, it encompasses volatile fatty acids as well as more resistant compounds like fulvic-like and humic-like substances. Xenobiotic organic compounds (XOCs) encompass a diverse range of chemical substances, including aromatic hydrocarbons, phenols, chlorinated aliphatics, pesticides, and plastizers (Abdel-Shafy et al., 2023b; Propp et al., 2021; Wang & Wang, 2022).

The quantity and composition of waste are primarily influenced by waste properties and composition, site-specific factors, climatic conditions, fill volume and depth, decomposition rate, landfill age, surface capping, and percolate volume. Elevated levels of diverse pollutants are detected in close proximity to the initiation of leaching from recently established landfills, whereas the attenuation and microbial degradation of organic substances diminish the potency of leachate in more mature landfills. The temporal evolution of leachate composition exhibits a resemblance to the distinct phases observed in the process of fermentation (Manheim et al., 2021a).

In instances where leachate is permitted to discharge from a site, such as in the context of older landfills lacking an impermeable liner at the base, its potency will undergo attenuation within the underlying substratum of the landfill and become diluted by groundwater situated beneath the water table. The contamination of groundwater poses a potential concern for locations characterized by a high water table. Due to its inherent potency, landfill leachate undergoes physical, chemical, or biological treatment within a dedicated facility prior to being discharged or reintroduced into the landfill for the purpose of waste stabilization (Manheim et al., 2021a; Pazoki et al., 2020).

Although the composition of leachate is complex, it can be classified into four main categories of pollutants. During the initial acid phase, the leachate exhibits increased concentrations of all its components. However, in the subsequent methanogenic phase, the leachate undergoes a stabilization process, resulting in decreased concentrations and a lower BOD/COD ratio. During the acid phase, it is typically observed that the ratio between biochemical oxygen demand (BOD) and chemical oxygen demand (COD) is equal to or greater than 0.4. In the transient phase, the aforementioned ratio generally falls within the range of 0.2–0.4. Ultimately, during the methanogenic phase, it is generally observed that the ratio of biochemical oxygen demand (BOD) to chemical oxygen demand (COD) is commonly less than or equal to 0.2. In general, the detection of heavy metals is typically conducted at extremely low concentrations. The liberation of accumulated heavy metals would not transpire over a temporal scale encompassing thousands of years (Abdel-Shafy et al., 2023a; Aziz et al., 2020).

The generation of leachate in landfill sites exhibits a gradual decline over time, although certain factors may retain environmental significance for periods ranging from several decades to centuries. The primary components of landfill leachate consist of dissolved methane, fatty acids, sulphate, nitrate, nitrite, phosphates, calcium, sodium, chloride, magnesium, potassium, and trace metals such as chromium, manganese, iron, nickel, copper, zinc, cadmium, mercury, and lead. The migration of leachate through the soil to groundwater or surface water can result in significant challenges for the rehabilitation of aquifers, primarily due to the lack of appropriate liner systems or the presence of damaged liners. It can be retained by soil, leading to negative consequences for the ecosystem (Iravanian & Ravari, 2020; Propp et al., 2021).

The uptake of metals by plants from the soil represents a significant pathway for the introduction of metals into the food chain. The deposition of trace metals in plants has the potential to significantly impact crop growth and productivity, while also presenting a heightened risk to animal health. Metals such as lead, zinc, and cadmium exhibit varying degrees of mobility across vegetation and invertebrate trophic levels, necessitating individualized assessments on a case-by-case basis. The combination of leachate and surface water containing elevated levels of nitrate and phosphates poses a significant risk, primarily in the form of eutrophication. The presence of eutrophic conditions in open sectors of lakes consistently leads to the overproduction of planktonic algae and cyanobacteria, which subsequently has detrimental effects on fish species. The ammonia that is produced as a result of leachate within landfills has the ability to migrate through the various layers of soil horizons. As it moves, it undergoes a process known as nitrification, where it is gradually converted into nitrite and nitrate. This transformation contributes to the occurrence of eutrophication issues (Olowoyo et al., 2012).

Landfills have the potential to emit mercury as a result of the disposal of batteries and paint residues within their confines. It is a hazardous metallic element that can be readily assimilated by various organisms, including animals, fish, and shellfish, thereby becoming incorporated into the food chain. Alloway's research findings indicate that the transformation of chromide into chromate within landfills holds considerable environmental importance, given that chromate exhibits greater toxicity towards plants compared to chromide (Lins & Lins, 2020).

The leachate typically contains a wide variety of xenobiotic organic compounds (VOCs), although their overall effect is generally not a major long-term concern in most cases. In contrast, ammonia is generally recognized as a significant contaminant in leachates and is a major concern in terms of its potential long-term environmental impact. The detrimental effect of high concentrations of ammonia on chemical oxygen demand (COD) removal necessitates the implementation of pretreatment methods to selectively precipitate NH4. The presence of a high concentration of ammonia has a detrimental effect on chemical oxygen demand (COD) removal, necessitating the implementation of pretreatment methods to selectively precipitate NH_4 (Duan et al., 2021).

4.3.2 Landfill Gas

The composition of landfill gas consists of a diverse array of hundreds of distinct gases. In terms of volume, landfill gas generally consists of methane ranging from 45 to 60% and carbon dioxide ranging from 40 to 60%. Landfill gas comprises minor quantities of nitrogen, oxygen, ammonia, sulphides, hydrogen, carbon monoxide and nonmethane organic compounds (NMOCs) like trichloroethylene, benzene and vinyl chloride (Manheim et al., 2021a).

Table 4.1 presents a comprehensive compilation of landfill gases, along with their respective volumetric percentages and distinctive characteristics.

Theoretically, the biological decomposition of one metric tonne of municipal solid waste results in the generation of 442 m^3 of landfill gas. This gas consists of 55% methane and possesses a calorific value ranging from 15 to 21 MJ/m^3, which is roughly half the calorific value of natural gas. The major components of landfill gas are methane (CH_4) and carbon dioxide (CO_2), with a large number of other constituents at low concentrations such as ammonia, sulfide, and non-methane volatile organic compounds (VOCs) (Pujotomo & Azzahra, 2019b). Landfill sites contribute 20% of the total global anthropogenic methane emission. In addition, chemical and biochemical transformations within the landfill create new organic or inorganic substances, such as tri- and per-chlorethylene to vinylchloride, amino acids to methyl- and ethyl-mercaptans, or sulphur compounds to hydrogen sulphide (H_2S). The US EPA listed 94 non-methane organic compounds found in air emissions from municipal solid waste landfills, including benzene, toluene, chloroform, vinyl chloride, carbon tetrachloride, and 1,1,1trichloroethane (Danthurebandara et al., 2012; Pujotomo & Azzahra, 2019b).

The uncaptured gas poses an environmental threat, as methane is a greenhouse gas and is explosive when mixed with air at around 5–15%, and at higher concentrations it will displace oxygen to produce an asphyxiation risk and many of the VOCs are odorous and toxic. VOCs play a significant role in the formation of ground-level ozone, which inhibits photosynthesis, reduces growth, and depresses agricultural yields. The lateral migration of gas through the soil has been the cause of hazardous explosions due to methane being inflammable and explosive when mixed with sufficient air. An unpleasant odor can be caused by trace elements present in landfill gas, such as organic fatty acids from the acid phase and H_2S and other sulfur-containing compounds. Hydrogen sulfide in the landfill gas creates odor and may also cause asphyxiation. Gas migration on landfills is governed by meteorological conditions, waste factors, and soil properties (Mor & Ravindra, 2023).

Table 4.1 Components of a typical landfill gas

Component	Percent by volume	Characteristics	References
Carbon monoxide	0–0.2	It is a colorless, odorless gas	Rosa et al. (2021)
Hydrogen	0–0.2	It is a colorless, odorless gas	Lam (2023)
Sulfides	0–1	Sulphides, such as hydrogen sulphide, dimethyl sulphide, and mercaptans, are naturally present gases that contribute to the unpleasant odour resembling rotten eggs in landfill gas. Even at extremely low	Chetri et al. (2020), Manheim et al. (2021a)
NMOCs (non-methane organic compounds)	0.01–0.6	Non-methane organic compounds (NMOCs) are a class of carbon-containing organic compounds. Methane, an organic compound, does not fall under the classification of non-methane organic compound (NMOC). Non-methane organic compounds (NMOCs) can arise through natural processes or be generated through synthetic chemical reactions. Non-methane organic compounds (NMOCs) that are frequently encountered in landfill environments encompass a range of chemical substances such as acrylonitrile, benzene, 1,1-dichloroethane, 1,2-cis dichloroethylene, dichloromethane, carbonyl sulphide, ethylbenzene, hexane, methyl ethyl ketone, tetrachloroethylene, toluene, trichloroethylene, vinyl chloride, and xylenes	Wang et al. (2021)
Oxygen	0.1–1	Approximately 21% of the atmosphere is oxygen. It is tasteless, odourless, and colourless	Reinhart et al. (2020), Wang et al. (2021)
Ammonia	0.1–1	It is a gaseous compound that lacks colour and possesses a strong, penetrating scent	Wang et al. (2021)
Nitrogen	2–5	Approximately 79% of the atmosphere is oxygen. It is tasteless, odourless, and colourless	Miao et al. (2019), Wang et al. (2021)

(continued)

Table 4.1 (continued)

Component	Percent by volume	Characteristics	References
Carbon dioxide	40–60	Carbon dioxide is present in the Earth's atmosphere at naturally occurring low concentrations, measuring approximately 0.03%. It has no colour or smell and is slightly acidic	Twaróg et al. (2020), Wang et al. (2021)
Methane	45–60	Methane is a gas that occurs naturally. The substance lacks colour and scent. Landfills represent the primary contributor to methane emissions of anthropogenic origin in the United States and India	Pujotomo and Azzahra (2019a)

The landfill gas problem persists as long as its emission and impact remain high. It is usually either vented or collected for energy production. Landfill odor is a sensitive issue, often becoming health concerns and provoking opposition from nearby residents against landfill operation and extension. The gas is generally controlled by installing vertical or horizontal wells within the landfill, which are either vented to the atmosphere or connected to a central blower system that pulls gas to a flare or treatment process. The Intergovernmental Panel on Climate Change (IPCC) report that the landfill gas collection efficiencies ranging from 9 to 90% and estimates an average of 20%. The gas production rate will peak after a few years, depending on the decomposition rate of the waste, which is under the influence of temperature and rainfall of the site. The methane concentration in the landfill gas will reach a maximum of 60% and become stable for some years. Control of these emissions at the source is necessary from an environmental protection viewpoint and to address the obligations under the Kyoto protocol. Gaseous pollutants have significant effects on plants, animals, and entire ecosystems. The lateral migration of gas through soil beyond landfill boundaries causes the displacement of oxygen from soil, leading to a decline in soil faunal populations, burrowing animals, and vegetation dieback. Acidic gaseous constituents contribute to acid rains and its secondary effects on the acidification of soils and ecosystems. Ammonia is a major acidic constituent found in landfill gas, which is a secondary acidifying agent following its atmospheric oxidation to nitric acid. It has effects on plants, causing loss of stomatal control, reduction in photosynthesis, enzyme inhibition, changes in synthetic pathways, and depressed growth and yield (Beguedou et al., 2023; Campbell, 2020).

Landfilling is widely employed as the primary method for waste disposal in both the United States and worldwide, resulting in the prevalence of landfill facilities. Landfills are prone to a range of natural and anthropogenic disasters as a result of their extensive geographical distribution. Landfills can sustain damage from different types of disasters, such as geophysical events like earthquakes, meteorological events like storms and extreme temperatures, hydrological events like floods and landslides, climatological events like wildfires, and accidents caused by various factors. Disaster events can have substantial effects on petrol emissions through two separate mechanisms: the impairment of cover systems, which leads to cracking and displacement, thereby causing an increase in emissions. The occurrence of gas extraction systems being damaged or experiencing power failures can lead to functional impairment, thereby causing an escalation in emissions. In the event of damage to the cover system, fully penetrating cracks indicate the release of emissions from an exposed scarp of a mass slide at a landfill. On the other hand, partially penetrating cracks indicate the release of emissions from a translational slide, where a partial thickness of the cover over a slope area is lost, along with fully or partially penetrating cracks occurring at any location on the surface of the landfill. The examination of various crack attributes enables the prioritization of post-disaster landfill repair and maintenance activities in order to minimize gas emissions (Xiu et al., 2021).

4.3.2.1 Landfill Gas Production

Landfill gas is formed through three distinct processes, namely bacterial decomposition, volatilization, and chemical reactions (Yaashikaa et al., 2022).

- Bacterial decomposition is the primary process responsible for the generation of landfill gas. This process occurs when bacteria naturally present in both the waste and the soil covering the landfill break down organic waste. Organic waste encompasses various materials such as food, garden waste, street sweepings, textiles, wood and paper products. The decomposition of organic waste by bacteria occurs in four distinct phases, with the composition of the gas evolving throughout each phase (Fig. 4.2).
- Volatilization: Landfill gases can be generated through the transformation of specific types of waste, particularly organic compounds, as they undergo a phase change from a liquid or solid state to a gaseous state. The phenomenon under discussion is commonly referred to as volatilization. Non-methane organic compounds (NMOCs) present in landfill gas can potentially arise from the volatilization of specific chemicals that have been deposited within the landfill.
- Chemical reactions which are fundamental processes in chemistry that involve the transformation of one or more substances into different substances. These

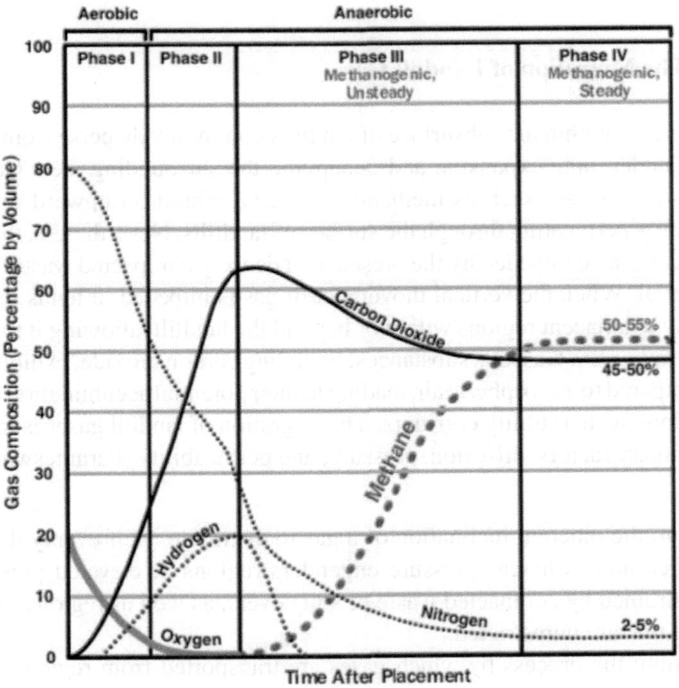

Fig. 4.2 Production phases of typical landfill gas (Vaverková, 2019)

reactions occur when the generation of landfill gas, which encompasses non-methane organic compounds (NMOCs), can occur as a result of chemical reactions involving specific substances found within waste materials. In the event that chlorine bleach and ammonia are exposed to one another within the confines of a landfill, the resultant outcome is the generation of a noxious gas.

Landfill gas production is influenced by various factors, including the presence of nutrients in organic waste, the presence of oxygen in landfills, the content of water in landfills, temperature, and the time of refusal. Bacteria can grow and multiply in organic waste, producing more methane and other gases. However, some wastes contain harmful compounds, leading to reduced gas production. Oxygen is crucial for aerobic bacteria to decompose waste, and more oxygen is available if waste is buried shallowly or regularly disturbed. High oxygen levels can slow methane production, and atmospheric oxygen can affect bacterial activity. Water content in landfills also contributes to methane production, with a moisture content of 40% or higher promoting bacterial growth and distributing nutrients. Temperature is a significant factor in landfill gas production, with temperatures ranging from 77 to 113 °F. The time of refusal from recently buried waste is higher than that from older waste, with most emissions occurring within the first 20 years after waste is deposited. The duration of gas production depends on the waste's organic content (Duan et al., 2021; Vaverková, 2019).

4.3.2.2 The Migration of Landfill Gas

Gases generated within the subsurface of landfills commonly disperse from the landfill site by undergoing expansion and occupying the surrounding area. Gases with lower density than air, such as methane, generally exhibit an upward movement, predominantly permeating through the surface of landfills. Nevertheless, the upward displacement can be impeded by the presence of densely compacted waste or landfill cover material. When the vertical movement of gas is impeded, it tends to disperse horizontally to adjacent regions within or beyond the landfill, allowing it to continue its upward trajectory. Gaseous substances, including carbon dioxide, exhibit a higher density compared to atmospheric air, leading to their potential accumulation in subterranean regions such as utility corridors. The migration of landfill gases is influenced by three primary factors: diffusion, pressure, and permeability (Parameswaran et al., 2023).

- Diffusion: the inherent inclination of a gas to distribute itself evenly throughout a specified area, whereas pressure engenders regions of elevated pressure that are constrained by compacted waste or soil covers, as well as regions of reduced pressure that are unrestricted.
- Convection: the process by which gases are transported from regions of higher pressure to regions of lower pressure, resulting in the movement of gases towards the surrounding ambient or indoor air.

- Permeability: the ability of gases and liquids to traverse interconnected voids or pores within waste materials and soils. Gaseous substances exhibit a propensity to migrate preferentially through regions characterized by elevated permeability, such as sand or gravel, as opposed to regions characterized by diminished permeability, such as clay or silt.

It is also influenced by various factors, such as the type of landfill cover, both natural and man-made pathways, wind speed and direction, moisture levels, groundwater levels, temperature, as well as barometric and soil gas pressure. Permeable landfill covers, such as those composed of gravel or sand, facilitate the upward migration of gas, whereas less permeable covers, such as silts and clays, induce horizontal underground gas migration. The field of geology encompasses the study of natural underground conduits that facilitate the movement of gases. The presence of wind speed and direction is a significant factor in the determination of gas concentration, which exhibits considerable variability on a daily basis (Kamaruddin et al., 2021).

However, the prediction of landfill gas migration distance is challenging due to the influence of numerous factors; nevertheless, observations have indicated travel distances exceeding 1,500 feet. Computer models that utilize data pertaining to the landfill and its surrounding soil conditions have the capability to forecast the approximate patterns of migration originating from pre-existing landfills (Li et al., 2022).

4.4 Greenhouse Gas Emissions from Engineered Landfills

Emissions from engineered landfills and their impacts on climate change and sustainable development are significant concerns for environmental scientists and policymakers. Landfills produce methane, a potent greenhouse gas that contributes to global warming. Additionally, the leachate from landfills can contaminate nearby water sources, posing risks to both human health and ecosystems. Therefore, finding sustainable waste management solutions is crucial to mitigate these emissions and promote a more environmentally friendly approach to landfilling (Ghosh et al., 2023).

Methane (CH_4) and carbon dioxide (CO_2) are classified as greenhouse gases, constituting the primary landfill gases. Carbon dioxide emissions from landfills are excluded from gas emission inventories because they are classified as biogenic in nature and are therefore not accounted for in this context. The trace components nitrous oxide (N_2O), carbon monoxide (CO), and hydrochlorofluorocarbons (CFCs, HCFCs, and HFCs, collectively referred to as F-gases in this context) are highly effective greenhouse gases. It is worth noting that N_2O and certain species within the F-gas chemical group are also recognized as substances that contribute to the depletion of the ozone layer. The emission of landfill gas can result in both direct and indirect radiative forcing (RF) impacts on the Earth's atmosphere (Manheim et al., 2021a).

Methane exhibits both direct and indirect radiative forcing effects. The process of absorbing outgoing radiation is facilitated, and as methane decomposes, it results in the production of greenhouse gases such as carbon dioxide, water vapor, and ozone. Moreover, carbon monoxide (CO) exerts both direct and indirect radiative forcing mechanisms. This is achieved through its absorption and emission of reflected infrared radiation, as well as its chemical influence on the levels of methane, ozone, and carbon dioxide. Both nitrous oxide and F-gases have significant climate impacts as a result of their high global warming potentials. Numerous non-methane volatile organic compounds (NMVOCs) exhibit a high level of reactivity and possess relatively short atmospheric half-lives, typically ranging from a few hours to a few days. As a consequence, these compounds have a significant impact on air quality at both local and regional levels. Non-methane volatile organic compounds (NMVOCs) serve as precursors for the generation of tropospheric ozone, photochemical smog, and secondary organic aerosol (SOA) in the Earth's atmosphere (Manheim et al., 2021b).

Non-methane volatile organic compounds (NMVOCs) play significant roles in the formation and degradation of ozone and secondary organic aerosols (SOA) in the Earth's atmosphere. As a result, they contribute to global climate change both indirectly and directly. Furthermore, certain non-methane volatile organic compounds (NMVOCs), such as benzene and other aromatic or halogenated hydrocarbons, have the potential to cause acute and/or chronic adverse effects on human health. Alkanes and alkenes are the primary categories of secondary organic aerosols (SOA) (Dave et al., 2020).

Substances that contribute to the formation of non-methane volatile organic compounds (NMVOCs) in the atmosphere are commonly referred to as NMVOC precursors. These precursors pose risks to human health and the environment, and as a result, they are classified as hazardous air pollutants (HAPs). The emission of reduced Sulphur compounds poses olfactory challenges and creates nuisances for communities residing in close proximity to these sources. The range of values is between 48 and 50. Certain non-methane volatile organic compounds (NMVOCs) that have undergone chlorination or bromination, such as chloroform or bromoform, have been identified as substances that contribute to the depletion of stratospheric ozone. The primary chemical families that contribute to the formation of tropospheric ozone are aromatics, alkenes, and aldehydes. Secondary organic aerosols (SOAs) refer to particles that are either in a liquid or solid state and are present in the atmosphere. These particles have an indirect impact on the Earth's energy balance and also pose a direct risk to human health. Oxygenated compounds, aromatics, alkanes, carbonyls, and alkenes are the major classes of SOA NMVOC precursors (Duan et al., 2022; Manheim et al., 2021a).

Consequently; landfills play a substantial role in anthropogenic climate change and serve as one of the main generators of global greenhouse gas (GHG) emissions, particularly methane (CH_4). These emissions from landfills contribute to approximately 1–5% of the overall GHG emissions. Another significant source of methane (CH_4) and nitrous oxide (N_2O) emissions in the waste sector is the leachates generated from landfill sites. The greenhouse gas (GHG) emissions from landfills are primarily influenced by factors such as the volume of the landfill, the age of the waste that is disposed of, the temperature, and the moisture content (Górka et al., 2021). Furthermore, the fate of carbon in waste, including carbon sequestered in landfills and CO_2 emissions resulting from collection and decomposition, has been extensively studied. Various countries have reported different rates of CH_4 emission flux from landfills, such as 31.0 mg m^2 min1 in South Africa, 0.21–266 mg m^2 min1 in Japan, 37.5 mg m^2 min1 in Florida, 0.38–89.5 mg m^2 min1 in Taiwan, 1030.6 mg m^2 min1 in Thailand, and 0–1112 mg m^2 min1 in Mexico and Malaysia. These emissions, Hence, the emissions of greenhouse gases (GHGs) from landfill sites are influenced by various stages of municipal solid waste (MSW) management. According to the report, methane (CH_4) emissions from landfills in Europe, the USA, and South Africa were found to range between 0.1 and 1.0 t CH_4 ha^{-1} day^{-1}. The emissions of greenhouse gases from landfills in the United Kingdom have exhibited a downward trend since 2009. A significant decrease of approximately 50% in greenhouse gas (GHG) emissions was observed in the United Kingdom between 2009 (29 million metric tonnes (MMt) of carbon dioxide equivalent (CO_2e) per year) and 2018 (14.1 MMt CO_2e per year). The year 2016 marked the lowest recorded emissions, with a value of 13.9 MMt CO2e per year (Manheim et al., 2021a; Zamri et al., 2021).

4.4.1 Methane Flux

When MSW is first deposited in a landfill, it undergoes an aerobic (with oxygen) decomposition stage when little methane is generated. Then, typically within less than 1 year, anaerobic conditions are established and methane-producing bacteria begin to decompose the waste and generate methane (Shammas et al., 2020). Figure 4.3 provides a comprehensive overview of methane flux data on a global scale, categorized by continent and country (Saunois et al., 2020).

The literature has reported both negative fluxes, representing the ingress of substances from the atmosphere into landfills, and positive fluxes, representing the releases of substances from landfills into the atmosphere. In general, the methane fluxes observed y varied between – 4.50 and 4.15 × 10^4 g/m^2 day, with an average of 7.81 × 10 g/m^2 (95% confidence intervals: 7.07–8.55 × 10 g/m^2). According to Saunois et al. (2020); Pan et al., 2023), the continent of South America exhibited the highest methane fluxes, with Asia, Europe, and North America following in descending order. Australia exhibited the lowest levels of methane fluxes. The phenomenon of methane uptake into soil covers, which entails negative fluxes, has been observed in the regions of North America, Africa, and Asia. The methane fluxes that were reported exhibited minimal variation, as indicated by the 95% confidence intervals which closely clustered around the mean values across all continents. The reviewed municipal solid waste landfills exhibited a significant elevation in methane flux, ranging from 10 to 10^2 grammes per square metre per day. The global mean flux experienced an increase from an average of 10 g/m^2 per day during the time period of 1988–1992 to approximately 10^2 g/m^2 in subsequent periods.

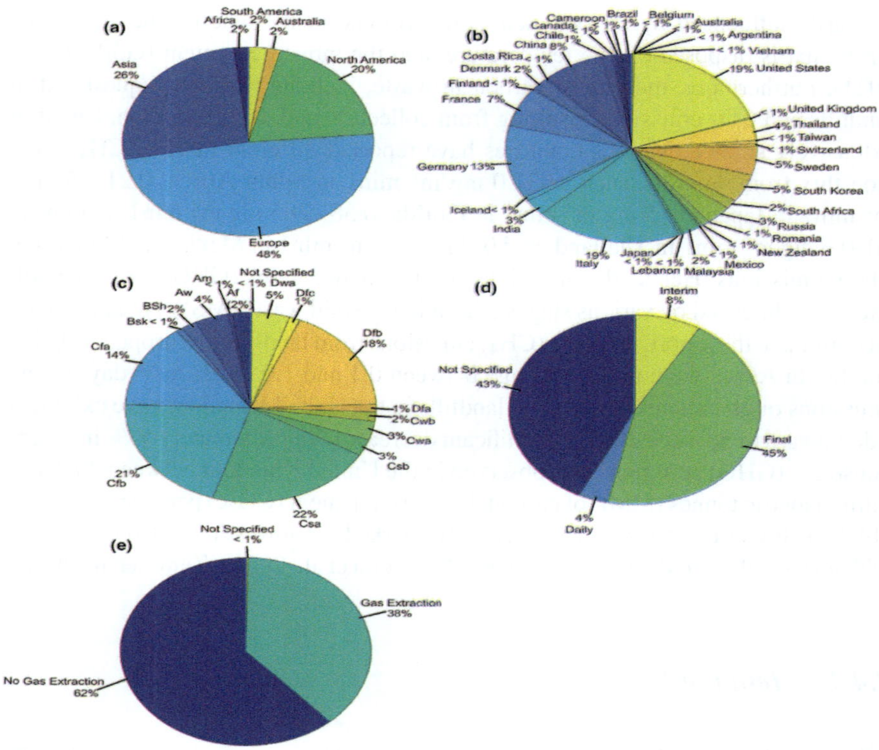

Fig. 4.3 Methane flux data as a function of **a** continent, **b** country, **c** climate zone, d cover category, and e presence of a gas extraction system (Manheim et al., 2021a)

The methane fluxes exhibited significant variation spanning eight orders of magnitude across different countries. The countries with the highest mean methane fluxes were Mexico, South Korea, Cameroon, Costa Rica, Chile, Malaysia, Germany, France, Italy, and the United States. The countries of Switzerland, New Zealand, Canada, and Lebanon exhibited the highest degree of variability in fluxes, specifically in terms of box length. The levels of flux variation were found to be relatively low in the countries of the United States, France, Germany, Italy, Malaysia, and India. However, it was consistently observed that the flux remained at an elevated level of 10^2 grammes per square metre per day. Furthermore, there was significant variation in methane fluxes, commonly referred to as "whiskers," across multiple countries including South Africa, the United States, Vietnam, Iceland, Germany, and Malaysia. Differences in methane emissions across nations can be attributed to factors such as the quantity and composition of waste, as well as the operational and climatic conditions of landfills. The elevated levels of emissions observed in developing countries can be attributed to the presence of organic waste and unregulated disposal practices. Conversely, in developed countries, the heightened emissions can be linked to the significant amount of waste disposed of per unit area through landfilling. The figure illustrates the variation in methane flux in relation to the organic waste fraction (Manheim et al., 2021a, 2021b; Saunois et al., 2020).

4.4.2 *Nitrous Oxide Flux*

Nitrous oxide fluxes ranged from − 2.50 to 3.75 × 10^{-3} g/m^2 day, with a mean of 6.50 × 10^{-2} g/m^2 day (95% confidence intervals: − 1.26 to 1.39 × 10 g/m^2). Africa reported the highest fluxes, followed by Europe, Asia, North America, and South America. Australia was not reported, and North America showed significant variation.

Finland, Cameroon, and India had the highest mean nitrous oxide fluxes, while Argentina and Malaysia had significantly lower fluxes. High flux variations were found in the U.S., Japan, and Thailand. Mean nitrous oxide fluxes varied by five orders of magnitude across countries. Variations in nitrous oxide fluxes between countries are likely due to waste amount, composition, and landfill operational and climatic conditions (Fig. 4.4). High emissions in developing countries stem from organic waste and uncontrolled disposal, while high emissions in developed countries may be due to high waste disposal per unit area of landfilling. The amount of organic matter and available nitrogen in waste streams affects nitrous oxide production through anaerobic microbial processes (denitrification) and cooxidation of bioavailable N and methane in cover soils with oxygen.

More climate-based variations were observed in nitrous oxide flux compared to methane flux. Tropical, temperate, and cold climates (A-, C-, and D-climates) had higher flux data, while arid climates (B-climate) had lower flux data. High temperatures and precipitation in most climate zones likely facilitated nitrous oxide formation and flux, while low precipitation, low waste mass moisture, and cold waste and cover temperatures likely reduced flux. Long-term cold or dry landfill conditions hinder efficient bacterial nitrogen cycling and nitrous oxide production year-round. Nitrous oxide flux shows an increase with rising average precipitation and temperature (Nemitz et al., 2018; USEPA, 2020).

4.4.3 *NMVOC Flux*

An Overall, NMVOC fluxes ranged from − 2.00 × 10^{-3} to 7.32 × 10^{-1} g/m^2 day, with a mean of 1.99 × 10^{-2} g/m^2 (95% confidence intervals: 1.50 × 10^{-2}–2.48 × 10^{-2} g/m^2). Alcohols had the highest mean fluxes, followed by aromatics, monoterpenes, ketones, and alkanes (Fig. 4.5). Ethanol, m-xylene, alpha-pinene, acetone, and propane had the highest average fluxes within these chemical families. Alcohols, monoterpenes, and F-gases had the greatest NMVOC emission variation (Fig. 4.5). High fluxes of oxygenated compounds (alcohols, ketones), aromatics (BTEX),

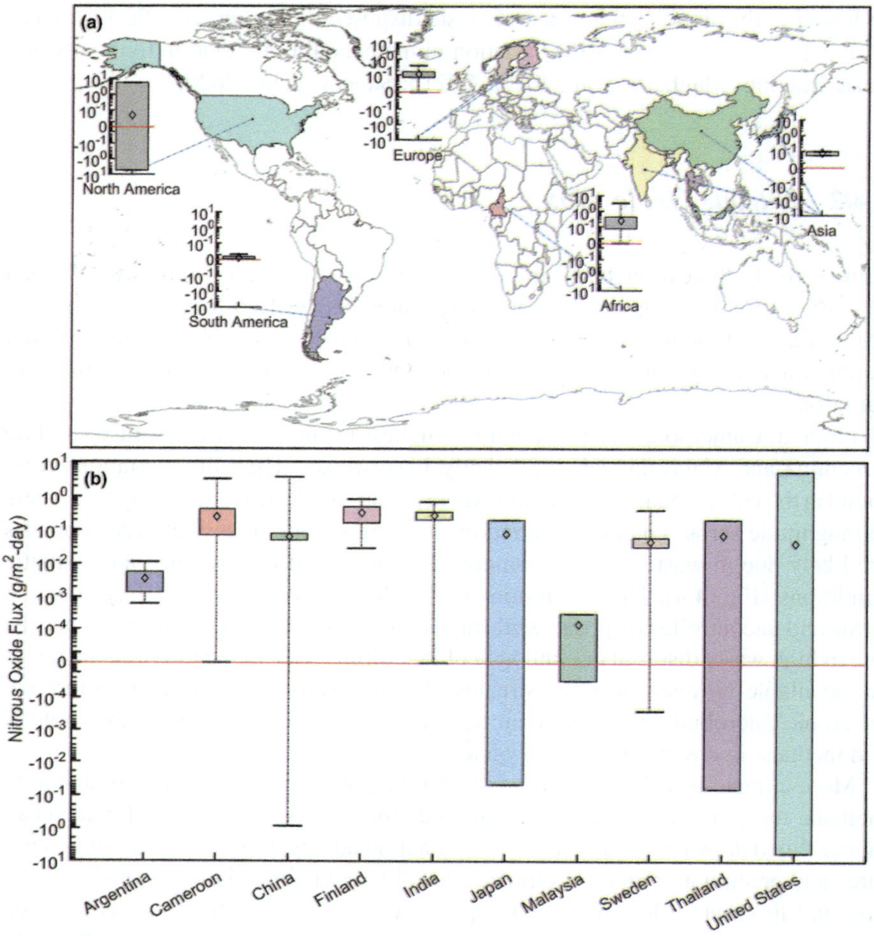

Fig. 4.4 Nitrous oxide flux (g/m^2 day) for MSW Landfills by **a** continent and **b** country. The diamonds, boxes, and whiskers are for the mean, 95% confidence intervals, and min/max fluxes (Manheim et al., 2021a)

monoterpenes, and aliphatic hydrocarbons indicate biotic and abiotic NMVOC generation in MSW landfills. The lowest NMVOC fluxes were for organic nitrates. Early anaerobic decomposition of waste materials (fresh wastes and younger landfills) produces oxygenated NMVOC compounds like alcohols, organic acids, and ketones. Solvents, oils, degreasers, cooking fuels, and other human-made sources generate aromatic and aliphatic hydrocarbons. NMVOC flux was more affected by climate than methane flux. Tropical and temperate climates (A- and C-climates) had higher flux data, while cold climates (D-climates) had lower flux data. High temperature and precipitation in the A- and C-climate zones likely facilitated the transport of volatile, moderate-to-high water solubility compounds (Manheim et al., 2021a).

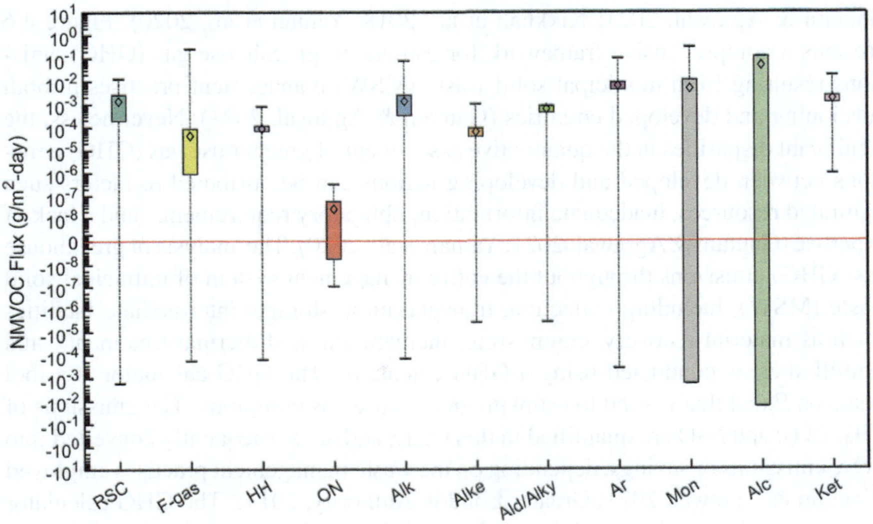

Fig. 4.5 NMVOC flux for MSW landfills by chemical family. The diamonds, boxes, and whiskers are for the mean, 95% confidence intervals, and min/max fluxes (Manheim et al., 2021a)

4.4.4 Estimating GHG Emissions: Approaches and Techniques

The inventory conducted by the Intergovernmental Panel on Climate Change (IPCC) includes a comprehensive list of calculation methods used to quantify greenhouse gas (GHG) emissions resulting from waste disposal (Xin et al., 2020). The Intergovernmental Panel on Climate Change (IPCC) classifies greenhouse gases (GHGs) into four primary categories, which include national accounting, corporate-level accounting, lifecycle assessment, and carbon trading methodologies (Yaman et al., 2020). Numerous scholars have adhered to the guidelines set forth by the Intergovernmental Panel on Climate Change (IPCC) when computing greenhouse gas (GHG) emissions related to municipal solid waste (MSW) management (Xin et al., 2020; Yaman et al., 2020). The carbon factor utilized in the calculation is subject to variation based on the specific type of vehicles and treatment procedures involved. The factors employed in the computation of greenhouse gas emissions are derived from the Waste and Resources Assessment Tool (WRATE) version 2 and data provided by the Department of Energy and Climate Change (Gautam & Agrawal, 2021). The Land GEM modelling approach is commonly employed for the analysis of greenhouse gas (GHG) emissions originating from landfill sites (Gautam & Agrawal, 2021; Xin et al., 2020). A comprehensive review of the available scholarly works pertaining to the quantification of greenhouse gas emissions (GHGs) in municipal solid waste (MSW) management was conducted, encompassing studies conducted in both developed and developing nations. The primary emphasis of this review was on the utilisation of life cycle assessment methodology (Fang et al., 2023;

Gautam & Agrawal, 2021; Nikkhah et al., 2018; Yaman et al., 2020). Figure 4.6 presents a comprehensive framework for estimating greenhouse gas (GHG) emissions resulting from municipal solid waste (MSW) management practises in both developing and developed countries (Gautam & Agrawal, 2021). Nevertheless, the significant disparities in the quantitative assessment of greenhouse gas (GHG) emissions between developed and developing nations can be attributed to factors such as limited resources, inadequate information, obligatory requirements, and a lack of expertise (Gautam & Agrawal, 2021; Yaman et al., 2020). The analysis of greenhouse gas (GHG) emissions throughout the entire management system of municipal solid waste (MSW), including collection, transportation, storage, intermediate facilities (such as material recovery, composting, incineration, and thermal treatment), and landfill sites, is conducted using a GHG calculator. The GHG calculator is a tool based on Excel that is used to estimate greenhouse gas emissions. The emissions of CH_4, CO_2, and N_2O are quantified in this study, and are subsequently converted into CO_2e emissions or savings, depending on the waste management practises employed (Gautam & Agrawal, 2021; Greater London Authority, 2014). The GHG calculator comprises various carbon modules, each of which provides data on the carbon dioxide equivalent (CO_2e) performance of distinct waste management systems.

Landfill gas (LFG) generation models are employed for the purpose of estimating the production of gas within landfill sites, with the primary objective of attaining optimal precision. The models under consideration encompass zero order, first order, second order, and/or multiphase configurations. Nevertheless, zero-order models lack reliability as they exhibit significant levels of inaccuracy, whereas higher order

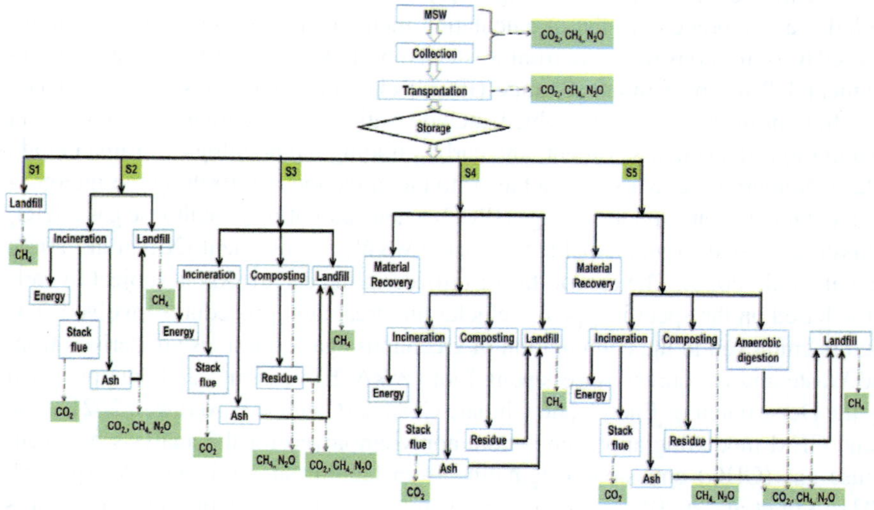

Fig. 4.6 Scenarios in life cycle assessment of municipal solid waste management (from collection to landfill wastes) with respect to major greenhouse gas emissions (CO_2, CH_4, and N_2O). CO_2: carbon dioxide; CH_4: methane; N_2O: nitric oxide; S1: scenario 1; S: scenario 2; S3: scenario 3; S4: scenario 4; S5: scenario 5 (Gautam & Agrawal, 2021)

models demonstrate reduced inaccuracies. The majority of users exhibit a preference for utilising a first order model (Nikkhah et al., 2018; Toha & Rahman, 2023).

Numerical models and their associated parameters have been developed, yet their practical application is limited by their lack of reliability. Several simplified empirical models have been developed in order to enhance their user-friendliness. Certain models are equipped with specialized computer software programmes in order to enhance their usability. Several examples of models that are commonly used in the field include the Intergovernmental Panel on Climate Change (IPCC) model, the Emissions and Land-use Planning System (E-PLUS), the French Agency for Ecological Transition (ADEME) models, the Dutch Multiphase (AMPM) model, and the United States Environmental Protection Agency (EPA) Landfill Gas Emissions Model (Land GEM).

The Land GEM model developed by the United States Environmental Protection Agency (US EPA) is utilized to estimate the emission rates of methane (CH_4), non-methane organic compounds (NMOCs), carbon dioxide (CO_2), and various other toxic air pollutants originating from landfill sites. The Triangular model posits a two-phase waste decomposition process, whereas the E-PLUS model assesses the economic and environmental impacts of landfill gas (LFG) collection projects, including the estimation of methane (CH_4) flow, non-methane organic compounds (NMOCs) emissions, and LFG flow. The ENCON model enables users to input precise waste environmental parameters, including temperature, moisture input, and extraction efficiency. The Intergovernmental Panel on Climate Change (IPCC) and landfill gas (LFG) generation models developed by the United States Environmental Protection Agency (EPA) have been extensively utilized by professionals involved in the assessment, design, and operation of landfills. Nevertheless, the primary parameters are consistently defined across all of these models. Numerous scholars have conducted investigations pertaining to the precision of LFG models in estimating gas generation rates originating from landfill sites. The researchers aimed to assess the accuracy of model predictions by comparing them to the empirical data obtained from LFG records. The findings indicated that the TNO and German EPER models exhibited an overestimation of the production rate of landfill gas (LFG) by a factor of two to six. Conversely, the Land GEM version 2.01 and the Belgium models demonstrated a standard error of less than 100% in their estimations (Bosetti, 2021; Skea et al., 2021; Srivastava & Chakma, 2020).

4.5 Ecological and Sociological Risks of GHG Emissions

The release of greenhouse gases (GHGs) has substantial consequences for both the natural environment and socioeconomic factors. The primary reason for their ability to contribute to global warming and climate change is responsible for impacting ecological balance, modifying biodiversity patterns, causing atmospheric pollution, and influencing the well-being of plants and animals.

4.5.1 Global Warming

CO_2 is the most important anthropogenic gas whose atmospheric concentration has increased since preindustrial times. Global atmospheric CO_2 concentration has increased from 354.39 to 414 ppm in the past decade, along with CH_4 and N_2O concentrations, which absorb outgoing radiant energy and cause "Global Warming" by 1.1 °C. Climate change potential (GWP) of CO_2, CH_4, and N_2O is 1, 28, and 265 over 100 years. It is the energy absorbed by 1 tonne of petrol emissions over a given time compared to 1 tonne of CO_2. The gas absorbs more energy with higher GWP. GHG concentrations raise atmospheric temperatures, warming the earth (Costa et al., 2021; Elemam & Eldeeb, 2023; Zheng et al., 2022).

4.5.2 Climate Change

The rise in greenhouse gas concentrations and subsequent increase in temperature are contributing to global changes in climatic patterns, commonly referred to as "forcing climate change." This phenomenon aims to restore equilibrium between incoming and outgoing solar radiation. The anthropogenic influence on climate change has resulted in alterations to cloud cover, wind speed, snowfall and rainfall patterns, as well as the duration of sunshine hours. Furthermore, it has had an impact on the typical weather patterns. Collectively, the alteration has resulted in a significant transformation in global climatic and weather patterns. The imposition of climate change has significant consequences for both terrestrial and aquatic ecosystems, encompassing the world's oceans and their respective ecological systems (Elemam & Eldeeb, 2023; Kumari et al., 2020).

4.5.3 Carbon Cycle

Carbon dioxide (CO_2), methane (CH_4), and nitrous oxide (N_2O) gases participate in a natural cycle. The exchange of carbon dioxide (CO_2) between the atmosphere and the Earth's surface encompasses various processes, including respiration, photosynthesis, decomposition, and combustion. Maintaining a consistent level of carbon dioxide (CO_2) is crucial for sustaining a balanced carbon cycle in the natural environment. However, the heightened concentration of CO_2 resulting from human activities poses a significant threat to the integrity of this natural cycling process (Vestin et al., 2020).

4.5.4 Effects on the Aquatic Ecosystem

Research findings indicate that, commencing from 1955, a significant proportion exceeding 80% of the heat has been absorbed by the Earth's oceans. From 1961 to 2003, there has been an observed increase in the global average temperature of the ocean up to a depth of 700 m by 0.10 °C. Within aquatic environments, numerous organisms exhibit optimal functioning within specific temperature ranges. However, any elevation in the ambient temperature can have detrimental effects on their physiological processes, metabolic activities, reproductive patterns, and overall survival. It was documented that the primary driver behind the global extinction of various marine organisms, such as phytoplankton, zooplankton, plants, animals, and coral bleaching, is the rise in sea temperature and ocean acidification (Hornbach et al., 2020). In the year 2016, a study revealed that 16% of corals experienced complete bleaching, while 54% experienced partial bleaching. These bleaching events were primarily attributed to the increase in temperature and the emission of carbon dioxide (CO_2) (Hédouin et al., 2020). The phenomenon of ocean warming leads to the expansion of water, while the melting of ice on land contributes to the increase in sea levels. The increase in greenhouse gas (GHG) emissions and subsequent climate change have resulted in a substantial elevation of the sea surface level between 2010 (54.5 mm) and 2020 (95 mm). An increase in the sea surface level leads to various adverse effects such as erosion, flooding, and submergence of coastal areas with low elevation. It also contributes to heightened storm-surge flooding and the inward movement of seawater into estuaries and aquifers. Consequently, these phenomena result in the destruction of habitats and loss of biodiversity (Elemam & Eldeeb, 2023).

4.5.5 Effects on the Terrestrial Ecosystem

The phenomenon of global warming and the resulting alteration of climatic patterns can be attributed to the heightened emissions of greenhouse gases (GHGs) and the release of other detrimental substances such as ammonia (NH_3) and volatile organic compounds (VOCs). These factors possess the capacity to significantly influence the functioning of terrestrial ecosystems. The release of detrimental components, such as volatile organic compounds (VOCs) and ammonia (NH_3), during various phases of municipal solid waste (MSW) management leads to the contamination of the atmosphere, water bodies, and soil. Greenhouse gas (GHG) emissions have a significant impact on various ecological processes, including evapotranspiration, net carbon storage, biodiversity patterns, species composition, nutrient cycling, and soil dynamics. (Lu et al., 2022).

Wetland and terrestrial plants are recognized as significant carbon sinks. However, both species are also susceptible to climatic fluctuations attributed to the impacts of global warming. The rise in greenhouse gas (GHG) concentrations in the atmosphere has been observed to have both positive and negative impacts on different aspects of a plant's carbon budget. These impacts include changes in photosynthesis, respiration, biomass accumulation, allocation, metabolic functioning, decomposition, growth, and reproductive development (Xue et al., 2021).

4.5.6 Socioeconomic Impacts of GHGs

The greenhouse gas emissions resulting from intermediate processes and landfills have been identified as significant contributors to various socioeconomic impacts. These impacts include adverse effects on human health, primarily caused by exposure to noxious gases. Additionally, the contamination of ground and surface water by leachates further exacerbates these issues. While advanced waste management systems in developed countries are effectively designed to minimize emissions, there remains a significant concern regarding the health of rag pickers and workers in close proximity to waste collection points and landfill sites (Gautam & Agrawal, 2021). Exposure to emissions, whether through direct contact, inhalation, or ingestion of contaminated food and water, has been linked to various diseases, including heart-related ailments, respiratory problems, skin irritation, metabolic dysfunction, congenital malformations, nonchromosomal birth defects, reduced birth weight, premature birth, improper child growth, and cancers affecting the lung, stomach, liver, bile ducts, cervix, and prostate (Danthurebandara et al., 2012; Ramos et al., 2020). The economic implications of greenhouse gas emissions from waste management systems are significant. In high-income countries, the operational expenses associated with waste management, spanning from the collection stage to the ultimate disposal, typically fall within the range of US$ 100–170 per metric tonne. Conversely, in low to middle income countries, this range fluctuates between US$ 20 and 50 per metric tonne. The impacts of global warming and climate change resulting from increased greenhouse gas emissions have substantial consequences for various sectors, including agriculture, fisheries, horticulture, forestry, animal welfare, and human health. These effects impose a significant strain on the global economy (Danthurebandara et al., 2012; Ramos et al., 2020).

4.6 Mitigation of Greenhouse Gases (GHG) Emissions and the Sustainable Development

4.6.1 Combating Strategies for Greenhouse Gases (GHG) Emissions from Municipal Solid Waste Management

In developing nations, the majority of waste is disposed of in an unscientific manner following treatment, resulting in various environmental issues including leaching, surface runoff, greenhouse gas emissions, and subsequent impacts on living organisms. Developed nations, including the United States, Japan, and various European countries, have implemented scientifically engineered landfills for the purpose of waste and residue disposal. The sanitary landfill is equipped with both side and bottom liners, which serve as protective barriers. Within this landfill, stabilized and unrecoverable waste is deposited in layers, adhering to established guidelines for solid waste management. The waste is subjected to compression in order to optimize spatial utilisation, and subsequently enveloped by an inert layer equipped with vents to facilitate the recovery of gases. Additionally, a drainage network is installed at the bottom to gather leachates (USEPA, 2020; Yaman et al., 2020).

4.6.1.1 Improved Landfill Practices

Current waste management practices have the potential to mitigate greenhouse gas (GHG) emissions in the waste sector. A diverse array of well-established, sustainable, and ecologically sound technologies have the capacity to mitigate greenhouse gas (GHG) emissions while concurrently fostering environmental preservation, safeguarding public health, and advancing sustainable development. These technologies have the potential to mitigate greenhouse gas (GHG) emissions by implementing effective landfill practices, recovering gas emissions, and implementing efficient wastewater management systems. Additionally, they can also prevent substantial GHG generation by employing advanced incineration techniques, promoting composting practices, and improving waste collection methods. Furthermore, the implementation of waste reduction, reutilization, and recycling strategies plays a crucial role in the indirect reduction of greenhouse gas (GHG) emissions. The aforementioned outcomes are a result of the efforts made to preserve natural resources, reduce reliance on non-renewable energy sources, and enhance the effectiveness and efficiency of energy and resource utilization. Waste management policies in developed nations such as Japan and the European Union are intricately interconnected and harmonized with climate policies established by international organizations like the Intergovernmental Panel on Climate Change (IPCC), the Environmental Protection Agency (EPA), and the World Bank. Current waste management practices have the potential to mitigate greenhouse gas (GHG) emissions in the waste sector. There is a wide array of sustainable technologies available that can effectively mitigate greenhouse gas emissions while simultaneously fostering environmental preservation, safeguarding public health, and facilitating sustainable development. When these technologies are integrated, they have the potential to mitigate greenhouse gas (GHG) emissions by improving land management, recovering gas, and managing wastewater. Additionally, they can also effectively prevent the substantial generation of GHGs through advanced incineration, composting, and waste collection methods. Furthermore, the implementation of waste reduction, reutilization, and recycling practices plays a crucial role in indirectly mitigating greenhouse gas (GHG) emissions. The aforementioned outcomes are a result of the efforts to preserve primary resources, reduce reliance on non-renewable energy sources, and enhance the efficacy of energy and resource utilization. Waste management policies in developed nations such as Japan and the European Union (EU) exhibit a strong interconnection and integration with climate policies established by the Intergovernmental Panel on Climate Change (IPCC), the Environmental Protection Agency (EPA), and the World Bank (Fang et al., 2023; USEPA, 2020; Yaman et al., 2020).

4.6.1.2 Waste Segregation Process

The management and environmental impact of municipal solid waste (MSW) are contingent upon the practice of waste segregation. The process of domestic waste segregation involves the categorization of waste materials into two main groups: organic/biodegradable and nondegradable components. This segregation allows for the separation of waste into distinct categories, including biodegradable, combustible, recyclable, hazardous, infectious, and inert wastes. The practice of segregating waste at facilities has been found to result in energy conservation and reduced greenhouse gas emissions. Developed nations such as the United States, Japan, and countries in Europe have made advancements in landfill practices in order to effectively manage and handle waste disposal (Roy & Tarafdar, 2022).

4.6.1.3 Controlled Composting and Anaerobic Digestion

The utilization of controlled composting and anaerobic digestion techniques has been shown to be a highly effective means of mitigating methane emissions originating from landfills. The mitigation of methane (CH_4) emissions in landfill generation models in the United States is achieved through the utilization of regulatory frameworks such as the Clean Air Act (CAA) and the New Source Performance Standards (NSPS). Based on the provisions outlined in the European Union landfill directive (1999/31/EC), a notable decrease of 35% was observed in the disposal of biodegradable wastes within landfills across the European Union. The utilization of increased landfill taxes and the establishment of regular landfill methane programs provide significant technical resources for the efficient control of landfill gases (Troncoso, 2022). The growing emphasis on the utilization of municipal solid waste (MSW) compost as a type of organic fertilizer can be ascribed to various factors, such as the constrained availability of land for waste disposal, the considerable expenses linked to waste management, and the environmental concerns associated with conventional waste disposal techniques. The incorporation of municipal solid waste (MSW) compost into agricultural and horticultural practices is regarded as a financially efficient and exceedingly encouraging approach for the management of MSW. This approach effectively addresses the adverse effects on both society and the environment, while simultaneously enhancing the nutritional content of the soil and plants (Cao et al., 2023; Lee et al., 2020).

4.6.1.4 Modern Controlled Incineration

Modern incineration techniques involve the application of regulated heat in order to extract usable energy from post-consumer waste. The utilization of incinerators has been restricted in many developing nations due to resource constraints associated with their installation. The implementation of subsidies for incinerators and the establishment of energy efficiency standards are imperative. Furthermore, financial support has been allocated for the development of incinerators, combustion chambers, and mechanical biological treatment infrastructure (Lee et al., 2020).

4.6.1.5 Implementing the 5-R Principle

The resource value of municipal solid waste (MSW) is emphasized as one of the guiding principles of the 5-R hierarchy (Fig. 4.7), which is recognized internationally. It is an extremely important factor in the decrease of greenhouse gas emissions and helps relieve pressure on the waste management system (Table 4.2). The hierarchy was established to facilitate management of the MSW on multiple levels: Reduction of wastes at source; reutilization of wastes wherever possible; recycling of the end products towards their useful life; recovery of the energy or materials from the wastes; and managing the residuals in an environmentally comprehensive way (Alsheyab, 2022).

4.6.1.6 U.S. EPA's Landfill Methane Outreach Program (LMOP)

The Landfill Methane Outreach Program (LMOP) of the United States Environmental Protection Agency encourages the use of landfill gas as a renewable and eco-friendly source of energy. Methane and carbon dioxide make up the bulk of landfill gas, which is produced as a natural by-product when solid waste breaks down in landfills. Landfill gas is also known as digester gas. Through the formation of partnerships with communities, landfill owners, utilities, power marketers, states, project developers, tribes, and non-profit organizations, LMOP assists project developers in overcoming

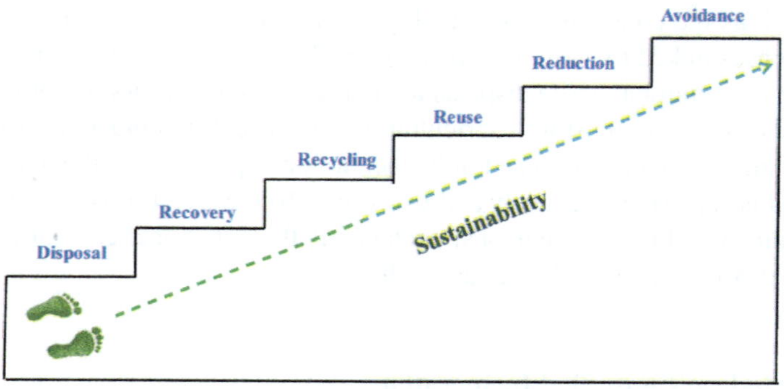

Fig. 4.7 The relationship between waste hierarchy and sustainable development (Alsheyab, 2022)

obstacles that stand in the way of project development by assisting them in determining whether or not a project is feasible, locating financing options, and informing the community about the benefits of project development (Hussein et al., 2022).

In a more prevalent manner, landfill gas is managed through active collection techniques, which involve the utilization of pipes to monitor the lateral migration of gas, or through extraction methods that involve the use of pumps connected to a network of horizontal gas trenches and vertical wells. The gas possessing a substantial energy content can be enhanced through the utilization of scrubbers, which serve to eliminate gaseous impurities including carbon dioxide, water vapor, and particulate matter. This process is employed in the context of power generation. The implementation of energy recycling practices is particularly feasible in the context of expansive landfill sites. The reclamation of landfill gas offers the potential for its utilization in two primary ways: direct use for heating purposes and indirect use for the generation of electricity. In instances where the rate of gas extraction is insufficient to warrant power generation, it is possible to incinerate the gas at a flare stack. The management of odor emanating from landfills can be effectively achieved through the implementation of certain practices. These practices include the maintenance of a small working face during routine operations and the subsequent covering of the waste spread after daily operations on active landfills. Improved land use planning and site selection, incorporating considerations such as the positioning of vulnerable populations and the influence of wind patterns and topography, has the potential to effectively mitigate odor-related grievances. In the context of post-closure landfills, the management of odor emissions can be efficiently achieved through the implementation of horizontal collectors for the collection of landfill gas, coupled with the strategic covering of manholes (Hadjidimoulas, 2018).

Table 4.2 Management strategies to combat greenhouse gas (GHGs) emissions from municipal solid waste management and subsequent benefit to environment and socioeconomic facets (Gautam & Agrawal, 2021)

Policies and preventive measures	Environmental and socio-economic benefit	Minimization of GHGs
5-Rs (Refuse, Reduce, Reuse, Repurpose, and Recycle)	Reduction in use of raw materials, energy, and economic acquisition Offers job creation for more people with implementation of health and safety provisions Reduce the burden on waste management system and landfill sites	CO_2, CH_4, and N_2O
Expanded sanitation coverage	Minimize un managed disposal of waste in rural and urban areas Offers job to many people Minimize the detrimental effects on environment and human health Maximize material and energy recovery from wastes and economic acquisition	NA
Prohibition on uncontrolled disposal and open burning of waste	Inculcation of alternate low-cost technological innervations Minimize the detrimental effects on environment and human health Inhibit pathogen growth and disease vectors	CO_2, CH_4, and N_2O
Use of natural gas in place of petrol or diesel-based vehicles and handcart system	Minimizes environmental pollution Diminishes emission of fluorinated gases Regulation economic incentive Creates job opportunities	CO_2, CH_4, and N_2O
Efficient waste segregation	Regulation economic incentive Reduce the burden on waste management system and landfill sites	CO_2, CH_4, and N_2O
Reduction in biodegradable waste which is landfilled	Prevent microbial decomposition of organic matter and subsequent emission of CH_4 and N_2O Regulation economic incentive Reduces foul smell	CO_2, CH_4, and N_2O
Aerobic biological treatment of wastes	Although reduces CO_2 emissions but can emit N_2O and CH4underreducedaerationoranaerobicconditions Reduces volume stabilizes organic C, and destroys pathogens Compost generation which can be used as fertilizer Foul odor reduction	CO_2
Anaerobic biological treatment of wastes with energy recovery system	Even though minimizes CO_2 generation but if not properly managed, leads to enormous CH_4 emission CH_4 in biogas can be used as fuels for heat or electricity generation Produce bio solid under controlled conditions Foul odor reduction	CH_4, and N_2O

(continued)

Table 4.2 (continued)

Policies and preventive measures	Environmental and socio-economic benefit	Minimization of GHGs
Mandate in installation of incineration and thermal processing for waste-to-energy	Reduce environmental pollution such as heavy metal contamination, atmospheric release of acid gases, dioxins and other noxious substituents Potential odor reduction Provide significant mitigation potential for the waste sector Regulation economic incentive offers replacement of fossil fuels Offers job creation for more people	CO_2, CH_4, and N_2O
Standards for combustion and landfill performance to harness gas with or without energy recovery	Regulation economic incentive Waste to energy recovery reduce the burden of electricity generation and fuel combustion Efficient and proper management of MSW with less adverse impact on environment and human health	CH_4 (predominant), N_2O, andCO_2

4.6.2 Sustainable Development

The concept of Sustainable Development (SD) was first introduced in 1987 through the publication of the Brundtland Report, which subsequently garnered significant global recognition and acceptance. The primary objective of this initiative is to enhance the quality of life in a manner that is fair and equitable across different generations, both within and between generations. This goal is pursued through various means, including fostering community engagement and implementing transformative measures that address social, economic, and environmental dimensions. In 2015, the Paris Agreement was established as a response to the pressing issue of climate change. Concurrently, the World Sustainable Development Summit endorsed the Agenda 2030, encompassing a comprehensive framework consisting of 17 Sustainable Development Goals (SDGs). The management of solid waste poses a considerable obstacle in the pursuit of sustainable development (SD), as insufficient disposal methods result in environmental pollution, depletion of resources, substandard living conditions, public health issues, and climate change. Landfill mining is considered a strategic alternative for sustainable development (SD) and effective waste management, offering various social, economic, and environmental advantages (Borowy, 2021). Landfill mining (LM) has the potential to mitigate environmental consequences and prolong the operational lifespan of landfills. Additionally, it enables the retrieval of valuable waste materials such as metals and rare-earth elements, thereby fostering economic expansion and facilitating the advancement of environmentally sustainable technologies. Furthermore, the implementation of LM contributes to the mitigation of greenhouse gas emissions originating from landfills and dumpsites by means of organic compound treatment and recovery. Another important strategic alternative is sustainable landfill which can be defined as a landfill site that maintains a state of equilibrium in the amount of waste it contains, while also ensuring that the release of emissions remains at a level that is considered environmentally acceptable within the specific local context. Furthermore, the management of emissions is accomplished by implementing measures that are uncomplicated, robust, and ecologically sustainable. Moreover, it can be considered as a repository for the extended retention of carbon and nitrogen.

The prioritization of landfill capacity optimization and the mitigation of dependence on new landfill sites are of paramount significance. The effective utilization of landfills allows for the establishment of a symbiotic relationship with the environment, thereby facilitating the restoration of land for ecological purposes or the implementation of new utilitarian functions. (Bello et al., 2022).

4.7 Conclusions

Landfills are crucial in waste management strategies, especially in developing countries, where they are the primary method for disposing of biodegradable waste. However, recent research has shown that landfills are a significant source of greenhouse gas (GHG) emissions, with methane and carbon dioxide being the main components of landfill gas. High-income nations produce approximately 683 million metric tonnes of municipal solid waste (MSW) annually, with the Middle East, North Africa, South Asia, and Sub-Saharan Africa projected to experience the most rapid waste generation by 2050. Regulation and enforcement of landfills have been established since the mid-twentieth century, but there have been disparities in regulation and enforcement among different landfill locations. Landfills contribute to global warming by emitting gases like CO_2 and methane into the atmosphere, contributing to environmental pollution. The degradation process is responsible for the generation of these gases, which subsequently contribute to environmental pollution.

There exist multiple strategies for mitigating greenhouse gas emissions originating from landfills; however, it is important to note that none of these approaches can operate independently without external support or resources. Optimal outcomes can be attained by employing diverse levels of intervention and approaches. In the case of expansive landfills characterized by a substantial presence of organic matter, an optimal approach could involve the integration of gas utilization techniques for energy generation alongside the implementation of a landfill biocover specifically engineered to facilitate efficient oxidation of CH_4. Nevertheless, the utilization of biogas for combustion encounters challenges, including the significant expenses associated with gas recovery and burning systems. These obstacles impede the widespread adoption of biogas technology in developing nations. According to projections, the global emissions stemming from landfills are anticipated to witness a substantial rise, escalating from 340 Tg CO_2eq in 1990 to 1500 Tg CO_2eq by 2030, and further surging to 2900 Tg CO_2eq by 2050. The estimated overall mitigation potential is projected to reach a maximum of 30% by the year 2030, and further increase to 50% by the year 2050, under the assumption of implementing incineration, recycling, and landfill gas recovery methods. Therefore, the establishment and enforcement of consistent waste landfilling standards are of utmost importance in order to mitigate the environmental consequences.

References

Abdel-Shafy, H. I., Ibrahim, A. M., Al-Sulaiman, A. M., & Okasha, R. A. (2023). Landfill leachate: Sources, nature, organic composition, and treatment: An environmental overview. *Ain Shams Engineering Journal*, 102293.

Abdel-Shafy, H. I., Ibrahim, A. M., Al-Sulaiman, A. M., & Okasha, R. A. (2023). Landfill leachate: Sources, nature, organic composition, and treatment: An environmental overview. *Ain Shams Engineering Journal*, 102293.

Al Raisi, S. A. H. (2022). The generation, composition and fate of landfill leachate: A review. *European Journal of Environment and Earth Sciences, 3*(2), 105–109.

Alsheyab, M. A. T. (2022). Recycling of construction and demolition waste and its impact on climate change and sustainable development. *International Journal of Environmental Science and Technology, 19*, 2129–2138.

Aziz, H. A., Rosli, N. A., & Hung, Y. T. (2020). Landfill methane emissions. In *Handbook of environment and waste management: Acid rain and greenhouse gas pollution control* (pp. 397–454).

Banias, G., Batsioula, M., Achillas, C., Patsios, S. I., Kontogiannopoulos, K. N., Bochtis, D., & Moussiopoulos, N. (2020). A life cycle analysis approach for the evaluation of municipal solid waste management practices: The case study of the region of central Macedonia, Greece. *Sustainability, 12*(19), 8221.

Barati, B., Zafar, F. F., & Wang, S. (2022). Different waste management methods, applications, and limitations. *Waste-to-energy: Recent developments and future perspectives towards circular economy* (pp. 21–58). Springer.

Batista, M., Caiado, R. G. G., Quelhas, O. L. G., Lima, G. B. A., Leal Filho, W., & Yparraguirre, I. T. R. (2021). A framework for sustainable and integrated municipal solid waste management: Barriers and critical factors to developing countries. *Journal of Cleaner Production, 312*, Article 127516.

Beguedou, E., Narra, S., Agboka, K., Kongnine, D. M., & Afrakoma Armoo, E. (2023, July). Review of Togolese policies and institutional framework for industrial and sustainable waste management. In *Waste* (Vol. 1, No. 3, pp. 654–671). MDPI.

Bello, A. S., Al-Ghouti, M. A., & Abu-Dieyeh, M. H. (2022). Sustainable and long-term management of municipal solid waste: A review. *Bioresource Technology Reports, 18*, Article 101067.

Borowy, I. (2021). The social dimension of sustainable development at the UN: From Brundtland to the SDGs. In *The struggle for social sustainability* (pp. 89–108). Policy Press.

Bosetti, V. (2021). *Integrated assessment models for climate change*. Oxford University Press.

Campbell, D. J. (2020). 3.2 Explosion and fire hazards associated with landfill gas. In *Landfilling of waste: Biogas* (p. 133).

Cao, X., Williams, P. N., Zhan, Y., Coughlin, S. A., McGrath, J. W., Chin, J., & Xu, Y. (2023). Municipal solid waste compost: Global trends and biogeochemical cycling. *Soil & Environmental Health*, 100038.

Chaturvedi, A., Singh, G. S. P., & Sharma, S. K. (2023). Stabilization of mine waste dumps through bio-engineering. *Journal of The Institution of Engineers (India): Series D*, 1–12

Chetri, J. K., Reddy, K. R., & Grubb, D. G. (2020). Carbon-dioxide and hydrogen-sulfide removal from simulated landfill gas using steel slag. *Journal of Environmental Engineering, 146*(12), 04020139.

Costa, C., Wironen, M., Racette, K., & Wollenberg, E. K. (2021). Global Warming Potential*(GWP*): Understanding the implications for mitigating methane emissions in agriculture.

Danthurebandara, M., Van Passel, S., Nelen, D., Tielemans, Y., & Van Acker, K. (2012). Environmental and socio-economic impacts of landfills. *Linnaeus Eco-Tech, 2012*, 40–52.

Dave, P. N., Sahu, L. K., Tripathi, N., Bajaj, S., Yadav, R., & Patel, K. (2020). Emissions of non-methane volatile organic compounds from a landfill site in a major city of India: impact on local air quality. *Heliyon, 6*(7).

Dimiskovska, B., & Berisha, A. (2021). Calculation of methane emissions from municipal solid waste landfill Germova using IPCC method. *Geo-Information, 26*, 26–36.

Duan, C., Liao, H., Wang, K., & Ren, Y. (2022). The research hotspots and trends of volatile organic compound emissions from anthropogenic and natural sources: A systematic quantitative review. *Environmental Research*, 114386.

Duan, Z., Scheutz, C., & Kjeldsen, P. (2021). Trace gas emissions from municipal solid waste landfills: A review. *Waste Management, 119*, 39–62.

Elemam, D. A., & Eldeeb, A. R. (2023). Climate change in the coastal areas: Consequences, adaptations, and projections for the Northern Coastal Area, Egypt. *Scientific Journal for Damietta Faculty of Science, 12*(2), 19–29.

US EPA. (2013). *Municipal solid waste.* United State Environmental Protection Agency. https://archive.epa.gov/epawaste/nonhaz/municipal/web/html/

US EPA. (2017). *Yard trimmings: material-specific data.* Available at https://www.epa.gov/facts-and-figures-about-materials-waste-and-recycling/yard-trimmings-material-specific-data

US EPA. (2022). *Basic information about landfills.* United State Environmental Protection Agency. Available at https://www.epa.gov/landfills/basic-information-about-landfills

Fang, W., Ding, Y., Geng, J., Liu, Z., Huang, Y., Yang, J., Ma, Z., Liu, J., Bi, J., Liu, M., & Qu, Z. (2023). High potential of coupling the source-separation and incineration promotion to reduce costs based on city-level cost-benefit analysis of municipal solid waste management strategies in China. *Resources, Conservation and Recycling, 197*, Article 107099.

Firestone, D. M. (2020). Utility Solid Waste Activities Group v. EPA and the EPA's Path toward Regulating Coal Combustion Residuals. *Energy LJ, 41*, 127.

Gautam, M., & Agrawal, M. (2021). Greenhouse gas emissions from municipal solid waste management: a review of global scenario. *Carbon footprint case studies: municipal solid waste management, sustainable road transport and carbon sequestration* (pp. 123–160).

Gautam, P., Kumar, S., & Lokhandwala, S. (2019). Advanced oxidation processes for treatment of leachate from hazardous waste landfill: A critical review. *Journal of Cleaner Production, 237*, Article 117639.

Ghasemzade, R., & Pazoki, M. (2017). Estimation and modeling of gas emissions in municipal landfill (case study: Landfill of Jiroft City). *Pollution, 3*(4), 689–700.

Ghosh, A., Kumar, S., & Das, J. (2023). Impact of leachate and landfill gas on the ecosystem and health: Research trends and the way forward towards sustainability. *Journal of Environmental Management, 336*, Article 117708.

Górka, M., Bezyk, Y., & Sówka, I. (2021). Assessment of GHG interactions in the vicinity of the municipal waste landfill site—Case study. *Energies, 14*(24), 8259.

Greater London Authority. (2014). *Greenhouse gas calculator for municipal waste user manual.* http://www.london.gov.uk/sites/default/files/archives/GreenhouseGasCalcUserManual.pdf

Guo, H., Xu, H., Liu, J., Nie, X., Li, X., Shu, T., Bai, B., Ma, X., & Yao, Y. (2022). Greenhouse gas emissions in the process of landfill disposal in China. *Energies, 15*(18), 6711.

Hadjidimoulas, C. (2018). Methane gas emissions: Methods of improving the efficiency of the biggest landfill gas waste to energy project in the middle east installed in Amman, Jordan. *Open Access Library Journal, 5*(8), 1–35.

Hassan, S. H., Aziz, H. A., Johari, I., & Hung, Y. T. (2022). Construction and demolition (C&D) waste management and disposal. *Solid waste engineering and management* (Vol. 2, pp. 165–216). Springer.

He, K., Zhang, J., & Zeng, Y. (2019). Knowledge domain and emerging trends of agricultural waste management in the field of social science: A scientometric review. *Science of the Total Environment, 670*, 236–244.

Hédouin, L., Rouzé, H., Berthe, C., Perez-Rosales, G., Martinez, E., Chancerelle, Y., Galand, P. E., Lerouvreur, F., Nugues, M. M., Pochon, X., Siu, G., & Planes, S. (2020). Contrasting patterns of mortality in Polynesian coral reefs following the third global coral bleaching event in 2016. *Coral Reefs, 39*, 939–952

Hornbach, M. J., Harris, R. N., & Phrampus, B. J. (2020). Heat flow on the US Beaufort Margin, Arctic Ocean: Implications for ocean warming, methane hydrate stability, and regional tectonics. *Geochemistry, Geophysics, Geosystems, 21*(5), e2020GC008933.

Hussein, L., Uren, C., Rekik, F., & Hammami, Z. (2022). A review on waste management and compost production in the Middle East-North Africa region. *Waste Management & Research, 40*(8), 1110–1128.

Iravanian, A., & Ravari, S. O. (2020, December). Types of contamination in landfills and effects on the environment: a review study. In *IOP Conference Series: Earth and Environmental Science* (Vol. 614, No. 1, p. 012083). IOP Publishing.

Jagaba, A. H., Kutty, S. R. M., Lawal, I. M., Abubakar, S., Hassan, I., Zubairu, I., Umaru, I., Abdurrasheed, A.S., Adam, A. A., Ghaleb, A. A. S., Almahbashi, N. M. Y., & Noor, A. (2021). Sequencing batch reactor technology for landfill leachate treatment: A state-of-the-art review. *Journal of Environmental management, 282*, 111946

Kamaruddin, M. A., Norashiddin, F. A., Yusoff, M. S., Hanif, M. H. M., Wang, L. K., & Wang, M. H. S. (2021). Sanitary landfill operation and management. *Solid Waste Engineering and Management:, 1*, 525–575.

Kaur, P., Kaur, G. J., Routray, W., Rahimi, J., Nair, G. R., & Singh, A. (2021). Recent advances in utilization of municipal solid waste for production of bioproducts: A bibliometric analysis. *Case Studies in Chemical and Environmental Engineering, 4*, Article 100164.

Kaza, S., Yao, L., Bhada-Tata, P., & Van Woerden, F. (2018). *What a waste 2.0: a global snapshot of solid waste management to 2050*. World Bank Publications.

Khan, S., Anjum, R., Raza, S. T., Bazai, N. A., & Ihtisham, M. (2022). Technologies for municipal solid waste management: Current status, challenges, and future perspectives. *Chemosphere, 288*, Article 132403.

Kumari, B., Solanki, H., & Kumar, A. (2020). Climate Change: A Burning Issue for the World. *Medicine, 35*, 501–507.

Lam, D. (2023). *Investigating the potential of amine-grafted silica materials for simultaneous removal of carbon dioxide, water vapor, and hydrogen sulfide from landfill gas* (Doctoral dissertation). Florida Atlantic University.

Lee, R. P., Meyer, B., Huang, Q., & Voss, R. (2020). Sustainable waste management for zero waste cities in China: Potential, challenges and opportunities. *Clean Energy, 4*(3), 169–201.

Li, G., Zhan, L., Zhang, S., Feng, S., & Zhang, Z. (2022). A dual-porosity model for coupled rainwater and landfill gas transport through capillary barrier covers. *Computers and Geotechnics, 151*, Article 104966.

Lino, F. A., Ismail, K. A., & Castañeda-Ayarza, J. A. (2023). Municipal solid waste treatment in Brazil: A comprehensive review. *Energy Nexus*, 100232.

Lins, E. A. M., & Lins, A. (2020). An analysis of the aspects and impacts to human health caused by effluents from a solid waste landfill: Case study. *International Journal of Advanced Engineering and Technology, 4*, 2.

Lu, Z. Q., Wu, C. G., Wu, N. Y., Lu, H. L., Wang, T., Xiao, R., Liu, H., & Wu, X. H. (2022). Change trend of natural gas hydrates in permafrost on the Qinghai-Tibet Plateau (1960–2050) under the background of global warming and their impacts on carbon emissions. *China Geology, 5*(3), 475–509.

Maalouf, A., Mavropoulos, A., & El-Fadel, M. (2020). Global municipal solid waste infrastructure: Delivery and forecast of uncontrolled disposal. *Waste Management & Research, 38*(9), 1028–1036.

Madon, I., Drev, D., & Likar, J. (2019). Long-term risk assessments comparing environmental performance of different types of sanitary landfills. *Waste Management, 96*, 96–107.

Malav, L. C., Yadav, K. K., Gupta, N., Kumar, S., Sharma, G. K., Krishnan, S., Rezania, S., Kamyab, H., Pham, Q. B., Yadav, S., & Bhattacharyya S., & Bach, Q. V. (2020). A review on municipal solid waste as a renewable source for waste-to-energy project in India: Current practices, challenges, and future opportunities. *Journal of Cleaner Production, 277*, 123227.

Manheim, D. C., Yeşiller, N., & Hanson, J. L. (2021a). Gas emissions from municipal solid waste landfills: A comprehensive review and analysis of global data. *Journal of the Indian Institute of Science, 101*(4), 625–657.

Manheim, D. C., Yesiller, N., & Hanson, J. L. (2021b). Climate change effects of gases from municipal solid waste landfills. *Japanese Geotechnical Society Special Publication, 9*(4), 142–147.

Mazzei, H. G., & Specchia, S. (2023). Latest insights on technologies for the treatment of solid medical waste: A review. *Journal of Environmental Chemical Engineering*, 109309.

Miao, L., Yang, G., Tao, T., & Peng, Y. (2019). Recent advances in nitrogen removal from landfill leachate using biological treatments–A review. *Journal of Environmental Management, 235*, 178–185.

Milyutina, N. O., Smyatskaya, Y. A., Politaeva, N. A., & Zelenkovsky, P. S. (2020). MSW landfill leachate: Formation, characterization and treatment. *Butler Communications, 61*, 79–85.

Mondal, T., Choudhury, M., Kundu, D., Dutta, D., & Samanta, P. (2023). Landfill: An eclectic review on structure, reactions and remediation approach. *Waste Management, 164*, 127–142.

Mor, S., & Ravindra, K. (2023). Municipal solid waste landfills in lower-and middle-income countries: environmental impacts, challenges and sustainable management practices. *Process Safety and Environmental Protection.*

Nanda, S., & Berruti, F. (2021). Municipal solid waste management and landfilling technologies: A review. *Environmental Chemistry Letters, 19*, 1433–1456.

Nemitz, E., Mammarella, I., Ibrom, A., Aurela, M., Burba, G. G., Dengel, S., Gielen, B., Grelle, A., Heinesch, B., Herbst, M., Hörtnagl, L., & Zahniser, M. (2018). Standardisation of eddy-covariance flux measurements of methane and nitrous oxide. *International Agrophysics, 32*(4), 517–549.

Nikkhah, A., Khojastehpour, M., & Abbaspour-Fard, M. H. (2018). Hybrid landfill gas emissions modeling and life cycle assessment for determining the appropriate period to install biogas system. *Journal of Cleaner Production, 185*, 772–780.

Olowoyo, J. O., Okedeyi, O. O., Mkolo, N. M., Lion, G. N., & Mdakane, S. T. R. (2012). Uptake and translocation of heavy metals by medicinal plants growing around a waste dump site in Pretoria, South Africa. *South African Journal of Botany, 78*, 116–121.

Ozbay, G., Jones, M., Gadde, M., Isah, S., & Attarwala, T. (2021). Design and operation of effective landfills with minimal effects on the environment and human health. *Journal of environmental and public health, 2021.*

Palaniandy, P., Aziz, H. A., Wang, L. K., Michael Terlecky Jr, P., & Hung, Y. T. (2022). Sanitary landfill types and design. In *Solid waste engineering and management: Volume 2* (pp. 543–597). Springer.

Pan, Q., Liu, Q. Y., Zheng, J., Li, Y. H., Xiang, S., Sun, X. J., & He, X. S. (2023). Volatile and semi-volatile organic compounds in landfill gas: Composition characteristics and health risks. *Environment International*, 107886.

Parameswaran, T. G., Nazeeh, K. M., Deekshith, P. K., GL, S. B., & TKK, C. D. (2023). Gas collection system design for a landfill via three-dimensional stochastic waste heterogeneity models and kriging. *Journal of Environmental Chemical Engineering, 11*(5), 110563.

Pazoki, M., Ghasemzadeh, R., Pazoki, M., & Ghasemzadeh, R. (2020). Landfilling. *Municipal Landfill Leachate Management*, 21–100

Pivato, A., Raga, R., Marzorati, S., Cerminara, G., Lavagnolo, M. C., & Schievano, A. (2022). Mitigating long-term emissions of landfill aftercare: Preliminary results from experiments combining microbial electrochemical technologies and in situ aeration. *Waste Management & Research, 40*(5), 596–606.

Propp, V. R., De Silva, A. O., Spencer, C., Brown, S. J., Catingan, S. D., Smith, J. E., & Roy, J. W. (2021). Organic contaminants of emerging concern in leachate of historic municipal landfills. *Environmental Pollution, 276*, Article 116474.

Pujotomo, I., & Azzahra, S. (2019). Electrical Energy Production Process from Landfill Gas. In *E3S Web of Conferences* (Vol. 125, p. 13002). EDP Sciences.

Pujotomo, I., & Azzahra, S. (2019). Electrical energy production process from landfill gas. In *E3S Web of Conferences* (Vol. 125, p. 13002). EDP Sciences.

Ramli, H., Aziz, H. A., & Hung, Y. T. (2021). Practices of solid waste processing and disposal. *Solid Waste Engineering and Management:, 1*, 625–673.

Ramos, A., Berzosa, J., Clarens, F., Marin, M., & Rouboa, A. (2020). Environmental and socio-economic assessment of cork waste gasification: Life cycle and cost analysis. *Journal of Cleaner Production, 249*, Article 119316.

Reinhart, D., Joslyn, R., & Emrich, C. T. (2020). Characterization of Florida, US landfills with elevated temperatures. *Waste Management, 118*, 55–61.

Rosa, L., Sanchez, D. L., & Mazzotti, M. (2021). Assessment of carbon dioxide removal potential via BECCS in a carbon-neutral Europe. *Energy & Environmental Science, 14*(5), 3086–3097.

Ross, D. E., & Law, H. J. (2023). Evolution of municipal solid waste management in North America. *Waste Management & Research*, 0734242X231173308.

Roy, D., & Tarafdar, A. (2022). Solid waste management and landfill in high-income countries. *Circular economy in municipal solid waste landfilling: Biomining & leachate treatment: Sustainable solid waste management: Waste to wealth* (pp. 1–23). Springer.

Saunois, M., Stavert, A. R., Poulter, B., Bousquet, P., Canadell, J. G., Jackson, R. B., Raymond, P. A., Dlugokencky, E. J., Houweling, S., Patra, P. K. Ciais, & Zhuang, Q. (2020). The global methane budget 2000–2017. *Earth system science data, 12*(3), 1561-1623.

Shammas, N. K., Wang, L. K., Wang, M. H. S., & Chen, S. L. (2020). Ecological impact and management of solid waste landfill gas. In *Handbook of environment and waste management: Acid rain and greenhouse gas pollution control* (pp. 455–482).

Shareefdeen, Z., & Bhojwani, J. (2022). Hazardous waste accidents: From the past to the present. *Hazardous waste management: Advances in chemical and industrial waste treatment and technologies* (pp. 27–56). Springer International Publishing.

Sharma, P., & Kumar, S. (2021). Characterization and phytotoxicity assessment of organic pollutants in old and fresh municipal solid wastes at open dump site: A case study. *Environmental Technology & Innovation, 24*, Article 101938.

Siddiqua, A., Hahladakis, J. N., & Al-Attiya, W. A. K. (2022). An overview of the environmental pollution and health effects associated with waste landfilling and open dumping. *Environmental Science and Pollution Research, 29*(39), 58514–58536.

Simon, C. E., & Ogunlowo, O. O. (2022). Assessment of leachate pollution potential of unlined landfill using leachate pollution index (LPI), Nigeria. *British Journal of Multidisciplinary and Advanced Studies, 3*(2), 67–76.

Singh, R. P., Tyagi, V. V., Allen, T., Ibrahim, M. H., & Kothari, R. (2011). An overview for exploring the possibilities of energy generation from municipal solid waste (MSW) in Indian scenario. *Renewable and Sustainable Energy Reviews, 15*(9), 4797–4808.

Skea, J., Shukla, P., Al Khourdajie, A., & McCollum, D. (2021). Intergovernmental Panel on Climate Change: Transparency and integrated assessment modeling. *Wiley Interdisciplinary Reviews: Climate Change, 12*(5), Article e727.

Sovacool, B. K., Griffiths, S., Kim, J., & Bazilian, M. (2021). Climate change and industrial F-gases: A critical and systematic review of developments, sociotechnical systems and policy options for reducing synthetic greenhouse gas emissions. *Renewable and Sustainable Energy Reviews, 141*, Article 110759.

Srivastava, A. N., & Chakma, S. (2020). Quantification of landfill gas generation and energy recovery estimation from the municipal solid waste landfill sites of Delhi, India. *Energy Sources, Part A: Recovery, Utilization, and Environmental Effects*, 1–14.

Tam, V. W. Y., & Lu, W. (2016). Construction waste management profiles, practices, and performance: A cross-jurisdictional analysis in four countries. *Sustainability, 8*(2), 190.

Tesseme, A. T., & Chakma, S. (2020). Small scale bioreactor studies for sustainable municipal solid waste landfilling management in developing countries. *International Journal of Environment and Waste Management, 25*(2), 194–219.

Tirkolaee, E. B., & Torkayesh, A. E. (2022). A cluster-based stratified hybrid decision support model under uncertainty: Sustainable healthcare landfill location selection. *Applied Intelligence, 52*(12), 13614–13633.

Toha, M., & Rahman, M. M. (2023). Estimation and prediction of methane gas generation from landfill sites in Dhaka city, Bangladesh. *Case Studies in Chemical and Environmental Engineering, 7*, Article 100302.

Townsend, T. G., & Anshassi, M. (2023). CDD Landfills. *Construction and demolition debris* (pp. 303–352). Springer.

Troncoso, S. L. (2022). Regulatory uncertainty and new source performance standards on methane. *Environmental Law Reporter, 52*, 10453.

Twaróg, A., Mamak, M., Sechman, H., Rusiniak, P., Kasprzak, E., & Stanek, K. (2020). Impact of the landfill of ashes from the smelter on the soil environment: Case study from the South Poland, Europe. *Environmental Geochemistry and Health, 42*(5), 1453–1467.

USEPA. (2020). Advancing sustainable materials management: 2016 and 2017 Tables and figures, assessing trends in material generation, recycling, composting, combustion with energy recovery and landfilling in the United States, November 2019.

Vaverková, M. D. (2019). Landfill impacts on the environment. *Geosciences, 9*(10), 431.

Vavilin, V. A., & Lokshina, L. Y. (2023). Carbon and hydrogen dynamic isotope equations are used to describe the dominant processes of waste biodegradation: Effect of aeration in methanogenic phase of the landfill. *Waste Management, 166*, 280–293.

Vestin, P., Mölder, M., Kljun, N., Cai, Z., Hasan, A., Holst, J., Klemedtsson, J., & Lindroth, A. (2020). Impacts of clear-cutting of a boreal forest on carbon dioxide, methane and nitrous oxide fluxes. *Forests, 11*(9), 961.

Wang, L. K., & Wang, M. H. S. (2022). Innovative bioreactor landfill and its leachate and landfill gas management. *Solid waste engineering and management* (Vol. 3, pp. 583–614). Springer.

Wang, Y., Levis, J. W., & Barlaz, M. A. (2021). Life-Cycle Assessment of a regulatory compliant US municipal solid waste landfill. *Environmental Science & Technology, 55*(20), 13583–13592.

Xin, C., Zhang, T., Tsai, S. B., Zhai, Y. M., & Wang, J. (2020). An empirical study on greenhouse gas emission calculations under different municipal solid waste management strategies. *Applied Sciences, 10*(5), 1673.

Xiu, W., Wang, S., Qi, W., Li, X., & Wang, C. (2021). Disaster chain analysis of landfill landslide: Scenario simulation and chain-cutting modeling. *Sustainability, 13*(9), 5032.

Xue, Z., Jiang, M., Zhang, Z., Wu, H., & Zhang, T. (2021). Simulating potential impacts of climate changes on distribution pattern and carbon storage function of high-latitude wetland plant communities in the Xing'anling Mountains, China. *Land Degradation & Development, 32*(9), 2704–2714.

Yaashikaa, P. R., Kumar, P. S., Nhung, T. C., Hemavathy, R. V., Jawahar, M. J., Neshaanthini, J. P., & Rangasamy, G. (2022). A review on landfill system for municipal solid wastes: Insight into leachate, gas emissions, environmental and economic analysis. *Chemosphere*, 136627.

Yaman, C., Anil, I., & Alagha, O. (2020). Potential for greenhouse gas reduction and energy recovery from MSW through different waste management technologies. *Journal of Cleaner Production, 264*, Article 121432.

Zamri, M. F. M. A., Hasmady, S., Akhiar, A., Ideris, F., Shamsuddin, A. H., Mofijur, M., Fattah, I. R., & Mahlia, T. M. I. (2021). A comprehensive review on anaerobic digestion of organic fraction of municipal solid waste. *Renewable and Sustainable Energy Reviews, 137*, Article 110637.

Zhao, X. Y., Yang, J. Y., Ning, N., & Yang, Z. S. (2022). Chemical stabilization of heavy metals in municipal solid waste incineration fly ash: A review. *Environmental Science and Pollution Research, 29*(27), 40384–40402.

Zheng, Y., Wu, S., Xiao, S., Yu, K., Fang, X., Xia, L., Wang, J., Liu, S., Freeman, C., & Zou, J. (2022). Global methane and nitrous oxide emissions from inland waters and estuaries. *Global Change Biology, 28*(15), 4713–4725.

Chapter 5
Emission Factors from Municipal Solid Waste (MSW) Landfills and Their Impact on Air Pollution

Mohammed Zari

Abstract Methane is one of the main greenhouse gases (GHGs) that contributes considerably to global warming, thus being a major source of concern. GHGs entering the atmosphere are the primary cause of climate change globally, which is manifested in violent storms, increasing sea levels, and deluges. Waste dumping is considered one of the main sources of these gases due to the increased waste accumulation in landfills. As a result, landfill gas generation continues to be released into the atmosphere which contains mainly methane and carbon dioxide. The potential of dust emission into the atmosphere is also significant because of different waste management activities. This review attempts to outline landfill emissions quantification methods and factors, as well as their potential control options by primarily focusing on methane and dust emissions.

Keywords Municipal solid waste · Emission factors · Air pollution · Landfill gas · Dust emission

5.1 Introduction

Rising living standards and the ongoing growth of the world economy have resulted in resource depletion and an increase in the amount of municipal solid waste (MSW) produced per person (Zhang et al., 2022). When improperly processed, managed, transported, and disposed of, this waste may constitute a major threat to human health or the environmental compartments (water, air, soil) (Soltanzadeh et al., 2022). The disposal and anaerobic biodegradation of wet waste in landfills accounts for the majority of greenhouse gas (GHG) emissions from waste management activities

M. Zari (✉)
Chemical and Environmental Engineering Department, Faculty of Engineering, University of Nottingham, Nottingham, UK
e-mail: Mohammed.Zari@nottingham.ac.uk

Department of Environment, Faculty of Environmental Sciences, King Abdulaziz University, Jeddah, Saudi Arabia

A. Anouzla et al. (eds.), *Handbook of the Effects of Landfill Emissions on Sustainable Development*, https://doi.org/10.1007/978-3-032-01951-6_5

(Couth et al., 2011). Due to this, MSW disposal in landfills produces landfill gas (LFG), the primary components of which are methane (CH_4) (55–60% v/v) and carbon dioxide (CO_2) (40–45% v/v) (Mønster et al., 2019). Methane is a potent greenhouse gas, having a global warming potential approximately 21 times that of carbon dioxide over a 100-year period (Purmessur & Surroop, 2019). It also has an atmospheric residence period of around 12 years (Kumar & Sharma, 2014). In the last 200 years, atmospheric CH_4 concentrations have risen to approximately 1000 ppb. Prior to that, they have only changed between 580 and 730 ppb over a 10,000-year period (Delgado et al., 2023). A significant portion of CH_4 emissions originate from landfills. On a global scale, the MSW Sector provides 11% of total yearly anthropogenic emissions as biogas (Maria et al., 2020). The emissions of LFG into the environment can have a number of environmental issues including contribution to climate change as a result of the release of methane in the atmosphere. Consequently, limiting fugitive methane emissions from landfill sites needs to be prioritized in waste management strategies (Scheutz & Kjeldsen, 2019).

Because of their climatic impact and potential for energy production, LFG emissions have received a lot of attention. However, particle matter (PM) has a significant impact on local air pollution as well as human health (Hroncová et al., 2020; Russell & Brunekreef, 2009; Zari et al., 2022). Operating MSW landfills emit a variety of air pollutants as a form of dust in a different size range (Koshy et al., 2009). The process of residues, mechanical recycling and composting, waste unloading and sorting, and waste transportation by trucks all contribute to the emissions of PM from landfills (Chalvatzaki et al., 2010). Based on health research, there is a strong association between airborne PM and adverse health effects, such as cardiovascular and respiratory diseases (Burnett et al., 2014). As a result, it is critical to understand how MSW landfill operations affect the level of PM in the environment.

This paper will put light on factors responsible for high and low MSW landfill emissions, as well as currently available methods for the estimation of emission rate in order to enable the minimisation of potential risk by mitigation measures.

5.2 Landfill Gas Emission Factors

There are five stages to the process of MSW waste biodegradation and the generation of LFG, which consists of a series of complex chemical and biological processes: (1) aerobic (hydrolysis and aerobic degradation converting readily degradable carbohydrates to simple sugars, CO_2, and H_2O); (2) acidogenic (hydrolysis and fermentation of simple sugars to soluble volatile acids); (3) acetogenic (conversion of soluble acids to acetic acids, H_2 and CO_2); (4) methanogenic (methane generation bacteria metabolize acetate and form CH_4 and CO_2); (5) aerobic (oxidation with the restoration of aerobic conditions) (Brown et al., 2018; Couth et al., 2011; Purmessur & Surroop, 2019).

There is a considerable spatial variability in LFG emissions throughout the surface of a landfill due to the heterogeneity of waste composition and the properties of the landfill surface covering system (Mønster et al., 2019). Additionally, other factors, such as weather and seasonal temperature variations, can have an impact on this variability (Czepiel et al., 2003). To limit methane emissions from landfills in the environment, regulators worldwide are putting waste management strategies and policies into practice (Di Bella et al., 2011). Numerous research has been conducted to determine the amount of gas produced by MSW landfills. In order to determine how successfully methane gas emissions could be reduced by enhanced sustainable waste management, it is crucial to model and predict the production rate of methane gas in a landfill (Fallahizadeh et al., 2019).

LFG emissions can be assessed using two different approaches: field measurements and mathematical modelling (Chiemchaisri & Visvanathan, 2008; Huang et al., 2022). Various LFG emissions estimation models have been developed over the previous few decades, and each one integrates new characteristics and factors to introduce distinct approaches or produce more accurate findings (Delgado et al., 2023). The construction of these models used for various landfill sites is developed by the significance of local elements, such as the waste composition, disposal, and protective mechanisms against potential impacts (Mohareb et al., 2008). GasSim, LandGEM, and IPCC (Landfill Gas Emission Model) models are examples of the methods that have been proposed in the literature to quantitatively describe methane generation from anaerobic waste degradation (Delgado et al., 2023; Scharff & Jacobs, 2006). In terms of field monitoring programs, a number of measurement techniques are currently available. However, none of these have been recognised as an international reference method (Mønster et al., 2019).

Critical factors including waste composition, age of refuse, moisture content, pH, temperature, and the presence of readily degradable organic components play an important role in determining the methane emission rate (Delgado et al., 2023; Mohsen et al., 2020). The temperature and pH are important for the survival and action of bacteria, and they impact the type of bacteria that prevail, as well as the rate of gas production (Machado et al., 2009). Moisture content is a factor that regulates methane production because it promotes microbial activity by improving interaction between soluble and insoluble substrates and bacteria (Mehta et al., 2002). Different waste components degrade at various rates over the years, depending on the nature of the waste. Typically, food waste and certain green waste (grass) are included in the components that are readily degradable. A portion of the paper waste and the rest of green waste are moderately degradable. The remaining portion of the paper waste (newsprint and coated paper), textiles, wood, and other materials are slowly degradable. Consequently, the carbon in these materials can be stored in landfills for many decades (Barlaz et al., 2002; Pheakdey et al., 2023). Inert materials such as plastic, glass, metal, concrete, and rubble are regarded as non-biodegradable (Machado et al., 2009).

5.3 Dust Emission Factors

Landfill related activities, particularly landfill mining (LFM) activities can generate dust emissions through excavation, screening, and shredding of waste materials during operational periods. These emissions are significant because the amount of fine fractions within a landfill body is often considerable, accounting for more than 50% of the total excavated MSW mass within the size range of < 10 mm to > 4 mm (Zari et al., 2022). Dust particles within MSW can contain toxic materials in a diverse size range (Dino et al., 2016). If these materials are not properly managed, they can be transported into the surrounding areas and have an impact on nearby communities (Pandey et al., 2014). Therefore, quantification of dust emission rate is significant for environmental risk assessment, and it can provide the foundation for analysing the environmental fate of these emissions (Richardson et al., 2019). Similar to LFG emissions assessment, dust emissions can be estimated using field monitoring networks or mathematical modelling. However, no model has been developed yet for the estimation of dust emission rate from various landfill activities. Thus, there is a pressing need for the development of mathematical modelling to estimate the dust that can be airborne during different landfill activities to understand factors responsible for higher emissions rate, determine where any pressure points of different landfill activities exist so as to mitigate potential risk, and enable air dispersion modelling for landfill environmental impact assessment studies of air pollution in the absence of actual monitoring data.

Dust emissions in the mining sector are impacted by a number of factors, including meteorological conditions, heavy equipment dimensions, emission control efficiency, and physical properties of materials (Chaulya et al., 2019). Surface mining activities generate dust from a range of activity sources, many of which are also employed in LFM activities. The distance travelled by dust emissions is determined by particle size, as well as wind speed and turbulence. Smaller dust particles remain in the air for a longer amount of time and travel over a larger region. Larger particles also remain in the air for extended periods of time if there are strong and turbulent winds. According to data gathered from quarries, finer dust particles (< 10 μm) can disperse up to 1 km, whereas coarser dust particles (> 30 μm) are primarily deposited within 100 m of the source. Intermediate particle (10–30 μm) can travel up to 500 m (Sairanen & Pursio, 2020). Ultrafine particles (2.5 m) are projected to go further or do not undergo deposition process (Hroncová et al., 2020).

The general equation for dust emissions estimation is the following (Kim et al., 2020):

$$E = A \times EF \times \left(\frac{1 - ER}{100} \right) \tag{5.1}$$

where E is the total emissions rate (e.g., kilograms of dust released per day), and A is the activity rate (e.g., tonnes of waste handled per day). EF represents the emission factor (e.g., kilograms of dust released per tonne of waste handled), and ER the overall emission reduction efficiency (%). The emission reduction only applies if there are any emission-controlling methods used (Abril et al., 2016). The total emissions can be calculated by summing the equation results for each source of an activity. This general equation can serve as a starting point for the formulation of different landfill activities' equations to estimate dust emissions.

5.4 Monitoring of Landfill Emissions and Potential Mitigation Measure

LFG in modern landfills is routinely collected or regulated to prevent its emissions into the atmosphere and migration through the surrounding soil. The recovered gas is sometimes flared, and can be used to generate energy due to its high calorific value, leading to efficient MSW management while minimising the associated negative environmental impacts (Purmessur & Surroop, 2019). Most landfills have installed gas collection systems in the last decade to reduce these emissions and use the generated biogas. However, a considerable portion of LFG is created at landfills that do not have a gas collection system (Fei et al., 2016) and consequently, LFG is emitted to the atmosphere. It is reported that between 50 and 90% of the LFG generated is captured by gas collection systems (Delgado et al., 2023).

In Europe, LFG is controlled through flaring, gas collection and energy use, or other techniques like mitigation that rely on the microbial oxidation of methane in cover soils or built biofilters (also known as "bio-mitigation technologies") (Scheutz & Kjeldsen, 2019).

Monitoring of methane emissions should be undertaken during both the active phase, when waste is being received at a site, and the aftercare phase, when waste is no longer being received. Even after the application of mitigation measures are terminated, gas monitoring should be conducted until no substantial methane emissions are detected.

In regard to dust mitigation measures, environmental licenses for landfill sites require that dusts to be sufficiently controlled using dust suppression methods, so that they do not pose a negative influence on public health. Monitoring of particles is also required for the environmental permit. To ensure that nuisance dust issues have the least impact on those living nearby, regulators should establish a nuisance criterion level (Environment Agency, 2004).

5.5 Conclusion

This article puts a spotlight on the significance of understanding factors that contribute to higher MSW landfill emissions to limit the impact of air pollution. Estimated methods for landfill emissions rate are presented using either quantitative computational models or actual field measurements with identification of a knowledge gap in literature. Furthermore, the current potential mitigation measures of landfill emissions are highlighted. A critical step to enhance the efficiency of specific measures to minimize landfill emissions is to make tools available that allow their impact to be adequately quantified or estimated.

References

Abril, G. A., Diez, S. C., Pignata, M. L., & Britch, J. (2016). Particulate matter concentrations originating from industrial and urban sources: Validation of atmospheric dispersion modeling results. *Atmospheric Pollution Research, 7*, 180–189.

Barlaz, M. A., Rooker, A. P., Kjeldsen, P., Gabr, M. A., & Borden, R. C. (2002). *Critical evaluation of factors required to terminate the postclosure monitoring period at solid waste landfills*. ACS Publications.

Brown, D., Shaw, S., Gergory, R., Robinson, H., & Robinson, T. (2018). *Landfill aftercare scoping study, UK*.

Burnett, R. T., Pope, C. A., III., Ezzati, M., Olives, C., Lim, S. S., Mehta, S., et al. (2014). An integrated risk function for estimating the global burden of disease attributable to ambient fine particulate matter exposure. *Environmental Health Perspectives, 122*, 397–403.

Chalvatzaki, E., Kopanakis, I., Kontaksakis, M., Glytsos, T., Kalogerakis, N., & Lazaridis, M. (2010). Measurements of particulate matter concentrations at a landfill site (Crete, Greece). *Waste Management, 30*, 2058–2064.

Chaulya, S., Trivedi, R., Kumar, A., Tiwary, R., Singh, R., Pandey, P., et al. (2019). Air quality modelling for prediction of dust concentrations in iron ore mines of Saranda region, Jharkhand. *India. Atmospheric Pollution Research, 10*, 675–688.

Chiemchaisri, C., & Visvanathan, C. (2008). Greenhouse gas emission potential of the municipal solid waste disposal sites in Thailand. *Journal of the Air & Waste Management Association, 58*, 629–635.

Couth, R., Trois, C., & Vaughan-Jones, S. (2011). Modelling of greenhouse gas emissions from municipal solid waste disposal in Africa. *International Journal of Greenhouse Gas Control, 5*, 1443–1453.

Czepiel, P., Shorter, J., Mosher, B., Allwine, E., McManus, J., Harriss, R., et al. (2003). The influence of atmospheric pressure on landfill methane emissions. *Waste Management, 23*, 593–598.

Delgado, M., López, A., Esteban-García, A. L., & Lobo, A. (2023). The importance of particularising the model to estimate landfill GHG emissions. *Journal of Environmental Management, 325*, 116600.

Di Bella, G., Di Trapani, D., & Viviani, G. (2011). Evaluation of methane emissions from Palermo municipal landfill: Comparison between field measurements and models. *Waste Management, 31*, 1820–1826.

Dino, G. A., Rossetti, P., Biglia, G., Coulon, F., Gomes, D., Wagland, S., et al. (2016). SMART GROUND Project: SMART data collection and inteGRation platform to enhance availability and accessibility of data and infOrmation in the EU territory on SecoNDary raw materials. *Energy Procedia, 97*, 15–22.

Environment Agency. (2004). *Model procedures for the management of land contamination*. Environment Agency Bristol.

Fallahizadeh, S., Rahmatinia, M., Mohammadi, Z., Vaezzadeh, M., Tajamiri, A., & Soleimani, H. (2019). Estimation of methane gas by LandGEM model from Yasuj municipal solid waste landfill. *Iran. Methodsx, 6*, 391–398.

Fei, X., Zekkos, D., & Raskin, L. (2016). Quantification of parameters influencing methane generation due to biodegradation of municipal solid waste in landfills and laboratory experiments. *Waste Management, 55*, 276–287.

Hroncová, E., Ladomerský, J., & Ladomerská, D. (2020). Landfill air pollution by ultrafine and microparticles in case of dry and windless weather conditions. *Detritus, 10*, 139.

Huang, D., Du, Y., Xu, Q., & Ko, J. H. (2022). Quantification and control of gaseous emissions from solid waste landfill surfaces. *Journal of Environmental Management, 302*, 114001.

Kim, H., Tae, S., & Yang, J. (2020). Calculation methods of emission factors and emissions of fugitive particulate matter in south Korean construction sites. *Sustainability, 12*, 9802.

Koshy, L., Jones, T., & BéruBé, K. (2009). Characterization and bioreactivity of respirable airborne particles from a municipal landfill. *Biomarkers, 14*, 49–53.

Kumar, A., & Sharma, M. (2014). Estimation of GHG emission and energy recovery potential from MSW landfill sites. *Sustainable Energy Technologies and Assessments, 5*, 50–61.

Machado, S. L., Carvalho, M. F., Gourc, J.-P., Vilar, O. M., & do Nascimento, J. C. (2009) Methane generation in tropical landfills: Simplified methods and field results. *Waste Management, 29*, 153–161.

Maria, C., Góis, J., & Leitão, A. (2020). Challenges and perspectives of greenhouse gases emissions from municipal solid waste management in Angola. *Energy Reports, 6*, 364–369.

Mehta, R., Barlaz, M. A., Yazdani, R., Augenstein, D., Bryars, M., & Sinderson, L. (2002). Refuse decomposition in the presence and absence of leachate recirculation. *Journal of Environmental Engineering, 128*, 228–236.

Mohareb, A. K., Warith, M. A., & Diaz, R. (2008). Modelling greenhouse gas emissions for municipal solid waste management strategies in Ottawa, Ontario, Canada. *Resources, Conservation and Recycling, 52*, 1241–1251.

Mohsen, R. A., Abbassi, B., & Zytner, R. (2020). Investigation of fugitive methane and gas collection efficiency in Halton landfill in Ontario. *Canada. Environmental Monitoring and Assessment, 192*, 1–12.

Mønster, J., Kjeldsen, P., & Scheutz, C. (2019). Methodologies for measuring fugitive methane emissions from landfills–A review. *Waste Management, 87*, 835–859.

Pandey, B., Agrawal, M., & Singh, S. (2014). Assessment of air pollution around coal mining area: Emphasizing on spatial distributions, seasonal variations and heavy metals, using cluster and principal component analysis. *Atmospheric Pollution Research, 5*, 79–86.

Pheakdey, D. V., Noudeng, V., & Xuan, T. D. (2023). Landfill biogas recovery and its contribution to greenhouse gas mitigation. *Energies, 16*, 4689.

Purmessur, B., & Surroop, D. (2019). Power generation using landfill gas generated from new cell at the existing landfill site. *Journal of Environmental Chemical Engineering, 7*, 103060.

Richardson, C., Rutherford, S., & Agranovski, I. (2019). Particulate emission rates for open surfaces in Australian open cut black coal mines. *Journal of Environmental Management, 232*, 537–544.

Russell, A. G., & Brunekreef, B. (2009) *A focus on particulate matter and health*. ACS Publications.

Sairanen, M., & Pursio, S. (2020). Near field modelling of dust emissions caused by drilling and crushing. *SN Applied Sciences, 2*, 1–15.

Scharff, H., & Jacobs, J. (2006). Applying guidance for methane emission estimation for landfills. *Waste Management, 26*, 417–429.

Scheutz, C., & Kjeldsen, P. (2019). Guidelines for landfill gas emission monitoring using the tracer gas dispersion method. *Waste Management, 85*, 351–360.

Soltanzadeh, F., Behbahani, A. E., Hosseinmostofi, K., & Teixeira, C. A. (2022). Assessment of the sustainability of fibre-reinforced concrete by considering both environmental and mechanical properties. *Sustainability, 14*, 6347.

Zari, M., Smith, R., Wright, C., & Ferrari, R. (2022). Health and environmental impact assessment of landfill mining activities: A case study in Norfolk, UK. *Heliyon*, e11594.

Zhang, Z., Malik, M. Z., Khan, A., Ali, N., Malik, S., & Bilal, M. (2022). Environmental impacts of hazardous waste, and management strategies to reconcile circular economy and eco-sustainability. *Science of the Total Environment, 807*, 150856.

Chapter 6
Impact of Technical Landfills on Air Pollution

Roukaya Bouyakhsass, Safaa khattabi Rifi, Idrissi Layla, Abdelaziz Madinzi, Khalid Digua, Younes Abrouki, Tonni Agustiono Kurniawan, Abdelkader Anouzla, and Salah Souabi

Abstract The waste burial was carried out precariously for a long time without any constraint or control of the different categories of waste stored and buried. The consequences are severe pollution, particularly of the air with olfactory or visual nuisances. Study the air quality around technical landfills involves, on the one hand, evaluating and locating as best as possible the atmospheric emissions generated by its operation and, on the other hand, addressing the question of potential or actual nuisances and / or risks for residents. This work aims to assess the impacts of technical landfill centers for household and similar waste on the air. The study showed that technical landfill centers have considerable impacts on the air, particularly the release of biogas produced during the fermentation of waste and leachate, the presence of pathogenic microorganisms in the air, etc. These pollutants harm air quality, human health, and environmental balance. Health effects, ranging from respiratory problems to cardiovascular diseases, highlight the urgency of mitigating this threat. The environmental impact of air pollution from landfills also extends to biodiversity, soils, and aquatic ecosystems, compromising the sustainability of our ecosystems. Furthermore, the economic fallout is considerable, increasing public health costs, hampering industrial development, and requiring investments to address climate change and environmental disasters.

Keywords Solid waste · Technical landfill · Air pollution · Impact · Human health

R. Bouyakhsass · S. Rifi · I. Layla · A. Madinzi · K. Digua · A. Anouzla (✉) · S. Souabi
Laboratory of Process Engineering and Environment, Faculty of Science and Technology, Hassan II University, Mohammedia, Morocco
e-mail: aanouzla@gmail.com

Y. Abrouki
Faculty of Science, Laboratory of Spectroscopy, Molecular Modeling, Materials, Nanomaterial, Water and Environment, CERNE2D, Mohammed V University in Rabat, Rabat, Morocco

T. A. Kurniawan
College of Ecology and the Environment, Xiamen University, Xiamen, Fujian, China

A. Anouzla et al. (eds.), *Handbook of the Effects of Landfill Emissions on Sustainable Development*, https://doi.org/10.1007/978-3-032-01951-6_6

6.1 Introduction

Air pollution remains one of the major concerns of our time, not only because of its major role in the climate change process but also because of its harmful impact on public and individual health. This precarious situation has direct and significant consequences, leading to a substantial increase in morbidity, i.e. illness, and a worrying rise in mortality rates. The fine particles and toxic gases in polluted air penetrate the respiratory tract and even reach the bloodstream, creating a wide range of health problems, from acute respiratory disorders to chronic cardiovascular problems. This puts considerable pressure on healthcare systems and generates substantial economic costs regarding medical care, days lost from work, and reduced productivity (Manisalidis et al., 2020). Air pollution is a significant cause of death today. In 2015, it caused between 4 and 6 million deaths worldwide, with 2 to 8 million linked to domestic air pollution and 2–4 million to ambient air pollution (Landrigan, 2017). Pollution from landfill sites is generally classified into different categories, including air emissions, effluent discharges into water, and deposits in soil (Siddiqua et al., 2022). Commonly used methods of waste disposal, such as illegal landfills, uncontrolled incinerators, and dumps, have harmful consequences, including air and water pollution, soil deterioration, the emission of gases such as methane and hazardous liquids, and climate change (Abubakar et al., 2022). Indeed, when waste is not treated correctly, such as when it is burnt, it can pollute the air with nasty substances like airborne dirt and ozone. It can also create heavy metals and chemicals that evaporate into the air (Kanhai et al., 2021). The uncontrolled open-air burning of municipal solid waste is a significant cause of air pollution, contributing 29% of human emissions of fine particles ($PM_{2.5}$) into the ambient air (Adetona et al., 2020). Due to the rapid increase in solid waste and the limited space available for landfill sites, waste treatment plants in China are becoming increasingly important. The transition to non-coal thermal energy sources also improves the regulation of polluting emissions from these facilities. As a safe waste treatment method and a source of low-carbon, green energy, these plants play a crucial role in China (Fu et al., 2022).

This chapter aims to provide an in-depth analysis of the various facets of the solid waste issue, examining its production, composition, and the degradation process to which it is subjected. In addition, the article aims to provide an understanding of the operation of landfill sites as waste management hotspots and to highlight their potential as significant sources of air pollution. The chapter aims to present a comprehensive assessment of the impacts of air pollution from landfill sites on human health, exploring the various respiratory and cardiovascular diseases that result. In addition, it aims to highlight the environmental consequences of this pollution on ecosystems while evaluating the economic spin-offs generated by this problem. Finally, this work proposes a set of viable techniques and strategies for mitigating the air pollution caused by landfill sites. By bringing these various elements together, the ultimate aim of this study is to contribute to understanding and taking concrete measures to resolve the challenges associated with air pollution from landfill sites.

6.2 Solid Waste Challenges

Solid waste management (SWM) is a crucial and complex municipal service offered to every resident, and it is often considered difficult to manage (Hannan et al., 2015). The collection and disposal of municipal waste is a major environmental problem worldwide (Abdel-Shafy & Mansour, 2018). Solutions for managing this waste must be economically viable, technically feasible, and socially and legally acceptable while preserving the environment. Inadequate collection and disposal of household waste means that large quantities of waste end up in open dumps. This jeopardizes the quality of surface and groundwater and encourages the proliferation of parasites. This situation is exacerbated by open burning and spontaneous combustion, which cause air pollution and odor nuisance, especially in informal settlements where the lack of containers exacerbates the problem (Mohammed & Elias, 2017). Municipal solid waste landfill sites generate emissions of solid and gaseous pollutants that can persist for decades or centuries after their closure. The landfill gases emitted, such as methane (CH_4), carbon dioxide (CO_2), and non-methane volatile organic compounds, contribute to greenhouse effects and environmental problems. Some compounds, such as chlorofluorocarbons (CFCs), can destroy the ozone layer and greenhouse effects (Tian et al., 2013).

Conventional waste management methods, such as uncontrolled landfills or incineration without adequate filtration, have the potential to considerably impact local air pollution, which, by extension, can affect the health of residents (Kanhai et al., 2021). The findings show that the incineration of municipal waste generated energy recovery of 7.394×10^9 kWh/year, leading to 340.15×10^9 kWh/year of greenhouse gas emissions between 2014 and 2015 (Chen, 2018). Among the various types of municipal waste, plastics stood out as generating the most significant quantity of GHGs, followed by paper and textiles. As for non-hazardous industrial waste (NHIW), rubber showed the highest potential for GHG emissions. When waste is not properly collected and treated, the repercussions extend beyond urban boundaries, creating socio-economic and environmental consequences. In ecological terms, these consequences are diverse and significant. One of the main concerns is the release of methane (CH_4), a potent greenhouse gas. To alleviate the problems associated with solid waste, reducing air pollution and improving public health is essential. Several measures are recommended to achieve this, including establishing more efficient waste collection and treatment systems, promoting recycling and reuse, and integrating greener technologies throughout the waste management process.

6.3 Waste Production

Population and economic growth have contributed significantly to the increasing production of municipal solid waste (MSW), with around 2.01 billion metric tonnes produced worldwide each year. The prediction is that this will double by 2050

(Gautam & Agrawal, 2021). As a result, every year, the world produces an average of 218 kg of municipal solid waste per person (Sipra et al., 2018). Indeed, domestic solid waste poses an environmental threat in Addis Ababa, where only 65% of the waste produced daily is collected, while 5% is recycled and 5% is composted. The remaining 25% is illegally dumped and not collected (Mohammed & Elias, 2017). Economic growth may have led to an increase in total waste, from 635 million tonnes (Mt) in 1965 to 1999 Mt in 2015, with a projection of 3539 Mt by 2050. Over 2015–2050, the share of organic waste will fall from 47 to 39%, while other categories of waste, notably paper, will increase (Chen et al., 2020). At the same time, the proportion of waste sent to landfills has fallen from 28 to 18% due to increased adoption of more sustainable treatment methods (Chen et al., 2020). Worldwide, management costs average USD 205.4 per tonne of waste, highlighting the scale of the economic challenges associated with waste management (Das et al., 2019). It is estimated that China will rank first in the world in terms of municipal solid waste production, with 1,397,755 tonnes per day by 2025, followed by India with 376,639 tonnes per day and Brazil with 330,960 tonnes per day (Nanda & Berruti, 2021). Malaysia currently has a per capita solid waste production of around 1.1 kg per day, with over 26,500 tonnes of solid waste disposed of mainly in 166 active landfill sites across the country (Kamaruddin et al., 2017). Generally speaking, SMEs generate 30.8 million tonnes of waste in England, more than households (Woodard, 2021).

6.4 Waste Composition

Solid waste is classified into different types depending on its source: residential, industrial, institutional, construction, demolition, and municipal waste. It can also be categorized according to its properties: corrosive, flammable, reactive, and toxic, which contain acids or bases that can corrode containers, can cause fires under certain conditions, are unstable by nature, and can cause explosions and toxic fumes when heated, and are harmful or even fatal if swallowed or absorbed (Alam & Ahmade, 2013). The estimated physical composition of solid waste is as follows: vegetables 4.2%, paper 2.5%, rubber/plastic 2.9%, wood 2.3%, bone 1.1%, textiles 2.9%, metals 0.9%, glass 0.5%, combustible foil 15.1%, non-combustible stone 2.5%, and all other fine particles represent 65% (Mohammed & Elias, 2017). However, the composition can vary according to socio-economic changes in society and seasonal variations, which influence the types of waste present in the waste stream (Gallardo et al., 2018). Indeed, low-income populations are mainly responsible for producing organic waste, while densely populated areas are more likely to generate materials such as metals, glass, and paper (Sharma et al., 2022). Solid waste is generally characterized by its high density and high moisture content in developing countries. Landfill gas is produced when this waste undergoes anaerobic decomposition in landfill sites. This gas is mainly composed of methane (CH_4) and carbon dioxide (CO_2), accompanied by small quantities of volatile organic compounds and other gases (Kumar & Prakash, 2020).

6.5 Waste Degradation Process

The degradation of organic waste in landfill sites mainly results from an anaerobic biodegradation process (Ogata et al., 2016). Anaerobic digestion (AD), a mechanism facilitated by enzymatic and bacterial activities, converts organic matter into biogas while minimizing sludge production. Organic waste, characterized by high levels of carbohydrates, proteins, and lipids, is an optimal substrate for this AD process (Golwala et al., 2021). This degradation process takes place in 5 stages (Luo & Wong, 2019; Wang et al., 2018): The initial **aerobic phase**, in which waste is deposited and compacted in successive horizontal layers in a cell to limit odor nuisance and flight. Inert materials then cover it. In other words, contact between the oxygen and the surface layers exists before they are protected. This is followed by the **hydrolysis phase**, during which the redox potential decreases and the organic load of the leachate increases. Anaerobic and facultative aerobic microorganisms degrade the biopolymers into low molecular weight oligomers and monomers. The result is the production of ammoniacal nitrogen and biogas containing mainly carbon dioxide. Hydrolysis is a prolonged process that continues even during the stages that follow it. The waste then undergoes the **acidogenesis phase**, in which the fatty acids, amino acids, and sugars produced during the previous step are fermented into volatile fatty acids, alcohol, and lactic acid by acid-producing microorganisms. The fourth phase is a**cetogenesis**, in which the main players are acetogenic microorganisms: Homo-acetogenic, hetero-acetogenic, and mixotrophic forms. These species produce acetic acid, hydrogen, and carbon dioxide from the volatile fatty acids and alcohols produced during the acidogenesis stage. This is a crucial stage in waste degradation, as any malfunction can accumulate VFAs and thus inhibit methanogenesis. The final step is **methanogenesis**, where all the reactions are carried out under strictly anaerobic conditions by methanogenic microorganisms: hydrogenophiles, acetoclasts, and methylotrophs. During the methanisation phase, the VFAs are consumed, which leads to an increase in pH and, consequently, the precipitation of inorganic elements.

Composting is an alternative way of managing solid waste, recycling organic matter into valuable products, and limiting the accumulation of waste (Kadir et al., 2016). This process is considered efficient, environmentally friendly, and beneficial for agriculture, as compost can be used as a soil conditioner, fertilizer, and nutrient supplier (Kadir et al., 2016). Composting involves the controlled degradation (microbial process) of plant and animal matter (organic waste) in an aerobic environment. During this process, organic matter is transformed into smaller biodegradable elements, which can be used to improve crop growth and yield (Koul et al., 2022). During this process, the organic matter present in the waste undergoes degradation, resulting in the production of methane (CH_4) and carbon dioxide (CO_2) (Kinyua et al., 2016).

Because of their complex polymeric structure, microplastics resist natural environmental degradation (Haedar et al., 2019). Abiotic degradation (physical and chemical) converts polymers into monomers. At the same time, biodegradation by microorganisms can transform them into biomass, carbon dioxide, or methane (Jin et al., 2017). The degradation of microplastics in landfills involves an initial aerobic environment, a prolonged anaerobic period, mechanical stress, and variations in pH, temperature, and humidity (Golwala et al., 2021). Plastics extracted from landfills show signs of increasing degradation over time, with those over ten years old degrading more rapidly (Canopoli et al., 2020).

6.6 Functioning of Landfill Sites

A public landfill, also known as a municipal landfill or sanitary landfill site, is a facility where solid waste is disposed of in a controlled manner (Meegoda et al., 2016). Before opening a landfill, careful planning is required. This includes the choice of location, considering environmental factors and proximity to populated areas. Solid waste from various sources, such as households, businesses, and industries, is collected and transported to landfills. Before being deposited, waste can be sorted into separate recyclable materials, hazardous waste, and ordinary waste (Hassan & Saleh, 2022; Iraguha et al., 2022). The waste is then disposed of in trenches or designated areas within the landfill (Chiguer et al., 2016). After each deposit, the waste is mechanically compacted to reduce its volume and maximize usable space. The waste is then leveled to create a uniform surface (Adamcová et al., 2020). In general, decomposing waste produces liquids called leachates, which can be contaminated with pollutants (Bouyakhsass et al., 2023). Modern landfills have leachate collection systems to prevent leachate from reaching the soil and groundwater. The leachate collected is generally treated to reduce its toxicity (Huang et al., 2022). Landfill sites can be equipped with gas collection systems to prevent gases from being released into the atmosphere. These gases can then be burnt or converted into energy (Saghir & Naimi, 2019). When the landfill reaches its maximum capacity, it is closed following the regulations (Serrano et al., 2023). Site remediation may include final covering with a material layer, planting vegetation to stabilize the soil, and implementing measures to minimize long-term impacts (Koda et al., 2023). Landfill sites can generate various types of pollutants that can contribute to air pollution (Iravanian & Ravari, 2020).

6.7 Landfill Sites: A Potential Source of Air Pollution

Air quality is a major issue for human health and the environment. Unfortunately, landfill sites, which receive large quantities of waste, are proving to be a worrying potential source of air pollution. The accumulation of organic and inorganic matter causes waste to decompose naturally under the action of microorganisms. However, due to the anaerobic conditions in these landfills, decomposition produces noxious gases (Siddiqua et al., 2022; Vaverková, 2019).

Methane formation in landfill sites results from the anaerobic decomposition of organic waste. As this waste decomposes without oxygen, microorganisms such as methanogenic bacteria come into play, producing methane as a by-product (Vaverková, 2019). This gas is then trapped in the porous structures of the soil or escapes into the atmosphere if collection mechanisms are not in place.

The global warming potential (GWP) of methane is of particular concern. Over a period of 100 years, the GWP of methane is around 28 times higher than that of CO_2 (Abdulmajeed et al., 2017). Cela signifie que, bien que le méthane reste moins abondant que le dioxyde de carbone dans l'atmosphère, il a une capacité beaucoup plus grande à absorber et à retenir la chaleur, contribuant ainsi de manière significative au réchauffement climatique.

On the other hand, public landfill sites receive a wide variety of waste, often including organic chemicals such as benzene, toluene, and xylene, which are classified as volatile organic compounds (VOCs) (Lakhouit & Alsulami, 2020; Pan et al., 2023). These substances are characterized by their volatility, meaning they tend to evaporate easily into the ambient air. When waste containing these VOCs is deposited in landfill sites, the compounds come into contact with the air and are gradually released into the atmosphere.

In the presence of sunlight and other atmospheric pollutants, VOCs can react chemically to form tropospheric ozone, also known as low-level ozone (Nair et al., 2019). Ground-level ozone is a major pollutant that poses severe risks to human health and the environment.

Moreover, when waste decomposes in landfill sites, it generates fine particles known as $PM_{2.5}$ and PM_{10} (Sharma et al., 2019). These fine particles are small solid or liquid particles suspended in the ambient air. Because of their minute size, they can penetrate deep into the respiratory tract when inhaled by humans, leading to worrying health problems (Feuyit et al., 2019).

The accumulation of these pollutants in the atmosphere above landfill sites can lead to a deterioration in the quality of the surrounding air, affecting the health of local populations and contributing to regional pollution problems.

6.8 Impact of Air Pollution from Landfill Sites on Human Health

The effect of air pollution from landfill sites on human health is a major concern. The decomposition of waste at these sites results in the emission of harmful gases and fine particles, which can devastate human health.

When released into the atmosphere, methane contributes to the greenhouse effect, causing global warming that can adversely affect human health. Climate change is associated with rising temperatures, more frequent heatwaves, and extreme weather events, leading to health problems such as heat stroke, respiratory issues, and heat-related infections.

On the other hand, exposure to tropospheric ozone, the result of chemical reactions of VOCs, can cause respiratory tract irritation, breathing difficulties, and asthma attacks and increase the risk of chronic respiratory diseases such as chronic bronchitis and emphysema (Njoku et al., 2019; Vinti et al., 2021). Moreover, Lakhouit and Alsulami (2020) mentioned in their work that benzene, toluene, ethylbenzene, and xylenes, VOCs released during the storage of waste in landfill sites, are individually and collectively carcinogenic to humans, even at low concentrations.

Fine particles ($PM_{2.5}$ and PM_{10}) are small solid or liquid particles that can penetrate deep into the respiratory tract when inhaled (Opara et al., 2021). These fine particles can cause airway inflammation, aggravate asthma symptoms, and increase the risk of respiratory problems such as bronchitis and pneumonia (Dalasile & Reddy, 2017; Hossain et al., 2019). In addition, fine particles can enter the bloodstream and cause cardiovascular problems such as heart disease, strokes, and cardiac arrhythmias (Jalali et al., 2021; Soile et al., 2018).

The most vulnerable populations, such as children, the elderly and people already suffering from health problems, are particularly exposed to the harmful effects of air pollution caused by landfill sites. In addition, people living near these sites are also more likely to suffer the consequences of this pollution on their health.

6.9 Environmental Impact of Air Pollution from Landfill Sites

Solid waste disposal sites generate a variety of pollutants that can have harmful consequences for ecosystems, biodiversity, and natural resources.

One of the main consequences of atmospheric pollution from landfill sites is the formation of tropospheric ozone. In the presence of sunlight and other atmospheric pollutants, the volatile organic compounds (VOCs) emitted by waste react to form low-level ozone. Tropospheric ozone can damage vegetation by impairing photosynthesis and damaging plant leaves (Siddiqua et al., 2022). This can lead to a reduction in agricultural productivity, affecting crops and harvests.

In addition to ozone, emissions of greenhouse gases (GHGs) such as methane (CH_4) from landfill sites contribute to global warming. Methane is a much more potent greenhouse gas than carbon dioxide (CO_2) in terms of global warming potential over 100 years (Abdulmajeed et al., 2017). Global warming can disrupt ecosystems, cause glaciers and ice caps to melt, raise sea levels, and lead to unpredictable climate change (Høyer et al., 2019; Kazaryan et al., 2021).

Air pollution from landfill sites can also affect air quality in surrounding areas. The fine particles ($PM_{2.5}$ and PM_{10}) emitted by landfill sites can be carried by the wind over long distances and adversely affect air quality in neighboring areas (Conte et al., 2018; Sharma et al., 2023).

In addition, air pollution from landfill sites can harm biodiversity and the ecological balance. Emissions of VOCs and fine particles can affect ecosystems and disrupt the life of animal and plant species, reducing biological diversity and impacting the food chain (Jakhar & Sachar, 2023).

6.10 Economic Impact of Air Pollution from Landfill Sites

The economic impact of air pollution from landfill sites is complex and can be significant. Although often seen as a low-cost solution to waste disposal, these solid waste disposal sites can have hidden costs and long-term economic consequences for local communities and governments.

First, air pollution from landfill sites can increase public health costs. Emissions of harmful gases, such as volatile organic compounds (VOCs) and fine particles, can aggravate respiratory, cardiovascular, and neurological problems in people near these sites (Dalasile & Reddy, 2017). This can lead to increased expenditure on healthcare, including medical treatment, hospitalization, and medication, putting pressure on local and national healthcare systems (Barwick et al., 2018; Croitoru & Sarraf, 2017).

Furthermore, global warming caused by greenhouse gas (GHG) emissions from landfills can have significant economic consequences. Extreme weather events, such as heatwaves, droughts and floods, can disrupt agriculture and supply chains, leading to crop losses and reduced productivity (Khan et al., 2018; Sharma et al., 2021; Wang et al., 2017). In addition, rising sea levels due to melting glaciers and ice caps can threaten coastal infrastructures, requiring major investment in coastal protection and the reconstruction of affected areas (Linquist et al., 2018).

6.11 Techniques for Reducing Atmospheric Pollution

Reducing air pollution is a significant concern if we are to preserve air quality, protect human health, and preserve the environment.

Methane is one of the main greenhouse gases emitted by landfill sites. Plaid landfill waste can be captured by installing biogas collection and treatment systems. This captured methane can then be used as a renewable energy source to power engines or generators, producing electricity and reducing the landfill's impact on the climate (Choe et al., 2022; Kumar et al., 2023).

On the other hand, rather than disposing of all organic waste in landfill sites, composting allows this material to decompose in a controlled, aerobic way (in the presence of oxygen) to produce compost (Ayilara et al., 2020). This approach reduces the amount of waste sent to landfills, limits methane production, and has an organic amendment beneficial to the soil (Abdel-Shafy & Mansour, 2018).

Public landfill sites must be appropriately managed to minimize the emission of atmospheric pollutants. The daily covering of waste with inert materials (such as earth) and the containment of landfill sites in suitable facilities reduce emissions of VOCs and fine particles (Majdinasab & Yuan, 2017; Randazzo et al., 2020).

By integrating more responsible waste management practices, we can effectively reduce the air pollution caused by landfill sites. This comprehensive approach will help to improve air quality, reduce the impact on human health, and preserve the environment for present and future generations.

6.12 Conclusion

In conclusion, solid waste management is a major global concern with far-reaching implications for human health, the environment, and the economy.

Decomposing waste in landfill sites generates emissions of greenhouse gases such as methane, volatile organic compounds, and fine particles. These pollutants harm air quality, human health, and the environmental balance. The effects on health, ranging from respiratory problems to cardiovascular disease, underline the urgency of mitigating this threat.

The environmental impact of air pollution from landfill sites also extends to biodiversity, soils, and aquatic ecosystems, compromising the sustainability of our ecosystems. Moreover, the economic fallout is substantial, adding to public health costs, hampering industrial development, and requiring investment to deal with climate change and environmental disasters.

However, there are promising solutions to these challenges. Air pollution abatement techniques, such as methane capture, waste-to-energy, and recycling, offer tangible approaches to mitigating the impacts of climate change.

References

Abdel-Shafy, H. I., & Mansour, M. S. M. (2018). Solid waste issue: Sources, composition, disposal, recycling, and valorization. *Egyptian Journal of Petroleum, 27*, 1275–1290. https://doi.org/10.1016/j.ejpe.2018.07.003

Abdulmajeed, A. M., Derby, S. R., Strickland, S. K., & Qaderi, M. M. (2017). Interactive effects of temperature and UVB radiation on methane emissions from different organs of pea plants grown in hydroponic system. *Journal of Photochemistry and Photobiology B: Biology, 166*, 193–201. https://doi.org/10.1016/j.jphotobiol.2016.11.019

Abubakar, I. R., Maniruzzaman, K. M., Dano, U. L., AlShihri, F. S., AlShammari, M. S., Ahmed, S. M. S., Al-Gehlani, W. A. G., & Alrawaf, T. I. (2022). Environmental sustainability impacts of solid waste management practices in the global south. *International Journal of Environmental Research and Public Health, 19*, 12717. https://doi.org/10.3390/ijerph191912717

Adamcová, D., Barton̆, S., Osinski, P., Pasternak, G., Podlasek, A., Vaverková, M. D., & Koda, E. (2020). Analytical modelling of MSW landfill surface displacement based on GNSS monitoring. *Sensors, 20*, 5998. https://doi.org/10.3390/s20215998

Adetona, O., Ozoh, O. B., Oluseyi, T., Uzoegwu, Q., Odei, J., & Lucas, M. (2020). An exploratory evaluation of the potential pulmonary, neurological and other health effects of chronic exposure to emissions from municipal solid waste fires at a large dumpsite in Olusosun, Lagos, Nigeria. *Environmental Science and Pollution Research International, 27*, 30885–30892. https://doi.org/10.1007/s11356-020-09701-4

Alam, P., & Ahmade, K. (2013). Impact of solid waste on health and the environment

Ayilara, M. S., Olanrewaju, O. S., Babalola, O. O., & Odeyemi, O. (2020). Waste management through composting: Challenges and potentials. *Sustainability, 12*, 4456. https://doi.org/10.3390/su12114456

Barwick, P. J., Li, S., Rao, D., & Zahur, N. B. (2018). The healthcare cost of air pollution: evidence from the world's largest payment network. In *Working Paper Series*. https://doi.org/10.3386/w24688

Bouyakhsass, R., Souabi, S., Rifi, S. K., Bouaouda, S., Taleb, A., Madinzi, A., Kurniawan, T. A., & Anouzla, A. (2023). Applicability of central composite design and response surface methodology for optimizing treatment of landfill leachate using coagulation-flocculation. *Chemical Engineering Research and Design, 197*, 669–684. https://doi.org/10.1016/j.cherd.2023.08.001

Canopoli, L., Coulon, F., & Wagland, S. T. (2020). Degradation of excavated polyethylene and polypropylene waste from landfill. *Science of the Total Environment, 698*, Article 134125. https://doi.org/10.1016/j.scitotenv.2019.134125

Chen, D.M.-C., Bodirsky, B. L., Krueger, T., Mishra, A., & Popp, A. (2020). The world's growing municipal solid waste: Trends and impacts. *Environmental Research Letters, 15*, Article 074021. https://doi.org/10.1088/1748-9326/ab8659

Chen, Y.-C. (2018). Evaluating greenhouse gas emissions and energy recovery from municipal and industrial solid waste using waste-to-energy technology. *Journal of Cleaner Production, 192*, 262–269. https://doi.org/10.1016/j.jclepro.2018.04.260

Chiguer, H., Khayyat, F., El rhaouat, O., Rifki, R., Bensaid, A., Kharrim, K., & Belghyti, D. (2016). Evaluation de la charge polluante des lixiviats de la décharge contrôlée de la ville d'Essaouira (MAROC) [Evaluation of the pollution load of leachates from the landfill in the city Essaouira (MOROCCO)]. *International Journal of Innovative and Applied Research, 14*, 863–874.

Choe, C., Byun, M., Lee, H., & Lim, H. (2022). Techno-economic and environmental assessments for sustainable bio-methanol production as landfill gas valorization. *Waste Management, 150*, 90–97. https://doi.org/10.1016/j.wasman.2022.06.040

Conte, M., Cagnazzo, V., Donateo, A., Cesari, D., Grasso, F. M., & Contini, D. (2018). A case study of municipal solid waste landfills impact on air pollution in south areas of Italy. *Open Atmospheric Science Journal, 12*, 1–3. https://doi.org/10.2174/1874282301812010001

Croitoru, L., & Sarraf, M. (2017). Estimating the health cost of air pollution: The case of Morocco. *Journal of Environmental Protection, 08*, 1087. https://doi.org/10.4236/jep.2017.810069

Dalasile, M., & Reddy, P. (2017). Respiratory health risks and exposure to particulate matter ($PM_{2.5}$) among informal waste pickers at a landfill site in Durban, South Africa. *African Journal for Physical Activity and Health Sciences AJPHES, 23*, 45–58. https://doi.org/10.10520/EJC-69faaca9c

Das, S., Lee, S.-H., Kumar, P., Kim, K.-H., Lee, S. S., & Bhattacharya, S. S. (2019). Solid waste management: Scope and the challenge of sustainability. *Journal of Cleaner Production, 228*, 658–678. https://doi.org/10.1016/j.jclepro.2019.04.323

Feuyit, G., Nzali, S., Lambi, J. N., & Laminsi, S. (2019). Air quality and human health risk assessment in the residential areas at the proximity of the Nkolfoulou landfill in Yaoundé Metropolis. *Cameroon Journal of Chemistry, 2019*, Article e3021894. https://doi.org/10.1155/2019/3021894

Fu, Z., Lin, S., Tian, H., Hao, Y., Wu, B., Liu, S., Luo, L., Bai, X., Guo, Z., & Lv, Y. (2022). A comprehensive emission inventory of hazardous air pollutants from municipal solid waste incineration in China. *Science of the Total Environment, 826*, Article 154212. https://doi.org/10.1016/j.scitotenv.2022.154212

Gallardo, A., Carlos, M., Colomer, F., & Edo-Alcón, N. (2018). Analysis of the waste selective collection at drop-off systems: Case study including the income level and the seasonal variation. *Waste Management and Research, 36*, 30–38. https://doi.org/10.1177/0734242X17733539

Gautam, M., & Agrawal, M. (2021). Greenhouse gas emissions from municipal solid waste management: A review of global scenario. In *Carbon Footprint Case Studies*. Springer.

Golwala, H., Zhang, X., Iskander, S. M., & Smith, A. L. (2021). Solid waste: An overlooked source of microplastics to the environment. *Science of the Total Environment, 769*, Article 144581. https://doi.org/10.1016/j.scitotenv.2020.144581

Haedar, N., Clara, T., Fahrudin, Abdullah, A., Fausiah, S., Rapak, & M. T. (2019). Selection of plastic degradation indigenous bacteria isolated from Tamangapa Landfill Macassar City. *Journal of Physics: Conference Series, 1341*, 022023. https://doi.org/10.1088/1742-6596/1341/2/022023

Hannan, M. A., Abdulla Al Mamun, M., Hussain, A., Basri, H., & Begum, R. A. (2015). A review on technologies and their usage in solid waste monitoring and management systems: Issues and challenges. *Waste Management, 43*, 509–523. https://doi.org/10.1016/j.wasman.2015.05.033

Hassan, A. I., & Saleh, H. M. (2022). Chapter 7—Toxicity and hazardous waste regulations. In D. Yadav, P. Kumar, P. Singh, D. A. Vallero (Eds.), *Hazardous waste management* (pp. 165–182). Elsevier. https://doi.org/10.1016/B978-0-12-824344-2.00012-4

Hossain, M. F., Kamal, A. M., Sikder, A. H. F., & Parveen, Z. (2019). *Air quality measurement at the solid waste disposal of matuail landfill site at Dhaka.*

Høyer, A.-S., Klint, K. E. S., Fiandaca, G., Maurya, P. K., Christiansen, A. V., Balbarini, N., Bjerg, P. L., Hansen, T. B., & Møller, I. (2019). Development of a high-resolution 3D geological model for landfill leachate risk assessment. *Engineering Geology, 249*, 45–59. https://doi.org/10.1016/j.enggeo.2018.12.015

Huang, F.-Y., Zhou, S.-Y.-D., Zhao, Y., Zhou, X.-Y., Li, H., Zhang, X., & Su, J.-Q. (2022). Dissemination of antibiotic resistance genes from landfill leachate to groundwater. *Journal of Hazardous Materials, 440*, Article 129763. https://doi.org/10.1016/j.jhazmat.2022.129763

Iraguha, F., Ramelan, A. H., & Setyono, P. (2022). Assessment of current solid waste management practices, community perceptions, and contributions in the City of Kigali, Rwanda. *IOP Conference Series: Earth and Environmental Science, 1016*, Article 012056. https://doi.org/10.1088/1755-1315/1016/1/012056

Iravanian, A., & Ravari, S. O. (2020). Types of contamination in landfills and effects on the environment: a review study. *IOP Conference Series: Earth and Environmental Science, 614*, Article 012083. https://doi.org/10.1088/1755-1315/614/1/012083

Jakhar, R., & Sachar, D. (2023). A study and analysis on waste fires in india and their corresponding impacts on environment and human health. *International Journal of Recent Technology and Engineering, 12*, 110–120. https://doi.org/10.35940/ijrte.B7655.0512223

Jalali, S., Karbakhsh, M., Momeni, M., Taheri, M., Amini, S., Mansourian, M., & Sarrafzadegan, N. (2021). Long-term exposure to $PM_{2.5}$ and cardiovascular disease incidence and mortality in an Eastern Mediterranean country: Findings based on a 15-year cohort study. *Environment Health, 20*, 112. https://doi.org/10.1186/s12940-021-00797-w

Jin, Z., Chang, F., Meng, F., Wang, C., Meng, Y., Liu, X., Wu, J., Zuo, J., & Wang, K. (2017). Sustainable pyrolytic sludge-char preparation on improvement of closed-loop sewage sludge treatment: Characterization and combined in-situ application. *Chemosphere, 184*, 1043–1053. https://doi.org/10.1016/j.chemosphere.2017.06.029

Kadir, A. A., Azhari, N. W., & Jamaludin, S. N. (2016). An overview of organic waste in composting. *MATEC Web Conference, 47*, 05025. https://doi.org/10.1051/matecconf/20164705025

Kamaruddin, M. A., Yusoff, M. S., Rui, L. M., Isa, A. M., Zawawi, M. H., & Alrozi, R. (2017). An overview of municipal solid waste management and landfill leachate treatment: Malaysia and Asian perspectives. *Environmental Science and Pollution Research, 24*, 26988–27020. https://doi.org/10.1007/s11356-017-0303-9

Kanhai, G., Fobil, J. N., Nartey, B. A., Spadaro, J. V., & Mudu, P. (2021). Urban municipal solid waste management: Modeling air pollution scenarios and health impacts in the case of Accra, Ghana. *Waste Management, 123*, 15–22. https://doi.org/10.1016/j.wasman.2021.01.005

Kazaryan, M., Semenishchev, E., & Voronin, V. (2021). Research of the impact of solid domestic and industrial waste on the snow and ice coverage of mountain territories of the Northern Caucasus using remote sensing technology. In *Earth observing systems XXVI. Presented at the Earth Observing Systems XXVI* (pp. 415–426) SPIE. https://doi.org/10.1117/12.2598314

Khan, M. T. I., Ali, Q., & Ashfaq, M. (2018). The nexus between greenhouse gas emission, electricity production, renewable energy and agriculture in Pakistan. *Renewable Energy, 118*, 437–451. https://doi.org/10.1016/j.renene.2017.11.043

Kinyua, M. N., Rowse, L. E., & Ergas, S. J. (2016). Review of small-scale tubular anaerobic digesters treating livestock waste in the developing world. *Renewable and Sustainable Energy Reviews, 58*, 896–910. https://doi.org/10.1016/j.rser.2015.12.324

Koda, E., Osiński, P., Podlasek, A., Markiewicz, A., Winkler, J., & Vaverková, M. D. (2023). Geoenvironmental approaches in an old municipal waste landfill reclamation process: Expectations vs reality. *Soils and Foundations, 63*, Article 101273. https://doi.org/10.1016/j.sandf.2023.101273

Koul, B., Yakoob, M., & Shah, M. P. (2022). Agricultural waste management strategies for environmental sustainability. *Environmental Research, 206*, Article 112285. https://doi.org/10.1016/j.envres.2021.112285

Kumar, M., & Prakash, V. (2020). A review on solid waste: Its impact on air and water quality. *Journal of Pollution Effects & Control, 8*, 1–3. https://doi.org/10.35248/2375-4397.20.8.252

Kumar, S. S., Kumar, A., Malyan, S. K., Ghosh, P., Kumar, M., Kapoor, R., Agrawal, A. K., Kumar, S., Kumar, V., & Singh, L. (2023). Landfill leachate valorization: A potential alternative to burden off resources and support energy systems. *Fuel, 331*, Article 125911. https://doi.org/10.1016/j.fuel.2022.125911

Lakhouit, A., & Alsulami, B. T. (2020). Evaluation of risk assessment of landfill emissions and their impacts on human health. *Arabian Journal of Geosciences, 13*, 1185. https://doi.org/10.1007/s12517-020-06218-5

Landrigan, P. J. (2017). Air pollution and health. *Lancet Public Health, 2*, e4–e5. https://doi.org/10.1016/S2468-2667(16)30023-8

Linquist, B. A., Marcos, M., Adviento-Borbe, M. A., Anders, M., Harrell, D., Linscombe, S., Reba, M. L., Runkle, B. R. K., Tarpley, L., & Thomson, A. (2018). Greenhouse gas emissions and management practices that affect emissions in US rice systems. *Journal of Environmental Quality, 47*, 395–409. https://doi.org/10.2134/jeq2017.11.0445

Luo, L., & Wong, J. W. C. (2019). Enhanced food waste degradation in integrated two-phase anaerobic digestion: Effect of leachate recirculation ratio. *Bioresource Technology, 291*, Article 121813. https://doi.org/10.1016/j.biortech.2019.121813

Majdinasab, A., & Yuan, Q. (2017). Performance of the biotic systems for reducing methane emissions from landfill sites: A review. *Ecological Engineering, 104*, 116–130. https://doi.org/10.1016/j.ecoleng.2017.04.015

Manisalidis, I., Stavropoulou, E., Stavropoulos, A., & Bezirtzoglou, E. (2020). *Environmental and health impacts of air pollution: A review* (p. 8). Public Health.

Meegoda, J. N., Hettiarachchi, H., & Hettiaratchi, P. (2016). Landfill design and operation. In *Sustainable solid waste management* (pp. 577–604). American Society of Civil Engineers (ASCE). https://doi.org/10.1061/9780784414101.ch18

Mohammed, A., & Elias, E. (2017). Domestic solid waste management and its environmental impacts in Addis Ababa city.

Nair, A. T., Senthilnathan, J., & Nagendra, S. M. S. (2019). Emerging perspectives on VOC emissions from landfill sites: Impact on tropospheric chemistry and local air quality. *Process Safety and Environment Protection, 121*, 143–154. https://doi.org/10.1016/j.psep.2018.10.026

Nanda, S., & Berruti, F. (2021). Municipal solid waste management and landfilling technologies: A review. *Environmental Chemistry Letters, 19*, 1433–1456. https://doi.org/10.1007/s10311-020-01100-y

Njoku, P. O., Edokpayi, J. N., & Odiyo, J. O. (2019). Health and environmental risks of residents living close to a landfill: A case study of Thohoyandou Landfill, Limpopo Province, South Africa. *International Journal of Environmental Research and Public Health, 16*, 2125. https://doi.org/10.3390/ijerph16122125

Ogata, Y., Ishigaki, T., Nakagawa, M., & Yamada, M. (2016). Effect of increasing salinity on biogas production in waste landfills with leachate recirculation: A lab-scale model study. *Biotechnology Reports, 10*, 111–116. https://doi.org/10.1016/j.btre.2016.04.004

Opara, A. I., Akaolisa, C. Z., Akakuru, C. O., Nkwoada, A. U., Ibe, F. C., Verla, A. W., & Chukwuemeka, I. C. (2021). Particulate matter exposure and non-cancerous inhalation health risk assessment of major dumpsites of Owerri metropolis, Nigeria. *Environmental Analysis, Health and Toxicology, 36*, Article e2021025. https://doi.org/10.5620/eaht.2021025

Pan, Q., Liu, Q.-Y., Zheng, J., Li, Y.-H., Xiang, S., Sun, X.-J., & He, X.-S. (2023). Volatile and semi-volatile organic compounds in landfill gas: Composition characteristics and health risks. *Environment International, 174*, Article 107886. https://doi.org/10.1016/j.envint.2023.107886

Randazzo, A., Asensio-Ramos, M., Melián, G.V., Venturi, S., Padrón, E., Hernández, P.A., Pérez, N. M., & Tassi, F. (2020). Volatile organic compounds (VOCs) in solid waste landfill cover soil: Chemical and isotopic composition versus degradation processes. *Science of the Total Environment, 726*, 138326. https://doi.org/10.1016/j.scitotenv.2020.138326

Saghir, M., & Naimi, Y. (2019). Energy recovery from waste in Fez city (Morocco). In 2019 *International Conference of Computer Science and Renewable Energies (ICCSRE). Presented at the 2019 International Conference of Computer Science and Renewable Energies (ICCSRE)* (pp. 1–6). https://doi.org/10.1109/ICCSRE.2019.8807644

Serrano, M. F., López, J. E., & Saldarriaga, J. F. (2023). Use of activated rice husk biochar for the removal of metals and microorganisms from treated leachates from landfills. *Journal of Material Cycles and Waste Management*. https://doi.org/10.1007/s10163-023-01762-0

Sharma G, Sinha B, Pallavi H, Chandra BP, Kumar A, Sinha V (2019) Gridded emissions of CO, NO_x, SO_2, CO_2, NH_3, HCl, CH_4, $PM_{2.5}$, PM_{10}, BC, and NMVOC from open municipal waste burning in India. *Environmental Science and Technology, 53*, 4765–4774 https://doi.org/10.1021/acs.est.8b07076

Sharma, G. D., Shah, M. I., Shahzad, U., Jain, M., & Chopra, R. (2021). Exploring the nexus between agriculture and greenhouse gas emissions in BIMSTEC region: The role of renewable energy and human capital as moderators. *Journal of Environmental Management, 297*, Article 113316. https://doi.org/10.1016/j.jenvman.2021.113316

Sharma, M., Khare, M., & Mishra, R. K. (2023). Air quality changes in Delhi due to open waste burning: An accidental fire in Bhalswa landfill. *International Journal of Environmental Science and Technology*. https://doi.org/10.1007/s13762-023-04921-w

Sharma, P., Soni, R., Srivastava, S. K., Singh, S. P. (2022). Modern landfilling approaches for waste disposal and management. In: D. C. Suyal, R. Soni (Eds.), *Bioremediation of environmental pollutants: Emerging trends and strategies* (pp. 239–252). Springer, Cham. https://doi.org/10.1007/978-3-030-86169-8_10

Siddiqua, A., Hahladakis, J. N., & Al-Attiya, W. A. K. A. (2022). An overview of the environmental pollution and health effects associated with waste landfilling and open dumping. *Environmental Science and Pollution Research, 29*, 58514–58536. https://doi.org/10.1007/s11356-022-21578-z

Sipra, A. T., Gao, N., & Sarwar, H. (2018). Municipal solid waste (MSW) pyrolysis for bio-fuel production: A review of effects of MSW components and catalysts. *Fuel Processing Technology, 175*, 131–147. https://doi.org/10.1016/j.fuproc.2018.02.012

Soile, O. B., Akiibinu, M. O., Bakare, T. E., Olaoye, G. O., Oyeyiola, F. A., Adeola, J. A., Alarape, B. B., Duduyemi, O. O., Anetor, J. I. (2018). Assessment of landfill-emitted gaseous pollutants and particulate matters in alimosho local government of Lagos state, Nigeria. *European Scientific Journal (ESJ), 14*, 272–272. https://doi.org/10.19044/esj.2018.v14n24p272

Tian, H., Gao, J., Hao, J., Lu, L., Zhu, C., & Qiu, P. (2013). Atmospheric pollution problems and control proposals associated with solid waste management in China: A review. *Journal of Hazardous Materials, 252–253*, 142–154. https://doi.org/10.1016/j.jhazmat.2013.02.013

Vaverková, M. D. (2019). Landfill Impacts on the Environment—Review. *Geosciences, 9*, 431. https://doi.org/10.3390/geosciences9100431

Vinti, G., Bauza, V., Clasen, T., Medlicott, K., Tudor, T., Zurbrügg, C., & Vaccari, M. (2021). Municipal solid waste management and adverse health outcomes: A systematic review. *International Journal of Environmental Research and Public Health, 18*, 4331. https://doi.org/10.3390/ijerph18084331

Wang, P., Wang, H., Qiu, Y., Ren, L., & Jiang, B. (2018). Microbial characteristics in anaerobic digestion process of food waste for methane production—A review. *Bioresource Technolnolgy. Bioconversion of Food Wastes, 248*, 29–36. https://doi.org/10.1016/j.biortech.2017.06.152

Wang, Z., Chen, J., Mao, S., Han, Y., Chen, F., Zhang, L., Li, Y., & Li, C. (2017). Comparison of greenhouse gas emissions of chemical fertilizer types in China's crop production. *Journal of Cleaner Production, 141*, 1267–1274. https://doi.org/10.1016/j.jclepro.2016.09.120

Woodard, R. (2021). Waste management in small and medium enterprises (SMEs): Compliance with duty of care and implications for the circular economy. *Journal of Cleaner Production, 278*, Article 123770. https://doi.org/10.1016/j.jclepro.2020.123770

Chapter 7
Biogas Generation and Ecological Implications: An Overview

Zakia Gueboudji

Abstract Biogas production is a sustainable and renewable process that converts organic waste into energy. This process involves the anaerobic digestion of organic materials such as agricultural waste, food waste, and sewage sludge, resulting in the production of biogas, which is mainly composed of methane and carbon dioxide. Biogas can be used as a source of renewable energy for electricity and heat production or as a transportation fuel. The production of biogas has various environmental benefits. Firstly, it reduces greenhouse gas emissions by capturing methane, which is a potent greenhouse gas, and using it as a fuel instead of releasing it into the atmosphere. Secondly, it provides an alternative to fossil fuels, reducing reliance on non-renewable energy sources. Thirdly, the production of biogas can reduce waste volumes and the need for landfill sites, leading to a reduction in environmental pollution. However, there are also potential negative environmental impacts associated with biogas production. For example, the production of biogas can lead to nutrient leaching and contamination of nearby water sources if not managed correctly. Additionally, the use of certain feedstocks, such as crops grown specifically for biogas production, can lead to land-use changes and biodiversity loss. Overall, biogas production has the potential to provide a sustainable source of energy while reducing greenhouse gas emissions and waste volumes. However, it is important to carefully manage the environmental impacts associated with biogas production to ensure its long-term sustainability.

Keywords Renewable energy · Biogas · Environmental sustainability · Anaerobic digestion · Methane emissions · Organic waste

Z. Gueboudji (✉)
Biotechnology, Water, Environment and Health Laboratory, Abbes Laghrour University of Khenchela, Khenchela, Algeria
e-mail: gueboudji.zakia@gmail.com; zakia.gueboudji@univ-batna2.dz

Department of Microbiology and Biochemistry, Faculty of Natural and Life Sciences, University of Batna 2 - Mostefa Ben Boulaid, Batna, Algeria

A. Anouzla et al. (eds.), *Handbook of the Effects of Landfill Emissions on Sustainable Development*, https://doi.org/10.1007/978-3-032-01951-6_7

7.1 Introduction

The global energy problem is one of the most pressing issues. The expanding population, finite fossil supplies, and increased environmental pollution have prompted energy experts and investors to consider alternate energies (Hashemi et al., 2019; Heidari-Maleni et al., 2023). The correlation between economic and environmental challenges has resulted in unique methods in the field of international environmental legislation, the most prominent of which is green economy (Raturi, 2019). The green economy is committed to lowering and mitigating greenhouse gas (GHG) emissions, as well as addressing the constraints and issues caused by the usage of fossil fuels and their depletion. Furthermore, even nations with abundant oil and gas reserves must embrace renewable energies to ensure the security of their fuel resources while keeping energy costs stable and environmental concerns in mind. As a result, long-term efforts have been undertaken to develop and promote the usage of renewable energy sources (Heidari-Maleni et al., 2023; Silva & Eus, 2019).

Biomass waste has great potential for use as energy source. An interesting aspect of the use of biomass is its renewable character, which guarantees that the sources will not be depleted (Ala'a et al., 2018; Tahie et al., 2020). Because biogas is mostly composed of organic waste and biomass, it is categorized as a renewable energy source, similar to wind and solar (Karlsson et al., 2017). Biomass is a sustainable energy source that is an organic leftover from agriculture, forestry, and urban development that may be utilized to benefit the environment (Mikusińska et al., 2023). Anaerobic digestion is the most common method for producing biogas. Microorganisms transform trash into energy in this biological process. Biogas is mostly composed of methane and carbon dioxide, but it also contains hydrogen sulfide and nitrogen (Avila-Lopez et al., 2023; Gueboudji, 2022).

At present, the quest for a sustainable energy resource for power generation, coupled with the challenges stemming from the creation and management of solid waste within urban areas, has prompted scholars to contemplate strategies for deriving energy from refuse. The technique of anaerobic digestion is employed to harness biogas from waste materials (Heidari-Maleni et al., 2023). Recent research findings indicate that the generation of biogas holds substantial promise in terms of substantially mitigating the release of greenhouse gases in contrast to the utilization of fossil fuels (Burg et al., 2018; Siddiki et al., 2021). The production of biogas has been documented as yielding noteworthy economic advantages through the reduction of domestic energy expenditures and expenses related to manure handling, in addition to generating revenue via the surplus biogas and digestate sales (Amoo et al., 2023).

Biogas constitutes a gas amalgam primarily composed of methane and carbon dioxide, accompanied by trace amounts of other gases including hydrogen sulphide (H_2S), ammonia (NH_3), oxygen (O_2), hydrogen (H_2), nitrogen (N_2), and carbon monoxide (CO) (Sun et al., 2015). This gaseous composition originates from a biochemical mechanism recognized as anaerobic digestion (AD). Anaerobic digestion involves the enzymatic degradation of organic materials into simpler compounds, giving rise to biogas and digestate, under the influence of microorganisms. The resulting digestate serves as a biofertilizer, contributing to the closing of nutrient deficiency in agricultural integrated systems (Rouhollahi et al., 2020). The technology of anaerobic digestion finds application in the treatment of diverse organic waste streams (Abdeshahian et al., 2016). Among these organic waste sources are animal manure, municipal solid waste, and agroindustrial residues, all of which function as substrates within the anaerobic digestion process (Nwokolo et al., 2020). Within the realm of biogas production, an array of factors holds substantial importance; nevertheless, the type of organic substrate employed has emerged as a pivotal determinant in both the output and constitution of the resultant biogas. Diverse biogas digesters operating across the globe harness an assortment of substrates, thereby giving rise to distinctive microbial communities and fluctuations in methane composition within these digesters. A number of investigations have leveraged sequencing technologies to scrutinize the composition of these microorganism communities participating in the anaerobic digestion (AD) process. This assembly of microorganisms encompasses bacteria, fungi, and archaea, collectively responsible for orchestrating the entire spectrum of reactions transpiring within the digester framework (Bücker et al., 2020; Wang et al., 2018).

This chapter book aims to provide a thorough overview of biogas generation processes and their ecological impacts. It explores the scientific principles of biogas production, delves into microbial digestion mechanisms, and examines the environmental benefits and challenges. The ecological footprint of different biogas production methods will be critically reviewed, considering factors like greenhouse gas mitigation, waste reduction, and circular economy integration. By offering insights into both technical and ecological aspects, this chapter aims to support informed decision-making for sustainable biogas development.

7.2 Biogas Production Technologies

7.2.1 Anaerobic Digestion Process

Anaerobic digestion (AD) systems have garnered significant attention within the scientific community due to their advantageous impacts on human health, the environment, the economy, and energy-related aspects (Van et al., 2020). The anaerobic digestion process, in which organic waste is broken down by microorganisms in the absence of oxygen, is at the heart of biogas generation. This section delves into the anaerobic digestion process, delving into its essential stages and processes. The microbial population involved in anaerobic digestion are essential for the stepwise breakdown of complex organic substrates into simple molecules, ultimately leading to biogas production. This section reviews the functional roles of key microbial groups throughout the digestion process. Furthermore, it examines the influence of critical operational parameters such as temperature, pH, hydraulic retention time

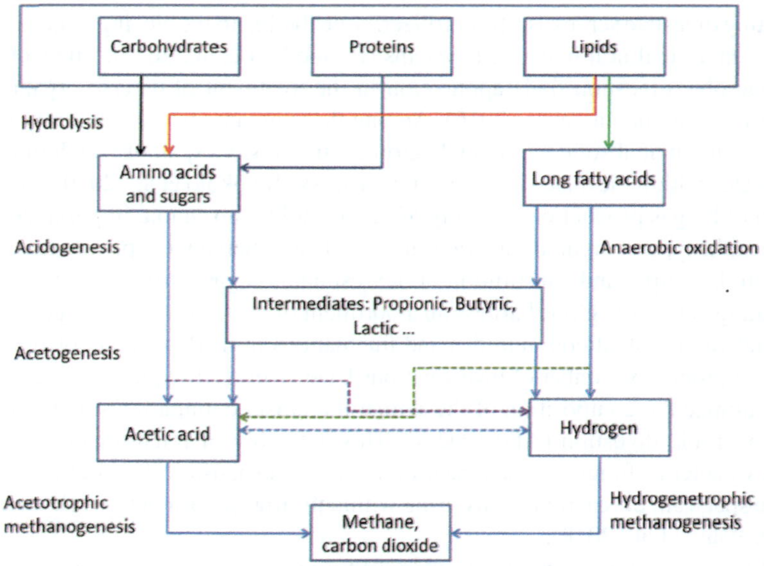

Fig. 7.1 Anaerobic digestion process (Tshemese et al., 2023)

(HRT), and organic loading rate (OLR) on the efficiency and stability of anaerobic digestion systems (Avila-Lopez et al., 2023; Sun et al., 2023).

The anaerobic biological degradation of organic substrates occurs through a sequential four-step process, namely hydrolysis, acidogenesis, acetogenesis, and methanogenesis. The ingenuity of this mechanism lies in its inherent interconnectedness, as the by-product generated in one phase seamlessly transitions into the substrate for the subsequent step, all within a unified system (Nwokolo et al., 2020). These stages of biochemical decomposition involve a cascade of chemical reactions, visually depicted in Fig. 7.1.

7.2.1.1 Hydrolysis

In the hydrolysis phase of anaerobic digestion, intricate biopolymeric compounds such as lipids, carbohydrates, and proteins undergo degradation into water-soluble compounds. This transformative process is facilitated by microorganisms including Bacteroides, Clostridia, and Bifidobacteria, and occasionally involves Streptococci and Enterobacteriaceae (Tshemese et al., 2023). This phase exhibits relatively sluggish kinetics, potentially impeding the overall digestion rate, particularly when solid materials serve as the substrate. As depicted in Eq. (7.1), cellulose ($C_6H_{10}O_5$) undergoes hydrolysis, yielding glucose ($C_6H_{12}O_6$) and hydrogen (H_2). This chemical transformation is facilitated by either homogeneous or heterogeneous acids, resulting in the formation of fermentable monosaccharides ($C_6H_{12}O_6$). Subsequently, these products ($C_6H_{12}O_6$ and H_2) are utilized by fermentative microorganisms in the subsequent stage to synthesize more complex organic compounds, including volatile fatty acids (Anukam et al., 2019; Sawyerr et al., 2019).

$$(C_6H_{10}O_5)_n + nH_2O \rightarrow n(C_6H_{12}) + nH_2 \tag{7.1}$$

7.2.1.2 Acidogenesis

The subsequent phase, recognized as the fermentation stage, involves the active participation of acidogenic bacteria, including Streptococcus, Escherichia, Staphylococcus, Pseudomonas, Bacillus, Sarcina, Desulfovibrio, Lactobacillus, and others (Deepanraj et al., 2014). Within this phase, these bacteria undertake the degradation of amino acids, lipids, and glucose, resulting in the formation of volatile fatty acids, organic acids, carbon dioxide, and hydrogen gas (as illustrated in Eqs. (7.2)–(7.7) below (Anukam et al., 2019)). Acetic acid (CH_3COOH) holds particular significance as the predominant organic acid generated during this stage, and it subsequently serves as the primary substrate for methanogenic microorganisms (Sarker et al., 2019). Notably, the production of volatile fatty acids is favored when the pH level exceeds 5, while ethanol production (CH_3CH_2OH) prevails under pH conditions lower than 5, with reactions ceasing at a pH level below 4 (Bajpai, 2017; Tshemese et al., 2023).

$$C_6H_{12}O_6 \leftrightarrow 2CH_3CH_2OH + 2CO_2 \tag{7.2}$$

$$C_6H_{12}O_6 + 2H_2 \leftrightarrow 2CH_3CH_2COOH + 2H_2O \tag{7.3}$$

$$C_6H_{12}O_6 \rightarrow 3CH_3COOH \tag{7.4}$$

$$C_3H_7O_{2N} + 2H_2O \rightarrow C_2H_4O_2 + NH_3 + CO_2 + 2H_2 + ATP \tag{7.5}$$

$$C_4H_9O_3 + NH_3 + CO_2 + H_2 + ATP \tag{7.6}$$

$$4CH_3COCOO^- + 4H_2O \rightarrow 5CH_3COO^- + 2HCO_3^- + 3H^+ \tag{7.7}$$

7.2.1.3 Acetogenesis

In this phase, the reactions are reversible and accompanied by hydrogen liberation. Volatile fatty acids, notably acetic acid and butyric acid, undergo transformation into carbon dioxide gas, hydrogen, and acetate, as illustrated in Eqs. (7.8)–(7.10). The bacteria actively involved in this phase encompass Clostridium, Syntrophomonas wolfeii, and Syntrophomonas wolinii (Deepanraj et al., 2014). This conversion of volatile fatty acids is facilitated by the presence of water molecules, acting as electron sources and originating from prior stages of anaerobic digestion. Equation (7.9) orchestrates the conversion of the intermediate product into acetate and hydrogen, which in turn become crucial components for the subsequent phase (Tshemese et al., 2023). This stage holds comparable significance, as it underpins the efficiency of biogas production, as about 70% of methane formation arises from the reduction of acetate ion. Acetate represents a central transitional product within this phase, constituting approximately 25% of the collective products generated, accompanied by around 11% hydrogen content (Tshemese et al., 2023).

$$CH_3CH_2COO^- + 3H_2O \leftrightarrow CH_3COO^- + H + HCO_3^- + 3H_2 \quad (7.8)$$

$$C_6H_{12}O_6 + 2H_2O \leftrightarrow 2CH_3COOH + 2CO_2 + 4H_2 \quad (7.9)$$

$$2CH_3CH_2OH + 2H_2O \leftrightarrow CH_3COO^- + 2H_2 + H^- \quad (7.10)$$

7.2.1.4 Methanogenesis

In the conclusive phase of the methanogenesis stage, acetic acid is transformed into methane and carbon dioxide by specialized microorganisms known as methanogens. These methanogens are anaerobic organisms highly susceptible to limited oxygen concentrations. Additionally, carbon dioxide, a by-product, participates in a reaction with hydrogen gas, leading to the further generation of methane. Conversely, ethanol experiences decarboxylation, resulting in the formation of methane. This stage encompasses two distinct categories of bacteria: the acetophilic methanogenic group, characterized by active species such as Methanosarcina and Methanosaeta, responsible for the decarboxylation of acetate into methane; and the hydrogenophilic methanogenic group, dominated by species like Methanospirilum, Methanobacterium formicicum, Methanoplanus, and Methanobrevibacterium, contributing to methane synthesis through a reaction involving carbon dioxide and hydrogen (Anukam et al., 2019; Deepanraj et al., 2014). The ultimate outcome of the anaerobic digestion process is the production of biogas, which comprises methane and carbon dioxide as its chief constituents.

$$CH_3COOH \rightarrow CH_4 + CO_2 \quad (7.11)$$

$$CO_2 + 4H_2 \rightarrow CH_4 + 2H_2O \quad (7.12)$$

$$2CH_3CH_2OH + CO_2 \rightarrow CH_4 + 2CH_3COOH \quad (7.13)$$

7.2.2 *Factors Influencing the Anaerobic Digestion Process and Biogas Generation*

Biogas generation is subject to various factors, including but not limited to substrate composition, temperature, pH level, organic loading rate, hydraulic retention time, among others (Avila-Lopez et al., 2023; Sun et al., 2023).

7.2.2.1 Substrate Type

A diverse array of biomass feedstocks can be harnessed for the purpose of biogas production, contingent upon their nutritional composition. In turn, these compositions exert influence over aspects such as biogas yield, methane content, degradation kinetics, and the biodegradability of the biomass. Essential nutritional elements within biomass substrates, namely carbohydrates, proteins, and fats, play a pivotal role in shaping their suitability for biogas generation. Theoretical assessments of the proportion of biogas and methane yield stemming from these nutrients have been documented in existing literature and are computed through the utilization of the Buswell formula (Atelge et al., 2020b).

Substrates rich in protein exhibit substantial potential for methane production, albeit their degradation results in the release of ammonium ions, contributing to an elevation in alkalinity within the anaerobic digestion process. This surge in alkalinity enhances the value of the digestate as a fertilizer, while concurrently tempering the activities of methanogenic microorganisms. The phenomenon of inhibition arises during the transition from ammonium to ammonia equilibrium, often occurring at variable concentrations. Furthermore, insights from literature indicate that microorganisms can acclimate to environments characterized by elevated ammonia levels, thereby effectively engaging in biogas production (Westerholm et al., 2016).

Substrates abundant in lipids, such as fats, exhibit significant potential for methane yield. These substrates encompass a range of sources including animal and plant tissue waste, biodegradable kitchen and canteen waste, grease and oil mixtures, and more (Diamantis et al., 2021). During their degradation, these substrates release long-chain fatty acids, which commonly possess toxicity toward the microbial milieu, consequently leading to a decline in pH (Rasit et al., 2015).

Beyond protein and lipid-rich substrates, there are several other types of biomass employed for biogas production. Examples include substrates characterized by a substantial lignocellulosic content, such as wheat straw, sorghum, and rice straw (Xu et al., 2019). This category of biomass poses challenges to degradation due to its recalcitrant nature, heterogeneous structure, and limited accessibility by enzymes targeting carbohydrate polymers (Dong et al., 2020). Nonetheless, pre-treatment methodologies can facilitate the breakdown of the intricate matrix, thus enhancing the porosity and surface area of lignocellulosic biomass, thereby augmenting biogas production.

The characterization of waste substrates is a pivotal step to ascertain the composition of each substrate. This typically involves evaluating the physical and chemical attributes, encompassing volatile solids, total solids, carbon-to-nitrogen (C/N) ratio, as well as elemental analysis covering carbon, nitrogen, hydrogen, and sulfur content (Cerón-Vivas et al., 2019). It is crucial to consider the origin (source) of the substrate during characterization, as waste composition is influenced by various factors, including weather conditions and the type of soil in which the source materials were cultivated (Chotovinský et al., 2016). The physical and chemical compositions are contingent upon the specific nature of the substrate; for instance, carbohydrates contain carbon and hydrogen, whereas proteins and lipids incorporate nitrogen within their composition (Schnürer, 2016). Analytical techniques of diverse nature are employed to scrutinize these substrate compositions.

7.2.2.2 Anaerobic Digestion pH

The operational pH plays a pivotal role in shaping both the progression of digestion and the resultant products within the anaerobic digestion (AD) process. Scholarly investigations indicate that the optimal encompassing pH range for AD falls within 4.0–8.5, a parameter that accommodates the requirements of fermentative bacteria, with a narrower favorable range of 6.5–7.2 emerging as conducive for the proliferation of methanogenic microorganisms (Tshemese et al., 2023). Notably, alterations in pH substantially influence the growth rate of microorganisms, with each microbial group exhibiting a distinct optimal pH value. This phenomenon is underscored by a notable increase in the abundance of microbial species, with counts rising from 6 to 14 as the pH shifts from 4.0 to 7.0. The dominance of bacterial populations undergoes shifts in accordance with varying pH levels. For instance, at pH 6.0, Clostridium butyricum asserts its dominance, while Propionibacterium spp. flourish during anaerobic acidogenesis at pH 8 (Tshemese et al., 2023).

Regulating the pH level to support optimal microorganism growth not only mitigates toxicity concerns, often arising from elevated levels of free ammonia (FA), but also has a pronounced impact on the composition of volatile fatty acids (VFAs). Within an anaerobic reactor, instability frequently leads to the accumulation of VFAs, culminating in a pH reduction and consequent acidification. However, the accumulation of VFAs does not always result in an immediate drop in pH, due to the buffering capacity exhibited by certain waste forms. Notably, in manure, an abundance of alkalinity exists, which implies that the expansion of VFAs must surpass a specific threshold before a significant pH alteration can be discerned (Akyol, 2020). Instances where the pH in the reactor diminishes often correspond to situations with potentially high concentrations of VFAs, potentially signifying prior process disruption. The buffering capacity is also influenced by compounds like hydrogen sulfide and phosphate. To rectify imbalances in pH, the introduction of a buffer solution into the bioreactor may be considered (Lattieff, 2016).

7.2.2.3 Temperature

Temperature stands as a pivotal parameter with far-reaching implications for anaerobic digestion (AD). Its influence extends to enzymatic and co-enzymatic activity, as well as to methane yield and the quality of the resulting digestate (effluent) (Vu et al., 2021). The growth of anaerobic bacteria is generally observed across three distinct temperature ranges: psychrophilic (10–30 °C), mesophilic (30–40 °C), and thermophilic (50–60°C). It is a common trend that AD performance is enhanced as temperature increases (Wei et al., 2022). Considerable attention has been directed toward the merits of thermophilic operation, characterized by heightened metabolic rates, accelerated pathogen elimination, and elevated specific growth rates, collectively contributing to increased biogas production (Wei et al., 2022).

Research conducted by Wei et al. (2022) revealed that biogas production at thermophilic conditions (55 °C) surpassed more than twofold the quantity generated under psychrophilic conditions (15 °C). Additionally, other investigations indicate that temperature elevation also correlates with an increase in phosphorus assimilation and organic nitrogen degradation. From a thermodynamic perspective, higher temperatures provide advantages to endergonic reactions, such as the breakdown of propionate into acetate, CO_2, and H_2, while they may not be as favorable for exergonic reactions, like hydrogenotrophic methanogenesis (Vu et al., 2021).

Furthermore, temperature can exert influence on the passive separation of solids, with notable improvements observed under thermophilic conditions compared to psychrophilic conditions. Nonetheless, there are certain drawbacks associated with thermophilic conditions, such as heightened sensitivity to environmental fluctuations when contrasted with the mesophilic process.

7.2.2.4 Organic Loading Rate (OLR)

Another pivotal operational parameter within the biogas production process is the Organic Loading Rate (OLR). This parameter quantifies the quantity of substrate introduced into a consistent digester system per unit volume per day. OLR is commonly expressed in units such as grams of total solids, chemical oxygen demand, or volatile solids per liter of digester volume per day (Lindmark et al., 2014). The calculation of this parameter is typically performed using Eq. (7.14):

$$OLR = COD_{feed} \times Q/V_r \tag{7.14}$$

In the provided equation, COD_{feed} represents the substrate strength measured in terms of Chemical Oxygen Demand (COD) concentration (mg/L), Q stands for the flow rate of the substrate (L/day), and Vr (L) denotes the working volume of the reactor.

Numerous literature studies have explored the influence of the Organic Loading Rate (OLR) on the biogas production process. For instance, Jiang et al. (2013) examined its effects on the acidogenesis of food waste. This research delved into the impact of OLR at different stages of the anaerobic digestion process, demonstrating that elevated OLR favored higher percentages of acetate and valerate, while propionate and butyrate percentages remained lower under the same OLR conditions (Meegoda et al., 2018; Roopnarain et al., 2021).

7.2.2.5 Hydraulic Retention Time (HRT)

Hydraulic Retention Time (HRT) is a fundamental parameter that gauges the average duration a liquid or dissolved component remains within a biogas study reactor. Calculated as the ratio of the tank volume to the influent flow rate, HRT serves as an approximation of the duration a substrate undergoes treatment within a process. The manner in which a digester is mixed significantly influences HRT, thereby exerting a substantial impact on biogas yield. The HRT is further subject to the influence of substrate type and distinct processes, yielding effects that span from several days to several months (Lindmark et al., 2014; Roopnarain et al., 2021).

Regarding its effects on anaerobic acidogenesis, there exists a diversity of data. Some researchers have found that acidification increases with an extended HRT. However, within the broader context of the AD process, the effect of HRT has been observed to be relatively consistent. Longer HRT durations are generally associated with higher methane content (Ma et al., 2013; Meegoda et al., 2018; Roopnarain et al., 2021).

7.2.3 Selection, Cultivation, and Inhibition of Microorganisms

In numerous scenarios, microorganisms have demonstrated superior cost-effectiveness compared to hydrolytic enzymes. These microorganisms possess the capability to convert high-molecular-weight compounds in substrates into lower-mass compounds through fermentation. This intricate process involves the synthesis of enzymes and the proliferation of decomposing microorganisms (Ferdes et al., 2020). Within this context, it becomes imperative to factor in the survival and growth conditions of valuable microorganisms, encompassing considerations such as nutrients, inhibitors, pH, temperature, oxygen concentration, and more. Adjustments to these parameters influence alterations in the population structure of microorganisms involved in substrate decomposition. Such adjustments can be tailored to meet the objectives and requirements of the biogas process (Nwokolo et al., 2020; Ferdes et al., 2020). However, it is worth noting that microorganisms often necessitate extended retention times, and there is a potential for the growth of undesired microorganisms, entailing more stringent operational conditions (Liu et al., 2025). Thus, the species-specific generation time under given conditions should be taken into account, with awareness that bacterial doubling time is generally shorter than that of fungi, necessitating careful selection based on prior investigation (Tshemese et al., 2023; Ferdes et al. 2020). Lastly, as proposed by Sawyerr et al. (2019), continuous research is indispensable for the evaluation of various biomass feedstocks and waste streams. These substrates play a pivotal role in the development of processes that induce kinetic reactions and amplify methane yield. This emphasis is critical because anaerobic digestion offers manifold advantages over alternative waste-management approaches. Its adaptability for both small and large scales, coupled with low operating costs, minimal energy consumption, and reduced environmental impact through the surplus digestate produced, positions AD as a promising technology (Sawyerr et al., 2019; Soccol et al., 2019). The resulting digestate can serve as a biofertilizer, as it is enriched with nitrogen, phosphorus, potassium, and traces of other elements and heavy metals. The fertilizer's value varies depending on the nutrient composition of the feedstock (Nwokolo et al., 2020; Tshemese et al., 2023).

7.2.4 Feedstock Selection and Preparation

The proper feedstock must be chosen for optimal biogas production. Anaerobic digestion systems utilize a wide range of feedstock sources, including agricultural residues, dedicated energy crops, food waste, and sewage sludge. Selection of appropriate feedstocks depends on multiple factors such as availability, physicochemical characteristics, and their potential for biogas yield. Moreover, feedstock pretreatment strategies such as shredding, homogenization, and various physical or chemical methods play a crucial role in enhancing substrate digestibility and improving overall biogas production efficiency (Mikusińska et al., 2023).

A diverse array of waste categories holds the potential to serve as primary materials for the generation of biogas through the utilization of anaerobic digestion (AD) technology. Substantial volumes of lignocellulosic waste are procured from agricultural, municipal, and assorted sources of waste generation. Prominent examples of waste streams commonly harnessed within the European energy sector encompass: excrement and liquid waste from animals, effluent from sewage systems, solid waste of municipal origin, and discarded food matter (Achinas et al., 2017). Table 7.1 showed a comparative analysis is presented therein, elucidating the output magnitude and energy potential pertaining to distinct raw materials exploitable for the purpose of biogas synthesis.

Biomass encompasses constituents such as carbohydrates, proteins, fats, cellulose, and hemicellulose, all of which are viable candidates for serving as raw materials in biogas production. In prevailing methodologies, co-substrates are commonly introduced to augment the organic content, thereby facilitating enhanced gas yield. Conventional co-substrates encompass organic byproducts originating from agricultural sectors, food waste, and/or biodegradable municipal waste sourced from residential environments. The constitution and output of biogas are contingent upon the specific characteristics of the primary feedstock and the accompanying co-substrate (Achinas et al., 2017).

Table 7.1 Comparative analysis of biogas yields and electricity generation from various potential substrates (Achinas et al., 2017; Stucki et al., 2011)

Type	Biogas yield per ton fresh matter (m^3)	Electricity produced per ton fresh matter [a] (kW·h)
Cattle dung	55–68	122.5
Chicken litter/dung	126	257.3
Fat	826–1200	1687.4
Food waste (disinfected)	110	224.6
Fruit wastes	74	151.6
Horse manure	56	114.3
Maize silage	200/220	409.6
Municipal solid waste	101.5	207.2
Sewage sludge	47	96.0

[a] 35% electrical efficiency combined heat power, heating value 21 MJ m^{-3}, 55% methane content, 3.6 MJ $(kW\ h)^{-1}$

7.2.5 *Digester Types and Configurations*

7.2.5.1 Digester Stage

In a general context, anaerobic digesters can be structured as single-stage, two-stage, or multi-stage reactors. Within these configurations, the steps of hydrolysis/acidogenesis and acetogenesis/methanogenesis can take place either jointly within the same digester or be segregated into different units (as depicted in Fig. 7.2). This separation of digesters simplifies process management, allowing for independent fine-tuning of operational and environmental parameters for the hydrolysis/acidogenesis and methanogenesis processes. This approach aims to optimize the overall reaction rate and enhance the yield of biogas through careful adjustment (Rabii et al., 2019).

Anaerobic digesters come in various types and configurations, each with its own advantages and limitations. This section presents a brief description of different digester types commonly used in biogas production, including batch digesters, continuous stirred tank reactors (CSTRs), plug flow digesters, and anaerobic sequencing batch reactors (ASBRs). It is explored the working principles, operational considerations, and suitability of each digester type for different feedstocks and scales of operation. Furthermore, it has been discussed the emerging trends in digester design, such as two-phase and multi-stage systems, aiming to improve process efficiency and stability (Van et al., 2020).

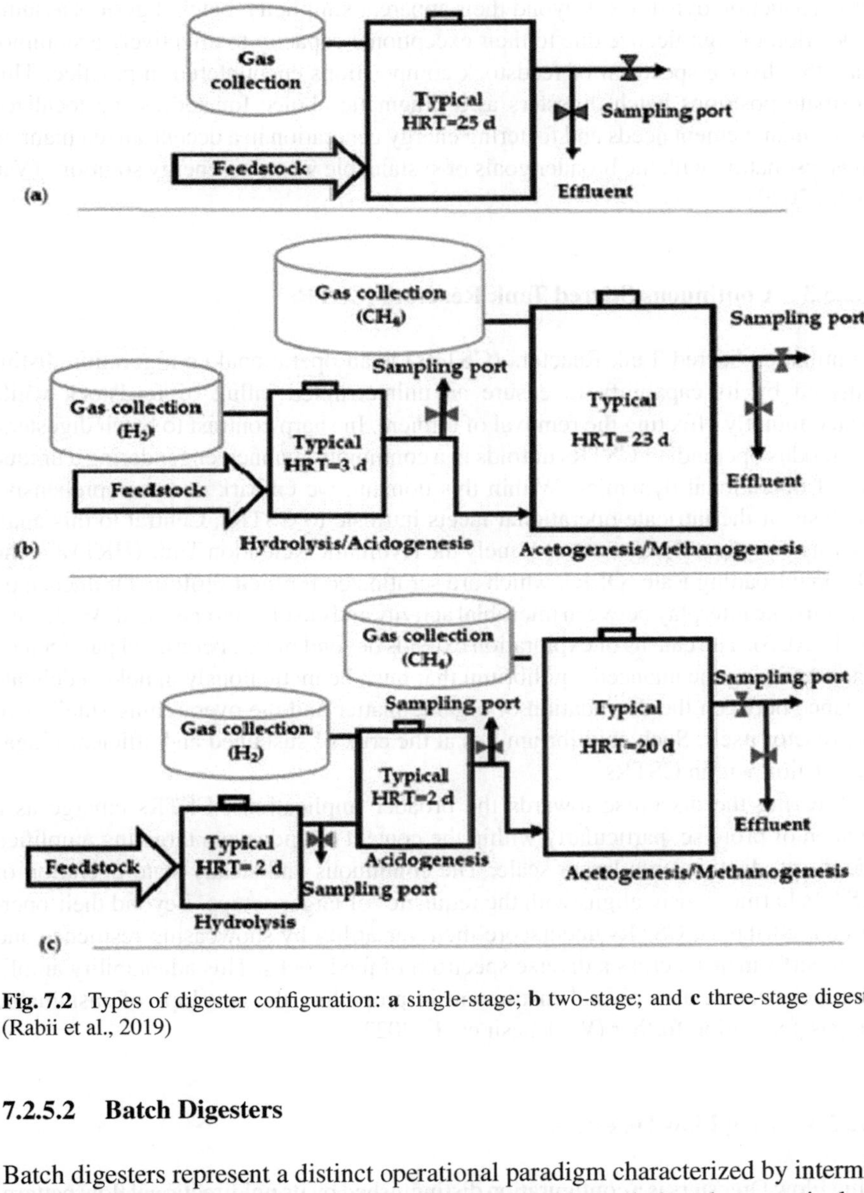

Fig. 7.2 Types of digester configuration: **a** single-stage; **b** two-stage; and **c** three-stage digester (Rabii et al., 2019)

7.2.5.2 Batch Digesters

Batch digesters represent a distinct operational paradigm characterized by intermittent feedstock loading and periodic effluent removal, rendering them particularly suitable for applications on a smaller scale. This subsection gives comprehensively into the nuanced working principles that underlie batch digesters, providing a meticulous exposition of the sequential stages that govern their operation, namely loading, digestion, and unloading. Moreover, it is meticulously scrutinized the operational subtleties intrinsic to batch digesters, including their remarkable adaptability to cope with the inherent variability of substrates and the consequential ramifications for gas production dynamics. Beyond their apparent simplicity, batch digesters assume a position of significance due to their exceptional capacity to effectively accommodate the diverse spectrum of feedstock compositions encountered in practice. This attribute positions batch digesters as a pragmatic choice for addressing localized waste management needs and fostering energy generation in a decentralized manner, thus resonating with the broader goals of sustainable waste-to-energy solutions (Van et al., 2020).

7.2.5.3 Continuous Stirred Tank Reactors (CSTRs)

Continuous Stirred Tank Reactors (CSTRs) is an operational configuration distinguished by its capability to ensure an uninterrupted influx of feedstock while concomitantly effecting the removal of effluent. In sharp contrast to batch digesters, the modus operandi of CSTRs unfolds in a continuous manner, engendering a unique set of operational dynamics. Within this domain, we embark on a comprehensive analysis of the intricate operational facets intrinsic to CSTRs. Central to this analysis are two pivotal parameters, namely the Hydraulic Retention Time (HRT) and the Organic Loading Rate (OLR), which are scrutinized for their profound influence on the intricate interplay between microbial activity and ensuing biogas yield (Wadchasit et al., 2023). The canvas of exploration extends beyond mere operational parameters, encompassing the nuanced equilibrium that must be meticulously upheld, a delicate balance between the degradation of organic matter and the overarching stability of the reactor itself. Such equilibrium lies at the crux of sustained and efficient biogas production within CSTRs.

Steering the discourse towards the broader implications, CSTRs emerge as a beacon of promise, particularly within the context of endeavors targeting amplified biogas production on a larger scale. The continuous and steady-state operation of CSTRs harmoniously aligns with the requisites of larger setups. Beyond their operational attributes, CSTRs underscore their versatility by showcasing resilience and high performance across a diverse spectrum of feedstocks. This adaptability amplifies their stature as a pivotal instrument in propelling the envelope of sustainable biogas generation further (Wadchasit et al., 2023).

7.2.5.4 Plug Flow Digesters

Plug Flow Digesters is a configuration distinguished by its unidirectional flow pattern, necessitating a meticulous and in-depth examination. This segment unveils the fundamental principles underlying hydrodynamics and substrate flow dynamics, which together elucidate their pivotal roles in optimizing the efficiency of anaerobic digestion. The discussion delves into the distinctive gradient of digestion stages that unfolds along the length of the digester, an aspect profoundly influencing the attainment of heightened degradation efficiencies. Significantly, the capacity of plug flow digesters to adeptly handle high-solid substrates emerges as a notable advantage, particularly in scenarios characterized by such feedstock compositions. Furthermore, this section underscores the broader applicability of plug flow digesters, highlighting their relevance within both agricultural and industrial contexts (Gómez et al., 2019).

7.2.5.5 Anaerobic Sequencing Batch Reactors (ASBRs)

Anaerobic Sequencing Batch Reactors (ASBRs) is representing a harmonious fusion of batch and continuous processes. This innovative approach unfolds through a series of sequential phases, encompassing feedstock loading, reaction, settling, and effluent removal. ASBRs' operational intricacies take center stage, offering insights into their profound implications. This multifaceted methodology not only unravels the orchestration of microbial consortia and the mitigation of washout phenomena but also delves into the intricacies of optimizing methane generation. Notably, ASBRs showcase their potential in effectively handling complex substrates, tackling inhibitory challenges, and dynamically adapting to load fluctuations, underscoring their operational flexibility and adaptability (Berhe & Leta, 2023).

Emerging design trends capture our attention, including the intriguing possibilities presented by two-phase and multi-stage systems. These configurations undergo scrutiny for their ability to enhance process efficiency, elevate gas yield, and ensure the stability of reactor performance. In examining these trends, it is uncover the underlying rationale driving their innovation and assess their applicability across a spectrum of operational scales and diverse feedstock compositions (Berhe & Leta, 2023).

7.2.5.6 Biogas Production Systems

As the quest for more efficient and effective biogas production intensifies, innovative configurations like two-phase and multi-stage systems have gained prominence, and their transformative potential warrants thorough investigation.

Two-Phase Systems

Within the framework of this evolving paradigm, two-phase systems stand out as a compelling innovation. These systems divide the digestion process into distinct phases, often segregating hydrolysis and methanogenesis. By creating tailored environments for each phase, two-phase systems can significantly enhance process efficiency. Hydrolysis, a rate-limiting step in anaerobic digestion, benefits from optimized conditions, resulting in improved substrate breakdown and subsequently higher methane yields. Moreover, this segregation minimizes the effects of inhibitory compounds, fostering a more stable microbial community and resilient system performance. The application of two-phase systems spans a broad spectrum of feedstock compositions and can prove especially valuable when dealing with complex or recalcitrant substrates (Yin et al., 2023).

Multi-stage Systems

Similarly, the emergence of multi-stage systems introduces a concept of sequential digestion stages, each optimized for specific microbial consortia and metabolic reactions. This approach allows for the exploitation of diverse microbial communities with distinct functions in various stages of degradation. By carefully orchestrating the progression of substrates through these stages, the overall efficiency of biogas production can be significantly enhanced. For instance, recalcitrant materials can be subjected to preliminary stages designed for efficient breakdown, followed by subsequent stages optimized for methane generation. The modular nature of multi-stage systems lends itself well to accommodating varying feedstock types and adapting to fluctuations in substrate availability (Rabii et al., 2019).

The underlying rationale driving these advancements is rooted in a profound understanding of microbial ecology, biochemistry, and process engineering. Designing systems that capitalize on specific microbial functions optimizes the degradation of complex organic matter and promotes the formation of methane-rich biogas. Furthermore, the robustness of such systems in the face of operational variations, substrate fluctuations, and inhibitory compounds is a testament to their potential for sustainable biogas production (Athar et al., 2022). The applicability of these configurations spans operational scales, from small-scale decentralized setups to large industrial installations. Their adaptability to varying feedstock compositions makes them versatile solutions for a range of organic waste streams, including agricultural residues, food waste, and industrial byproducts. By embracing these emerging trends, practitioners and researchers are poised to reshape the landscape of biogas production, inching closer to the goal of efficient waste valorization, renewable energy generation, and sustainable environmental practices (Thomson et al., 2020).

Figure 7.3 show a biogas and digestate production plan. The feedstock is first directed into the fermentation chamber. Biogas and digestate are produced as a result of anaerobic digestion. Biogas can provide heat, electricity, and be processed to make biomethane. Digestate is an important byproduct of biogas generation. This by-product's correct handling might be helpful to the environment. Digestate is a nutrient-rich compound that promotes plant development. Another advantage is the absence of odor in comparison to the initial feedstock, such as manure (Mikusińska et al., 2023). Organic components, macronutrients, and micronutrients make up its composition. Carbon, nitrogen, phosphorous, potassium, and sodium are the primary components of digestate. The amount of carbon in the digestate shows the anaerobic digestion process's effectiveness. Nitrogen is an important ingredient if the digestate is to be utilized as a fertilizer. Phosphorus is also an important component of the digestate. In their study, Wang & Lee (2021) found that the phosphorus concentration of silage and manure digestate was 1.35%, sewage sludge was 2.7%, and swine manure was 11.5%. Digestate may contain cellulose, hemicellulose, and lignin as

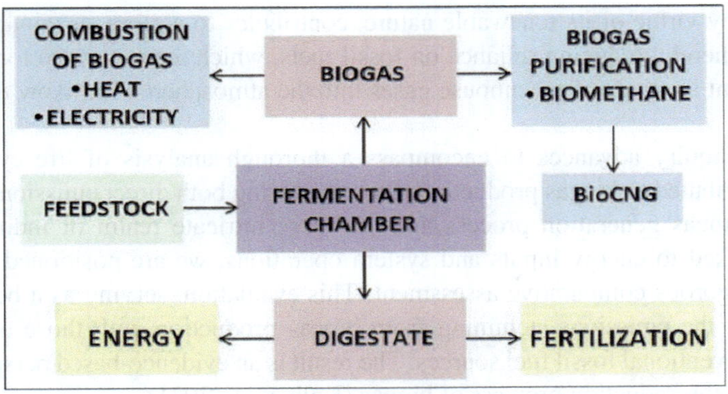

Fig. 7.3 Production of biogas and digestate (Mikusińska et al., 2023)

well. To generate additional biogas, digestate with a high concentration of cellulose and a trace of lignin is reintroduced to the reactor, where the anaerobic digestion process takes place (Mikusińska et al., 2023; Wang & Lee, 2021).

7.3 Environmental Impacts of Biogas Production

7.3.1 Greenhouse Gas Emissions

Biogas production plays a crucial role in mitigating greenhouse gas emissions. In this section, it is explored the environmental benefits of biogas as a renewable energy source. It is delved into the reduction of methane emissions through anaerobic digestion, as methane is a potent greenhouse gas with a significantly higher global warming potential than carbon dioxide. It is analyzed the life cycle emissions of biogas, considering both direct and indirect emissions, and compare them to conventional fossil fuel sources. Furthermore, we discuss the importance of proper management of biogas systems to prevent unintended methane leaks (Al-Ismail et al., 2023; Mohammed et al., 2021).

Central to the discourse is the recognition of methane, a potent greenhouse gas, as a primary target for reduction. By immersing ourselves in the mechanics of anaerobic digestion, we unearth the crux of this mitigation strategy. The anaerobic digestion process serves as an orchestrated stage where organic matter's controlled decomposition leads to the conversion of methane into biogas, substantially curtailing its release into the atmosphere. This phenomenon serves as a critical linchpin in our global strategy to counteract climate change (Li et al., 2019).

The exploration takes us deeper into the realm of biogas's environmental virtues as a renewable energy source. The section lays bare the multifaceted benefits that biogas offers in its capacity to mitigate greenhouse gas emissions. We elucidate how biogas, by virtue of its renewable nature, contributes to a more sustainable energy matrix, thereby reducing reliance on fossil fuels, which are notorious for releasing significant amounts of greenhouse gases into the atmosphere (Semikow & Sumera, 2019).

The inquiry advances to encompass a thorough analysis of life cycle emissions attributed to biogas production. By considering both direct emissions inherent to the biogas generation process and the more intricate realm of indirect emissions linked to energy inputs and system operations, we are positioned to undertake a rigorous comparative assessment. This evaluation, serving as a benchmark, contrasts the emissions stemming from biogas production with those originating from conventional fossil fuel sources. The result is an evidence-based perspective on the emission-reduction prowess of biogas (Kolb et al., 2021).

Even as biogas emerges as a formidable contender in the realm of emission mitigation, an essential caveat emerges, the importance of proper management to stave off unintended methane leaks. This discussion underscores the urgency of ensuring the integrity of biogas systems to prevent the inadvertent release of methane. By underlining the significance of vigilant system design, operation, and maintenance, we fortify the foundation upon which the environmental benefits of biogas production stand (Farghali et al., 2022).

7.3.2 Air Pollution and Odor Issues

Amidst the promise of biogas production's environmental advantages, it is imperative to recognize the potential challenges it may pose concerning air pollution and odor. This section delves deeply into these critical concerns, presenting a thorough investigation into the emissions linked to biogas production, and their associated implications. We navigate through a range of emissions, including volatile organic compounds (VOCs), ammonia, and hydrogen sulfide, that can cast a shadow over biogas's environmentally-friendly reputation (Zhang et al., 2019).

The focus is first directed towards emissions that can emanate from biogas production processes. VOCs, ammonia, and hydrogen sulfide are explored, with a particular emphasis on their environmental and potential health ramifications. The understanding of these emissions as potential contributors to air pollution and local discomfort underscores the complexity of balancing the benefits of biogas against its challenges (Zhang et al., 2019).

Through meticulous scrutiny, we uncover the potential consequences of these emissions for the environment and human health. The section delves into the far-reaching effects that these emissions can have on air quality, ecosystems, and public well-being. This holistic examination paints a comprehensive picture of the intricate interplay between biogas production and potential detrimental effects on the surroundings (Ali et al., 2023).

To address these challenges head-on, this section explores strategies for mitigating air pollution and controlling odor issues in the context of biogas production. Central to this exploration is the unveiling of advanced gas treatment technologies, which have emerged as potent solutions. The discussion encompasses the deployment of techniques such as scrubbers and biofilters. These technologies are evaluated for their capacity to intercept and neutralize harmful emissions before they are released into the atmosphere. By implementing these technologies, practitioners aim to not only minimize emissions but also ensure compliance with stringent air quality standards (Lu et al., 2023).

7.3.3 Water Pollution and Nutrient Management

In this section, it is analyzed the potential risks associated with the discharge of digestate, the residual material from anaerobic digestion. It is explored the nutrient content of digestate and its impact on water bodies, including eutrophication and contamination of groundwater sources. Additionally, we discuss best practices for digestate treatment and utilization as a nutrient-rich fertilizer, aiming to minimize environmental impacts and promote sustainable nutrient management practices (Wang et al., 2023).

The inquiry embarks on an exploration of the intricate dynamics surrounding digestate and its interaction with water bodies. By plumbing the depths of the nutrient content within digestate, we begin to appreciate its potential influence on water systems. A nuanced analysis ensues, unveiling the potential repercussions of introducing digestate into water bodies, including the contentious issues of eutrophication and the contamination of groundwater sources (Manyi-Loh et al., 2019).

A key focus of this section rests on eutrophication, a complex phenomenon triggered by excessive nutrient inputs, often culminating in an overgrowth of algae and other aquatic vegetation. The exploration extends into the mechanisms by which nutrient-rich digestate can contribute to this ecological imbalance, resulting in reduced oxygen levels and impaired aquatic ecosystems. By unpicking the intricacies of nutrient load and their far-reaching consequences, we underscore the urgent need for proactive nutrient management (Malhotra et al., 2022).

Venturing beyond the challenges, this section showcases a beacon of hope in the form of digestate's transformation into a nutrient-rich fertilizer. This approach aligns with the principles of circular economy, where waste is transformed into a valuable resource. This section delves into the best practices for treating and utilizing digestate as a potent agricultural input, thereby curbing the potential environmental impacts of its discharge. This not only optimizes nutrient utilization but also minimizes the risk of polluting water bodies (Lee et al., 2021).

At the heart of this exploration is the overarching commitment to sustainable nutrient management practices. By embracing the utilization of digestate as a fertilizer, practitioners can contribute to reducing the demand for synthetic fertilizers and the associated environmental toll. The discussion encompasses strategies that strive to balance nutrient requirements in agriculture with the preservation of water quality and ecosystem health (Lee et al., 2021; Manyi-Loh et al., 2019; Rouhollahi et al., 2020).

7.3.4 Case Studies on Environmental Impacts of Biogas

To provide practical insights into the environmental impacts of biogas production, this section presents a collection of case studies from diverse geographical locations and feedstock sources. These case studies examine the environmental challenges faced during biogas production and the strategies implemented to mitigate negative impacts (Ali et al., 2023).

- **Greenhouse Gas Emissions**

According to Paolini et al. (2018), a primary objective of the biogas industry is to decrease the consumption of fossil fuels, ultimately aiming to alleviate the effects of global warming. Nevertheless, the process of anaerobic digestion is linked to the generation of several greenhouse gases, notably carbon dioxide, methane, and nitrous oxide. Consequently, specific measures must be implemented to curtail these emissions. Pivotal approaches for enhancing the potential reduction of global warming in biogas plants encompass preventing methane discharge by utilizing flares, covering tanks, improving the efficiency of combined heat and power (CHP) units, refining strategies for electricity utilization, maximizing the utilization of thermal energy, and preventing leakages (Hijazi et al., 2016). Similar findings were deduced by Buratti et al. (2013) in their study concerning cereal crop-based biogas production in Umbria, Italy. The biomethane production chain surpasses the minimum greenhouse gas (GHG) savings requirement (35%), mainly attributed to the open storage of digestate. Common practices to heighten GHG reduction (up to 68.9%) entail utilizing heat and electricity generated by biogas CHP plants, along with the coverage of digestate storage tanks.

The evaluation of the impact of biogas plants on global warming necessitates a case specific examination. Bachmaier et al. (2010) calculated the GHG impact of ten agricultural biogas plants. GHG emissions resulting from electricity generation in these examined biogas plants ranged from − 85 to 251 g CO_2-eq/kWhel, while the GHG savings were 2.31–3.16 kWhfossil/kWhel. The findings also underscored that accurate estimates of GHG emissions related to biogas electricity production can only be formulated based on individualized monitoring data. These factors include reducing direct methane emissions and leakage, harnessing heat from cogeneration, evaluating input material quantity and characteristics, nitrous oxide emissions (e.g., from energy crop cultivation), and managing digestate.

In a case study involving an intensive dairy farm in Italy's Po Valley, Battini et al. (2014) computed a GHG emission reduction attributed to anaerobic digestion, which ranged between − 23.7 and − 36.5%, contingent on digestate management strategies. Optimizing all pertinent process parameters is crucial in terms of final environmental impact. For instance, a specific study on wastewater treatment indicated that process optimization could lead to emission reductions amounting to 1103 kg CO_2 eq/d for N_2O, 256 kg eq/d for CO_2, and 87 kg CO_2 eq/d for CH_4 (Mikosz, 2016).

- **Carbon dioxide emissions**

Harmful substances and airborne pollutants are introduced into the environment during the production and utilization of biogas, both through processes involving combustion and diffusive release. Regarding carbon dioxide specifically, the combustion of biogas results in the efficient conversion of methane into CO_2, occurring at a rate of 83.6 kg per gigajoule (based on biogas composed of 65% CH_4 and 35% CO_2 (Paolini et al., 2018)). Other instances of this pollutant being released are connected to the transportation and storage of biomass, as well as the use of digestate. In both the scenarios of biogas combustion and biomass/digestate emissions, the CO_2 is classified as biogenic and its impact on the climate is considered neutral in calculations. Factoring in the reduction in fossil fuel usage, it can be evidenced that global biogas production contributes to mitigating the anthropogenic greenhouse effect on the environment. Poeschl et al. (2012), in their research, examined the CO_2 emissions linked to biogas production from various raw materials, as well as the proportional contributions from factors such as feedstock provision, biogas plant operations and infrastructure, biogas utilization, and digestate management. According to their findings, the use of biogas leads to an overall negative CO_2 balance due to the fact that the capture of CO_2 outweighs the positive emissions arising from feedstock supply and biogas plant operations. As anticipated, the production of biogas from byproducts (e.g., food residues, pomace, slaughter waste, cattle manure, etc.) proves to be a more sustainable approach compared to the utilization of energy crops like whole-wheat plant silage. Furthermore, in the case of specific feedstocks such as municipal solid waste, the management of digestate significantly contributes to the overall reduction of emissions.

- **Methane Emissions**

The release of methane through biogas processes is generally not considered a significant health concern. While exposure to hydrocarbon mixtures can have certain negative impacts on human health, there is currently no evidence indicating significant interactions between methane and biological systems. Nevertheless, methane is classified as a potent greenhouse gas, with an estimated global warming potential 28 to 36 times greater than that of CO_2 over a 100-year period. Consequently, it stands as the second most prominent contributor among anthropogenic greenhouse substances. Consequently, when assessing the impact of the biogas industry on climate change, the emissions of methane emerge as a critically important factor (Paolini et al., 2018).

Methane can be discharged during the incomplete combustion of biogas. However, a substantial portion of methane emissions arises from diffusive releases linked to the storage of biomass and the management of digestate. Conversely, alternative strategies for biomass management need to be considered to mitigate emissions stemming from biogenic methane. The study by Poeschl et al. (2012), mentioned earlier, also addressed methane emissions. Across all the cases examined, the emission rates were found to be below 5 g per kilogram. In the context of cattle manure, notable reductions in methane emissions are linked to the processing and handling of digestate. This is particularly relevant, as this type of biomass tends to exhibit a high methane emission rate when spread in the field without any pre-treatment.

- **Nitrous Oxide Emissions**

In addition to carbon dioxide (CO_2) and methane (CH_4), nitrous oxide (N_2O) holds significance as another crucial greenhouse gas (GHG). Due to its notably high potential for causing a greenhouse effect, the emissions of N_2O resulting from biogas production processes can make a substantial contribution to the overall global warming impact (Paolini et al., 2018). The extent of the relative influence of nitrous oxide largely hinges on the specific climate metrics adopted. In fact, under certain metrics like the Global Temperature change Potential over a 100-year period (GTP-100), the impact of N2O emissions can even surpass that of CO_2 and CH_4 (Iordan et al., 2016).

The comprehensive greenhouse gas emissions associated with energy production from biogas are typically computed within a range of 0.10–0.40 kg of CO_2-equivalent per kilowatt-hour of electricity (kg CO_2-eq/kWhel). Notably, this range reflects a reduction of 22–75% in comparison to the greenhouse gas emissions stemming from the current energy mix prevalent in Germany (Meyer-Aurich et al., 2012). The significant variability in estimates concerning the potential for mitigating global warming is primarily attributable to uncertainties in the assessment of N_2O emission rates, as well as the handling and utilization of digestate as a fertilizer, as elaborated upon in the subsequent paragraphs (Paolini et al., 2018).

- **Emission Pathways in Biogas Production**

The environmental implications associated with biogas production are intricately interconnected with the pathways through which emissions of deleterious compounds are introduced into the atmosphere. A comprehensive comprehension of these emission pathways is of paramount importance for both the holistic evaluation of the ecological footprint of biogas processes and the formulation of efficacious strategies aimed at mitigating their environmental impact (Kodba et al. 2023).

The genesis of hazardous compounds during biogas production is a result of intricate mechanisms that encompass various stages of the process. These compounds span a spectrum of pollutants and greenhouse gases, each carrying unique implications for both the environment and human health. A pivotal step towards addressing the environmental consequences stemming from biogas production entails an exhaustive exploration of these intricate mechanisms, with the objective of identifying precise junctures amenable to intervention and optimization (Kodba et al. 2023).

Emissions originating from biogas production can be broadly categorized into two principal pathways: combustion and diffusive emissions. Combustion, involving the controlled ignition of biogas for the purpose of energy generation, leads to the release of compounds such as carbon dioxide (CO_2) and water vapor. This emission pathway is often characterized by a higher degree of manageability and controllability owing to its localized nature. In contrast, diffusive emissions encompass the uncontrolled release of gases like methane (CH_4) and nitrous oxide (N_2O) during various phases of biogas production, encompassing storage and transportation. Mitigating diffusive emissions presents a more intricate challenge due to their dispersed nature and their sensitivity to factors such as temperature, pressure, and storage methodologies (Beylot et al., 2013).

The incineration of biogas serves a dual purpose: energy generation and the abatement of methane emissions. Despite methane's potency as a greenhouse gas, its combustion within biogas systems facilitates the efficient conversion of methane into carbon dioxide and water, thereby substantially diminishing its global warming potential. This process of methane oxidation within biogas combustion contributes not only to energy production but also plays a pivotal role in alleviating the environmental impact of methane emissions. Furthermore, the conversion of methane to carbon dioxide assumes paramount significance, particularly considering the elevated warming potential of methane relative to carbon dioxide within abbreviated timeframes. The incineration of biogas, therefore, emerges as a viable avenue for converting a more potent greenhouse gas into one associated with comparatively reduced implications for global warming (Alrbai et al., 2023).

7.4 Mitigation Strategies and Best Practices

Methane capture and utilization involve capturing methane gas emissions from various sources, such as landfills, wastewater treatment plants, and agricultural activities, and utilizing it as a valuable energy resource rather than releasing it into the atmosphere. Some common methods for methane capture include anaerobic digestion, landfill gas recovery, and upgrading biogas for use in energy generation or other applications. By implementing these strategies, methane emissions can be significantly reduced, thereby mitigating climate change and promoting sustainable energy production (Chinea et al., 2023).

Digestate refers to the residual organic material left after the anaerobic digestion process. Effective management of digestate is crucial to maximize its value while minimizing environmental impacts. Best practices for digestate management include optimizing the anaerobic digestion process to enhance nutrient breakdown, separating solid and liquid fractions for appropriate treatment and utilization, and implementing nutrient recovery technologies to extract valuable nutrients like nitrogen and phosphorus. By efficiently managing digestate, we can reduce waste and pollution while promoting circular economy principles (Malhotra et al., 2022; Wang et al., 2023).

Odor control and air pollution abatement are essential aspects of sustainable waste management practices. Techniques such as biofiltration, activated carbon adsorption, and chemical scrubbing can be employed to mitigate odorous emissions from waste treatment facilities. Additionally, implementing proper ventilation systems, covering waste storage areas, and employing advanced air pollution control technologies can help reduce harmful pollutants, such as volatile organic compounds (VOCs), particulate matter, and greenhouse gases, emitted from various waste management operations (Lee et al., 2023).

To ensure the environmental sustainability of bioenergy and biogas production, it is important to select and manage feedstock resources wisely. Sustainable feedstock selection involves considering factors such as carbon intensity, land use impacts, water requirements, and biodiversity considerations. Utilizing agricultural residues, dedicated energy crops grown on marginal lands, and organic wastes for feedstock can help minimize competition with food production and avoid negative environmental impacts. Proper management practices, including optimizing feedstock preparation and storage, also play a crucial role in maximizing energy yields and reducing emissions during the conversion process (Rocha-Meneses et al., 2023).

7.5 Environmental Impact Assessment (EIA) for Biogas Projects

The EIA process is a systematic evaluation of the potential environmental impacts of a proposed project. It involves identifying and assessing the direct and indirect effects on various environmental components, including air, water, land, biodiversity, and human health. The process typically includes screening, scoping, impact assessment, mitigation planning, and monitoring stages.

During the EIA process, methodologies such as baseline data collection, predictive modeling, and impact assessment tools are employed. These may include environmental surveys, modeling software, and specialized assessment techniques to evaluate the potential impacts on specific environmental receptors. The methodologies used depend on the project's scope, scale, and location (Sechoala et al., 2023).

Life Cycle Assessment (LCA) is a valuable tool used to evaluate the environmental impacts of biogas production throughout its entire life cycle. LCA takes into account all stages, including feedstock production, transportation, anaerobic digestion, biogas upgrading, and utilization or disposal of by-products.

LCA assesses various environmental indicators, such as greenhouse gas emissions, energy consumption, water use, land use, and eutrophication potential. It helps identify hotspots and evaluate the environmental performance of different system configurations and management practices, aiding in decision-making for sustainable biogas projects (Bruno et al., 2023).

7.6 Policy and Regulatory Framework

National and international biogas policies play a crucial role in promoting the development and deployment of biogas projects. These policies aim to create a favorable regulatory framework, establish targets, provide financial incentives, and support research and development. Examples of national biogas policies include feed-in tariffs, renewable portfolio standards, tax incentives, grants, and subsidies for biogas production and utilization. International agreements and initiatives, such as the Paris Agreement and Sustainable Development Goals, also encourage countries to adopt sustainable energy practices, including biogas (Gustafsson & Anderberg, 2021).

Renewable energy targets and incentives are key drivers for biogas project implementation. Many countries have set renewable energy targets to increase the share of renewable energy in their energy mix. These targets provide a long-term vision and create a market demand for biogas. Incentives such as feed-in tariffs, power purchase agreements, tax credits, and renewable energy certificates encourage investment in biogas projects by providing financial benefits and reducing risks (Rocha-Meneses et al., 2023).

Standards and certification systems play a significant role in ensuring the quality, safety, and environmental performance of biogas projects. These standards cover various aspects, including feedstock requirements, digester design, gas quality, by-product management, and health and safety regulations. Certification schemes help establish credibility and market recognition for biogas producers, facilitating trade and enhancing consumer confidence (Lange et al., 2023). Examples of relevant standards and certifications include the International Organization for Standardization (ISO) standards (e.g., ISO 14001 for environmental management and ISO 9001 for quality management), the European Biogas Association (EBA) certification scheme, and regional or national standards specific to biogas production and utilization.

Financial support mechanisms are essential to incentivize and support biogas projects. These mechanisms can include grants, subsidies, loans, tax incentives, and venture capital funding. They aim to reduce the upfront costs, enhance project viability, and improve the return on investment for biogas producers. Government agencies, development banks, and private investors often provide financial support for biogas projects to promote renewable energy deployment, reduce greenhouse gas emissions, and enhance energy security (Rocha-Meneses et al., 2023).

7.7 Technological Advancements

Biogas production technologies have been advancing to improve efficiency, increase gas yield, and enhance the overall process. Some notable innovations include the development of high-solid anaerobic digestion systems that can handle a wider range of feedstocks, the use of advanced pre-treatment techniques to enhance the biodegradability of complex substrates, and the application of microbial consortia and synthetic biology approaches to optimize anaerobic digestion performance. Additionally, advancements in biogas upgrading technologies, such as membrane separation and pressure swing adsorption, enable the production of biomethane with higher purity for injection into natural gas grids or use as a transportation fuel (Baizharikova et al., 2023).

The integration of biogas systems with other renewable energy systems offers opportunities for enhanced energy generation and system flexibility. Biogas plants can be combined with solar and wind power installations to create hybrid renewable energy systems that utilize complementary energy sources. These integrated systems provide a more stable and reliable energy supply, as biogas production can be sustained even when solar or wind resources are intermittent. Furthermore, surplus electricity generated from biogas can be stored or used to support the electricity grid during peak demand periods (Dodo & Ashigwuike, 2023).

Wastewater treatment plants (WWTPs) present a significant opportunity for bioenergy generation through anaerobic digestion of sewage sludge. Advances in anaerobic digestion technologies, along with improved energy recovery systems, allow WWTPs to not only treat wastewater effectively but also produce biogas for heat and power generation. Additionally, the integration of nutrient recovery technologies in WWTPs enables the extraction of valuable nutrients from the sludge, contributing to resource efficiency and circular economy principles (Shen et al., 2015).

Biogas production plays a vital role in the circular economy by utilizing organic waste streams that would otherwise contribute to pollution or landfilling. Innovative approaches are being developed to maximize the valorization of biogas by optimizing the use of digestate as a nutrient-rich fertilizer, producing bio-based chemicals and materials from biogas-derived intermediates, and exploring new markets for biogas utilization, such as in the production of hydrogen or as a feedstock for synthetic natural gas (Kapoor et al., 2020).

7.8 Ecological Benefits of Biogas Technology

According to Omer (2016), simpler circumstances are evident when examining the environmental impacts of various pathways for the utilization of biogas. The fundamental assumptions guiding the comparison of distinct biogas utilization processes encompass:

- Biogas deployment in a heat-demand regulated gas engine, connected to the natural gas grid, with a capacity of 500 kWe. This entails an electrical efficiency of 37.5%, a thermal efficiency of 42.5%, and a methane loss of 0.01.
- Biogas utilization within a local gas engine, situated at the biogas plant also with a capacity of 500 kWe. Here, the electrical efficiency is 37.5%, thermal efficiency is 42.5%, and methane loss amounts to 0.5.

- Biogas production derived from maize silage employing a biogas plant with a covered storage tank, characterized by methane losses equaling 1% of the produced biogas.
- Biogas refinement involving a power consumption of 0.3 kWh/m3 of biogas, alongside methane losses of 0.5.

Diverse avenues exist for the conversion of biogas into energy. Predominantly, the method of combined heat and power (CHP) generation using a gas engine positioned at the site of biogas production is adopted. This preference rests upon two primary rationales. Firstly, biogas production operates on an almost continuous basis, rendering the regulation of anaerobic digesters in line with specific demand patterns relatively intricate or, in the immediate timeframe, even infeasible. Secondly, the impetus behind the advancement of renewable energies centers on electricity generation. As a result, operators of biogas plants receive a substantial portion of their income from secured feed-in tariffs designated for electricity. Summarizing the findings from the ecological assessments, it becomes evident that, not merely with the use of fossil fuels but also with the utilization of renewable sources such as biogas, the optimal strategy for combating climate change is combined heat and power cogeneration. From a technological vantage point, it can be inferred that the production of biogas, the conversion of renewable materials and organic waste into energy, epitomizes state-of-the-art technology.

7.9 Future Prospects

The process of biogas production entails the conversion of organic matter into methane-rich gas through a series of microbial reactions (Atelge et al., 2020a, 2020b). Nevertheless, the efficiency of this biogas production process is frequently constrained by the availability of organic materials accessible for microbial degradation (Nwokolo et al., 2020). Despite the progress made in various techniques for pre-treating substrates, there remains a discernible gap necessitating further investigation into more refined and innovative pre-treatment approaches. These approaches aim to deconstruct the intricate organic compounds within feedstocks into simpler constituents, thereby augmenting the yield of biogas (Stanley et al., 2022). An unexplored realm for extended research lies in the realm of emergent technologies, such as nanotechnology, which presents an avenue for novel possibilities (Nasrollahzadeh et al., 2019).

According to Roopnarain et al. (2021), biological biomass conversion technologies, particularly anaerobic digestion (AD), hold significant promise in the realm of renewable energy due to their eco-friendly nature and the diverse range of products they yield, which contribute to economic viability and social responsibility (Sarker et al., 2019). Abundant biomass resources are available for AD, capable of being converted into various energy sources with broad applications, offering potential for long-term global carbon footprint reduction.

Despite the positive outlook for AD, certain challenges hinder its progress within the global energy landscape. These obstacles can be surmounted through the development of process models grounded in knowledge and driven by information, resulting in a high degree of process automation. Automation would rely on real-time data from process monitoring to simulate and execute process controls, responding automatically and providing recommendations for plant operations. To achieve this, extensive research is required for real-time measurement of process parameters, nutrient composition, substrate utilization kinetics, microbial growth rates and kinetics, gene expression, metabolic profiling, and product yield (Theuerl et al., 2019a, 2019b). Automating the AD process would also mitigate the risk of process failure (Sahu et al., 2017).

Efforts must also be directed towards profiling the predictable functions of uncultivated microorganisms by combining genome/metagenome assembly and genome-based metabolic tools. Integrated multi-metaomics analyses could elucidate the AD microbiome, revealing microbial communities' presence, genetic potential, metabolic activities, and enzyme profiles. These meta-omic approaches need integration to provide a comprehensive view of participating organisms' identity, function, genetic components, and enzyme activities (Hassa et al., 2018; Theuerl et al., 2019a, 2019b). Optimization of the AD process involves refining process parameters and integrating novel technologies such as micro-aeration and bioaugmentation with microbial consortia or nanoparticles, while considering scalability, economy, and eco-friendliness.

Disruptive technologies like artificial intelligence, machine learning, and the Internet of Things are pertinent for predicting process disturbances, real-time monitoring, and enhancing process performance and maintenance (Logan et al., 2019; Theuerl et al., 2019a, 2019b). Machine learning algorithms aid in predicting non-linear interactions in AD datasets, unveiling the intricacies within the system. The complexity of these interactions necessitates data mining to extract valuable insights. While artificial intelligence learns from experimental data, it requires substantial data under varying conditions. Mathematical mechanistic models are crucial for acquiring a non-linear and systemic perspective. Models such as the Anaerobic Digestion Model No. 1 (ADM1) have been instrumental, and they continue to evolve, accommodating specific digester scenarios (Oliveros-Muñoz et al., 2021; Roopnarain et al., 2021). These models can harness the multi-metaomics information for data curation, optimizing AD and preventing system collapse while minimizing costs and time spent on optimization (Theuerl et al., 2019a, 2019b; Wade, 2020).

7.10 Conclusions

Biogas production offers a promising solution for addressing environmental challenges while simultaneously promoting renewable energy and sustainable waste management practices. Biogas production helps to mitigate greenhouse gas emissions by capturing and utilizing methane, a potent greenhouse gas, from organic waste sources. By implementing mitigation strategies such as methane capture, digestate management, odor control, and sustainable feedstock selection, the environmental impact of biogas production can be minimized. These strategies contribute to reducing air pollution, water contamination, and waste generation, resulting in improved environmental quality. The environmental impact assessment (EIA) process, supported by methodologies like life cycle assessment (LCA), provides a systematic approach to evaluate and manage the potential environmental effects of biogas projects. It ensures compliance with regulations, identifies areas for improvement, and facilitates stakeholder engagement. Policy and regulatory frameworks play a critical role in driving the development and deployment of biogas projects. National and international biogas policies, renewable energy targets, financial support mechanisms, and standards and certifications create an enabling environment for the growth of the biogas sector. Technological advancements in biogas production, integration with renewable energy systems, bioenergy from wastewater treatment plants, and circular economy approaches offer exciting prospects for the future. Continued research and exploration in these areas, along with addressing research gaps, will further enhance the efficiency, environmental performance, and sustainability of biogas technologies.

In summary, biogas production has the potential to make a positive environmental impact by reducing greenhouse gas emissions, promoting sustainable waste management, and contributing to the transition to a low-carbon energy system. By adopting best practices, implementing supportive policies, and advancing technological innovations, we can harness the benefits of biogas production while minimizing its environmental footprint.

Acknowledgements Many thanks to the leader editors Doctor Anouzla Abdelkader and Professor Salah Souabi, Hassan II University Casablanca, Faculty of Sciences and Techniques of Mohammedia, Morocco, for the suggestion to contribute to this book.

References

Abdeshahian, P., Lim, J. S., Ho, W. S., Hashim, H., & Lee, C. T. (2016). Potential of biogas production from farm animal waste in Malaysia. *Renewable and Sustainable Energy Reviews, 60*, 714–723.

Achinas, S., Achinas, V., & Euverink, G. J. W. (2017). A technological overview of biogas production from biowaste. *Engineering, 3*(3), 299–307.

Akyol, Ç. (2020). In search of the optimal inoculum to substrate ratio during anaerobic co-digestion of spent coffee grounds and cow manure. *Waste Management & Research, 38*(11), 1278–1283.

Ala'a, H., Jamil, F., Al-Haj, L., Myint, M. T. Z., Mahmoud, E., Ahmad, M. N., & Rafiq, S. (2018). Biodiesel production over a catalyst prepared from biomass-derived waste date pits. *Biotechnology Reports, 20*, e00284.

Ali, S., Yan, Q., Irfan, M., & Fahad, S. (2023). Relating biogas technology and environmental impact assessment: a roadmap towards clean energy for environmental sustainability. *Environmental Science and Pollution Research*, 1–22.

Al-Ismail, F. S., Alam, M. S., Shafiullah, M., Hossain, M. I., & Rahman, S. M. (2023). Impacts of renewable energy generation on greenhouse gas emissions in Saudi Arabia: A comprehensive review. *Sustainability, 15*(6), 5069.

Alrbai, M., Ahmad, A. D., Al-Dahidi, S., Abubaker, A. M., Al-Ghussain, L., Alahmer, A., & Akafuah, N. K. (2023). Performance and sensitivity analysis of raw biogas combustion under homogenous charge compression ignition conditions. *Energy, 283*, 128486.

Amoo, A. O., Ijanu, E. M., Haruna, A., Adeleye, A. O., & Sabo, A. (2023). Review on the pre-treatment advancements of biogas production barriers. *UMYU Journal of Microbiology Research (UJMR), 8*(1), 6–30.

Anukam, A., Mohammadi, A., Naqvi, M., & Granström, K. (2019). A review of the chemistry of anaerobic digestion: Methods of accelerating and optimizing process efficiency. *Processes, 7*(8), 504.

Atelge, M. R., Atabani, A. E., Banu, J. R., Krisa, D., Kaya, M., Eskicioglu, C., & Duman, F. A. T. İH. (2020a). A critical review of pretreatment technologies to enhance anaerobic digestion and energy recovery. *Fuel, 270*, 117494.

Atelge, M. R., Krisa, D., Kumar, G., Eskicioglu, C., Nguyen, D. D., Chang, S. W., & Unalan, S. (2020b). Biogas production from organic waste: Recent progress and perspectives. *Waste and Biomass Valorization, 11*, 1019–1040.

Athar, T., Pandey, A., Khan, M. K., Saqib, Z. A., Jabeen, M., Shahid, S., & Elinson, M. A. (2022). Potential role of beneficial microbes for sustainable treatment of sewage sludge and wastewater. *Sustainable Management and Utilization of Sewage Sludge,* 71–96.

Avila-Lopez, M., Robles-Rodriguez, C., Tiruta-Barna, L., & Ahmadi, A. (2023). Toward thermal autarky for large-scale biogas plants: Dynamic energy modeling for energy efficiency in anaerobic digesters with enhanced multimembrane gasholders. *Fuel, 339*, 126978.

Bachmaier, J., Effenberger, M., & Gronauer, A. (2010). Greenhouse gas balance and resource demand of biogas plants in agriculture. *Engineering in Life Sciences, 10*(6), 560–569.

Baizharikova, M., Tlebayeva, G., Tlebayev, M., Beksultanov, Z., & Tazhiyeva, R. (2023) Development of a new innovative technological process for continuous methane fermentation in a three-stage biogas plant. *Biomass Conversion and Biorefinery,* 1–9.

Bajpai, P. (2017). *Anaerobic technology in pulp and paper industry* (pp. 7–13). Springer.

Battini, F., Agostini, A., Boulamanti, A. K., Giuntoli, J., & Amaducci, S. (2014). Mitigating the environmental impacts of milk production via anaerobic digestion of manure: Case study of a dairy farm in the Po Valley. *Science of the Total Environment, 481*, 196–208.

Berhe, S., & Leta, S. (2023). Anaerobic co-digestion of agro-industrial wastes using anaerobic sequencing batch reactor for bio-energy recovery: Focus on process performance and stability of the methanogenic step. *Journal of Water Process Engineering, 54*, 103993.

Beylot, A., Villeneuve, J., & Bellenfant, G. (2013). Life cycle assessment of landfill biogas management: Sensitivity to diffuse and combustion air emissions. *Waste Management, 33*(2), 401–411.

Bruno, M., Marchi, M., Ermini, N., Niccolucci, V., & Pulselli, F. M. (2023). Life cycle assessment and cost-benefit analysis as combined economic-environmental assessment tools: Application to an anaerobic digestion plant. *Energies, 16*(9), 3686.

Bücker, F., Marder, M., Peiter, M. R., Lehn, D. N., Esquerdo, V. M., de Almeida Pinto, L. A., & Konrad, O. (2020). Fish waste: An efficient alternative to biogas and methane production in an anaerobic mono-digestion system. *Renewable Energy, 147*, 798–805.

Buratti, C., Barbanera, M., & Fantozzi, F. (2013). Assessment of GHG emissions of biomethane from energy cereal crops in Umbria; Italy. *Applied Energy, 108*, 128–136.

Burg, V., Bowman, G., Haubensak, M., Baier, U., & Thees, O. (2018). Valorization of an untapped resource: Energy and greenhouse gas emissions benefits of converting manure to biogas through anaerobic digestion. *Resources, Conservation and Recycling, 136*, 53–62.

Cerón-Vivas, A., Cáceres, K. T., Rincón, A., & Cajigas, Á. A. (2019). Influence of pH and the C/N ratio on the biogas production of wastewater. *Revista Facultad de Ingeniería Universidad de Antioquia, 92*, 70–79.

Chinea, L., Slopiecka, K., Bartocci, P., Park, A. H. A., Wang, S., Jiang, D., & Fantozzi, F. (2023). Methane enrichment of biogas using carbon capture materials. *Fuel, 334*, 126428.

Chotovinský, O., Altmann, V., & P˘rikryl, M, (2016). Influence of weather conditions on waste biomass production. *Research in Agricultural Engineering, 62*(2), 83–91.

Deepanraj, B., Sivasubramanian, V., & Jayaraj, S. (2014). Biogas generation through anaerobic digestion process-an overview. *Research Journal of Chemistry and Environment, 18*, 5.

Diamantis, V., Eftaxias, A., Stamatelatou, K., Noutsopoulos, C., Vlachokostas, C., & Aivasidis, A. (2021). Bioenergy in the era of circular economy: Anaerobic digestion technological solutions to produce biogas from lipid-rich wastes. *Renewable Energy, 168*, 438–447.

Dodo, U. A., & Ashigwuike, E. C. (2023). Techno-economic and environmental analysis of utility-scale hybrid renewable energy system integrating waste-to-energy plant to complement an unreliable grid operation. *Energy, Ecology and Environment*, 1–18.

Dong, L., Cao, G., Tian, Y., Wu, J., Zhou, C., Liu, B., & Ren, N. (2020). Improvement of biogas production in plug flow reactor using biogas slurry pretreated cornstalk. *Bioresource Technology Reports, 9*, 100378.

Farghali, M., Osman, A. I., Umetsu, K., & Rooney, D. W. (2022). Integration of biogas systems into a carbon zero and hydrogen economy: A review. *Environmental Chemistry Letters, 20*(5), 2853–2927.

Ferdeş, M., Dincă, M. N., Moiceanu, G., Zăbavă, B. Ş, & Paraschiv, G. (2020). Microorganisms and enzymes used in the biological pretreatment of the substrate to enhance biogas production: A review. *Sustainability, 12*(17), 7205.

Gómez, D., Ramos-Suárez, J. L., Fernández, B., Muñoz, E., Tey, L., Romero-Güiza, M., & Hansen, F. (2019). Development of a modified plug-flow anaerobic digester for biogas production from animal manures. *Energies, 12*(13), 2628.

Gueboudji, Z. (2022). *Effect of storage time on the quality of olive oil mill wastewaters obtained from the cold extraction of olive oil in the region of Khenchela and their biological properties.* Doctoral dissertation, Abbes Laghrour University-Khenchela, p250.

Gustafsson, M., & Anderberg, S. (2021). Dimensions and characteristics of biogas policies–Modelling the European policy landscape. *Renewable and Sustainable Energy Reviews, 135*, 110200.

Hashemi, S. S., Karimi, K., & Karimi, A. M. (2019). Ethanolic ammonia pretreatment for efficient biogas production from sugarcane bagasse. *Fuel, 248*, 196–204.

Hassa, J., Maus, I., Off, S., Pühler, A., Scherer, P., Klocke, M., & Schlüter, A. (2018). Metagenome, metatranscriptome, and metaproteome approaches unraveled compositions and functional relationships of microbial communities residing in biogas plants. *Applied Microbiology and Biotechnology, 102*, 5045–5063.

Heidari-Maleni, A., Taheri-Garavand, A., Rezaei, M., & Jahanbakhshi, A. (2023). Biogas production and electrical power potential, challenges and barriers from municipal solid waste (MSW) for developing countries: A review study in Iran. *Journal of Agriculture and Food Research*, 100668.

Hijazi, O., Munro, S., Zerhusen, B., & Effenberger, M. (2016). Review of life cycle assessment for biogas production in Europe. *Renewable Sustainable Energy Rev, 54*, 1291–1300.

Iordan, C., Lausselet, C., & Cherubini, F. (2016). Life-cycle assessment of a biogas power plant with application of different climate metrics and inclusion of near-term climate forcers. *Journal of Environmental Management, 184*, 517–527.

Jiang, J., Zhang, Y., Li, K., Wang, Q., Gong, C., & Li, M. (2013). Volatile fatty acids production from food waste: Effects of pH, temperature, and organic loading rate. *Bioresource Technology, 143*, 525–530.

Kapoor, R., Ghosh, P., Kumar, M., Sengupta, S., Gupta, A., Kumar, S. S., & Pant, D. (2020). Valorization of agricultural waste for biogas based circular economy in India: A research outlook. *Bioresource Technology, 304*, 123036.

Karlsson, N. P., Halila, F., Mattsson, M., & Hoveskog, M. (2017). Success factors for agricultural biogas production in Sweden: A case study of business model innovation. *Journal of Cleaner Production, 142*, 2925–2934.

Kodba, A., Pukšec, T., & Duić, N. (2023). Analysis of specific greenhouse gas emissions savings from biogas production based on agricultural residues and industrial by-products. *Energies, 16*(9), 3721.

Kolb, S., Plankenbühler, T., Hofmann, K., Bergerson, J., & Karl, J. (2021). Life cycle greenhouse gas emissions of renewable gas technologies: A comparative review. *Renewable and Sustainable Energy Reviews, 146*, 111147.

Lange, N., Moosmann, D., Majer, S., Meisel, K., Oehmichen, K., Rauh, S., & Thrän, D. (2023). Assessment of greenhouse gas emission reduction from biogas supply chains in Germany in context of a newly implemented sustainability certification. *Progress in life cycle assessment 2021* (pp. 85–101). Springer International Publishing.

Lattieff, F. A. (2016). A study of biogas production from date palm fruit wastes. *Journal of Cleaner Production, 139*, 1191–1195.

Lee, J., Lee, S., Lin, K. Y. A., Jung, S., & Kwon, E. E. (2023). Abatement of odor emissions from wastewater treatment plants using biochar: Review of the state-of-the-art approaches. *Environmental Pollution*, 122426.

Lee, M. E., Steiman, M. W., & Angelo, S. K. S. (2021). Biogas digestate as a renewable fertilizer: Effects of digestate application on crop growth and nutrient composition. *Renewable Agriculture and Food Systems, 36*(2), 173–181.

Li, Y., Chen, Y., & Wu, J. (2019). Enhancement of methane production in anaerobic digestion process: A review. *Applied Energy, 240*, 120–137.

Lindmark, J., Thorin, E., Fdhila, R. B., & Dahlquist, E. (2014). Effects of mixing on the result of anaerobic digestion. *Renewable and Sustainable Energy Reviews, 40*, 1030–1047.

Liu, L., Wang, F., Wang, L., Jiang, X., & Zhang, H. (2025). Addressing challenges in microbial manufacturing: Systematic microbial biotechnology. *The Innovation, 6*(6).

Logan, M., Safi, M., Lens, P., & Visvanathan, C. (2019). Investigating the performance of internet of things based anaerobic digestion of food waste. *Process Safety and Environmental Protection, 127*, 277–287.

Lu, H., Wang, H., Wu, Q., Luo, H., Zhao, Q., Liu, B., & Ren, N. (2023). Automatic control and optimal operation for greenhouse gas mitigation in sustainable wastewater treatment plants: A review. *Science of the Total Environment, 855*, 158849.

Ma, J., Zhao, B., Frear, C., Zhao, Q., Yu, L., Li, X., & Chen, S. (2013). Methanosarcina domination in anaerobic sequencing batch reactor at short hydraulic retention time. *Bioresource Technology, 137*, 41–50.

Malhotra, M., Aboudi, K., Pisharody, L., Singh, A., Banu, J. R., Bhatia, S. K., & Tyagi, V. K. (2022). Biorefinery of anaerobic digestate in a circular bioeconomy: Opportunities, challenges and perspectives. *Renewable and Sustainable Energy Reviews, 166*, 112642.

Manyi-Loh, C. E., Mamphweli, S. N., Meyer, E. L., & Okoh, A. I. (2019). Microbial anaerobic digestion: Process dynamics and implications from the renewable energy, environmental and agronomy perspectives. *International Journal of Environmental Science and Technology, 16*, 3913–3934.

Meegoda, J. N., Li, B., Patel, K., & Wang, L. B. (2018). A review of the processes, parameters, and optimization of anaerobic digestion. *International Journal of Environmental Research and Public Health, 15*(10), 2224.

Meyer-Aurich, A., Schattauer, A., Hellebrand, H. J., Klauss, H., Plöchl, M., & Berg, W. (2012). Impact of uncertainties on greenhouse gas mitigation potential of biogas production from agricultural resources. *Renewable Energy, 37*(1), 277–284.

Mikosz, J. (2016). Analysis of greenhouse gas emissions and the energy balance in a model municipal wastewater treatment plant. *Desalination Water Treat, 57*(59), 28551–28559.

Mikusińska, J., Kuźnia, M., Czerwińska, K., & Wilk, M. (2023). Hydrothermal carbonization of digestate produced in the biogas production process. *Energies, 16*(14), 5458.

Mohammed, S., Gill, A. R., Alsafadi, K., Hijazi, O., Yadav, K. K., Hasan, M. A., & Harsanyi, E. (2021). An overview of greenhouse gases emissions in Hungary. *Journal of Cleaner Production, 314*, 127865.

Nasrollahzadeh, M., Sajadi, S. M., Sajjadi, M., & Issaabadi, Z. (2019). An introduction to nanotechnology. In *Interface science and technology* (Vol. 28, pp. 1–27). Elsevier.

Nwokolo, N., Mukumba, P., Obileke, K., & Enebe, M. (2020). Waste to energy: A focus on the impact of substrate type in biogas production. *Processes, 8*(10), 1224.

Oliveros-Muñoz, J. M., Martínez-Villalba, J. A., Jiménez-Islas, H., Luna-Porres, M. Y., Escamilla-Alvarado, C., & Ríos-Fránquez, F. J. (2021). Luus-Jaakola method and ADM1 based optimization of hydrogen sulfide in anaerobic digestion of cow manure. *Biochemical Engineering Journal, 171*, 108012.

Omer, A. M. (2016). An overview of biomass and biogas for energy generation: Recent development and perspectives. *Agriculture Advances, 5*(2), 251–268.

Paolini, V., Petracchini, F., Segreto, M., Tomassetti, L., Naja, N., & Cecinato, A. (2018). Environmental impact of biogas: A short review of current knowledge. *Journal of Environmental Science and Health, Part A, 53*(10), 899–906.

Poeschl, M., Ward, S., & Owende, P. (2012). Environmental impacts of biogas deployment e part I: Life cycle inventory for evaluation of production process emissions to air. *Journal of Cleaner Production, 24*, 168–183.

Rabii, A., Aldin, S., Dahman, Y., & Elbeshbishy, E. (2019). A review on anaerobic co-digestion with a focus on the microbial populations and the effect of multi-stage digester configuration. *Energies, 12*(6), 1106.

Rasit, N., Idris, A., Harun, R., & Ghani, W. A. W. A. K. (2015). Effects of lipid inhibition on biogas production of anaerobic digestion from oily effluents and sludges: An overview. *Renewable and Sustainable Energy Reviews, 45*, 351–358.

Raturi, A. K. (2019). *Renewables 2019 global status report.*

Rocha-Meneses, L., Luna-delRisco, M., González, C. A., Moncada, S. V., Moreno, A., Sierra-Del Rio, J., & Castillo-Meza, L. E. (2023). An overview of the socio-economic, technological, and environmental opportunities and challenges for renewable energy generation from residual biomass: A case study of biogas production in Colombia. *Energies, 16*(16), 5901.

Roopnarain, A., Rama, H., Ndaba, B., Bello-Akinosho, M., Bamuza-Pemu, E., & Adeleke, R. (2021). Unravelling the anaerobic digestion 'black box': Biotechnological approaches for process optimization. *Renewable and Sustainable Energy Reviews, 152*, 111717.

Rouhollahi, Z., Ebrahimi-Nik, M., Ebrahimi, S. H., Abbaspour-Fard, M. H., Zeynali, R., & Bayati, M. R. (2020). Farm biogas plants, a sustainable waste to energy and bio-fertilizer opportunity for Iran. *Journal Cleaner Production, 253*, 119876.

Sahu, P., Sahu, S., Purohit, R., Warudkar, V., Arisutha, S., & Suresh, S. (2017). Automation in biogas plant for enhancement of efficiency and safety. *Materials Today: Proceedings, 4*(4), 5351–5356.

Sarker, S., Lamb, J. J., Hjelme, D. R., & Lien, K. M. (2019). A review of the role of critical parameters in the design and operation of biogas production plants. *Applied Sciences, 9*(9), 1915.

Sawyerr, N., Trois, C., Workneh, T., & Okudoh, V. (2019). An overview of biogas production: Fundamentals, applications and future research. *International Journal of Energy Economics and Policy, 9*(2), 105.

Schnürer, A. (2016). Biogas production: microbiology and technology. *Anaerobes in Biotechnology*, 195–234.

Sechoala, T., Popoola, O., & Ayodele, T. (2023). Economic and environmental assessment of electricity generation using biogas and heat energy from municipal solid waste: A case study of Lesotho. *AIMS Energy, 11*(2), 337–357.

Semikow, J., & Sumera, J. (2019). Some aspects of long-term testing of cogeneration set fueled with biogas. *Combustion Engines, 58*(4), 136–141.

Shen, Y., Linville, J. L., Urgun-Demirtas, M., Mintz, M. M., & Snyder, S. W. (2015). An overview of biogas production and utilization at full-scale wastewater treatment plants (WWTPs) in the United States: Challenges and opportunities towards energy-neutral WWTPs. *Renewable and Sustainable Energy Reviews, 50*, 346–362.

Siddiki, S. Y. A., Uddin, M. N., Mofijur, M., Fattah, I. M. R., Ong, H. C., Lam, S. S., & Ahmed, S. F. (2021). Theoretical calculation of biogas production and greenhouse gas emission reduction potential of livestock, poultry and slaughterhouse waste in Bangladesh. *Journal of Environmental Chemical Engineering, 9*(3), 105204.

Silva, V., & Eus, D. (2019). *Techno-economic analysis of a biomass gasification power plant dealing with forestry residues blends for electricity production in Portugal*, p. 212.

Soccol, C. R., Faraco, V., Karp, S. G., Vandenberghe, L. P., Thomaz-Soccol, V., Woiciechowski, A. L., & Pandey, A. (2019). Lignocellulosic bioethanol: current status and future perspectives. In *Biofuels: alternative feedstocks and conversion processes for the production of liquid and gaseous biofuels* (pp. 331–354).

Stanley, J. T., Thanarasu, A., Kumar, P. S., Periyasamy, K., Raghunandhakumar, S., Periyaraman, P., & Subramanian, S. (2022). Potential pre-treatment of lignocellulosic biomass for the enhancement of biomethane production through anaerobic digestion—A review. *Fuel, 318*, 123593.

Stucki, M., Jungbluth, N., & Leuenberger, M. (2011). Life cycle assessment of biogas production from different substrates. In *Final report. Bern: Federal department of environment, transport, energy and communications, federal office of energy.*

Sun, J., Xu, Y., Nairat, S., Zhou, J., & He, Z. (2023). Prediction of biogas production in anaerobic digestion of a full-scale wastewater treatment plant using ensembled machine learning models. *Water Environment Research*, e10893.

Sun, Q., Li, H., Yan, J., Liu, L., Yu, Z., & Yu, X. (2015). Selection of appropriate biogas upgrading technology—A review of biogas cleaning, upgrading and utilisation. *Renewable and Sustainable Energy Reviews, 51*, 521–532.

Tahir, A. H., Al-Obaidy, A. H. M., & Mohammed, F. H. (2020). Biochar from date palm waste, production, characteristics and use in the treatment of pollutants: A Review. In *IOP Conference Series: Materials Science and Engineering* (Vol. 737, No. 1, p. 012171). IOP Publishing.

Theuerl, S., Herrmann, C., Heiermann, M., Grundmann, P., Landwehr, N., Kreidenweis, U., & Prochnow, A. (2019a). The future agricultural biogas plant in Germany: A vision. *Energies, 12*(3), 396.

Theuerl, S., Klang, J., & Prochnow, A. (2019b). Process disturbances in agricultural biogas production—Causes, mechanisms and effects on the biogas microbiome: A review. *Energies, 12*(3), 365.

Thomson, R., Kwong, P., Ahmad, E., & Nigam, K. D. P. (2020). Clean syngas from small commercial biomass gasifiers; a review of gasifier development, recent advances and performance evaluation. *International Journal of Hydrogen Energy, 45*(41), 21087–21111.

Tshemese, Z., Deenadayalu, N., Linganiso, L. Z., & Chetty, M. (2023). An overview of biogas production from anaerobic digestion and the possibility of using sugarcane wastewater and municipal solid waste in a South African context. *Applied System Innovation, 6*(1), 13.

Van, D. P., Fujiwara, T., Tho, B. L., Toan, P. P. S., & Minh, G. H. (2020). A review of anaerobic digestion systems for biodegradable waste: Configurations, operating parameters, and current trends. *Environmental Engineering Research, 25*(1), 1–17.

Vu, M. T., Nguyen, L. N., Johir, M. A. H., Zhang, X., Nghiem, L. D., & Elimelech, M. (2021). Biogas sparging to control fouling and enhance resource recovery from anaerobically digested sludge centrate by forward osmosis. *Journal of Membrane Science, 625*, 119176.

Wadchasit, P., Rakmak, N., Sompong, O., Rattanasak, U., Imai, T., Jitpinit, S., & Nuithitikul, K. (2023). Improvement of biogas production and quality by addition of struvite precipitates derived from liquid anaerobic digestion effluents of palm oil wastes. *Journal of Environmental Chemical Engineering, 11*(1), 109081.

Wade, M. J. (2020). Not just numbers: Mathematical modelling and its contribution to anaerobic digestion processes. *Processes, 8*(8), 888.

Wang, P., Wang, H., Qiu, Y., Ren, L., & Jiang, B. (2018). Microbial characteristics in anaerobic digestion process of food waste for methane production—A review. *Bioresource Technology, 248*, 29–36.

Wang, W., Chang, J. S., & Lee, D. J. (2023). Anaerobic digestate valorization beyond agricultural application: Current status and prospects. *Bioresource Technology*, 128742.

Wang, W., & Lee, D. J. (2021). Valorization of anaerobic digestion digestate: A prospect review. *Bioresource Technology, 323*, 124626.

Wei, Y., Gao, J., Shi, Z., Li, X., Ma, W., & Yuan, H. (2022). Effect of hydrothermal pretreatment on two-stage anaerobic digestion of food waste and Enteromorpha: Digestion performance, bioenergy efficiency, and microbial community dynamics. *Fuel, 318*, 123639.

Westerholm, M., Moestedt, J., & Schnürer, A. (2016). Biogas production through syntrophic acetate oxidation and deliberate operating strategies for improved digester performance. *Applied Energy, 179*, 124–135.

Xu, N., Liu, S., Xin, F., Zhou, J., Jia, H., Xu, J., & Dong, W. (2019). Biomethane production from lignocellulose: Biomass recalcitrance and its impacts on anaerobic digestion. *Frontiers in Bioengineering and Biotechnology, 7*, 191.

Yin, Z., Zhou, S., Zhang, X., Li, X., Wang, Z., Wang, J., & Sun, C. (2023). A novel batched four-stage–two-phase anaerobic digestion system to facilitate methane production from rice straw and cow manure with low inoculum/substrate ratios. *Fermentation, 9*(6), 565.

Zhang, Y., Zhu, Z., Zheng, Y., Chen, Y., Yin, F., Zhang, W., ... & Xin, H. (2019). Characterization of volatile organic compound (VOC) emissions from swine manure biogas digestate storage. *Atmosphere, 10*(7), 411.

Chapter 8
Innovative Approaches of Landfill Gas Collection Systems in Landfill Sites for Efficient Methane Capture

Mohammed Abdillah Ahmad Farid, Mohd Nor Faiz Norrahim, Jacqueline Lease, and Yoshito Andou

Abstract Methane emissions from landfills are a significant contributor to greenhouse gas emissions and climate change. This article, hence, presents strategies for landfill gas collection systems aimed at improving methane recovery. A comparative analysis of gas collection system designs, including horizontal, vertical, and slotted wells, is presented. Additionally, the utilization of vacuum-assisted gas collection systems and gas permeable barriers is discussed, highlighting the potential benefits and drawbacks of each approach. Challenges and opportunities in implementing innovative landfill gas collection systems are discussed, including technical, regulatory, and financial aspects. Collaboration between stakeholders is emphasized as crucial for the successful implementation and operation of these systems. The potential of innovative landfill gas collection systems for reducing greenhouse gas emissions from landfills is highlighted, and a call to action for further research and

M. A. Ahmad Farid
Institute of Tropical Forestry and Forest Products (INTROP), Universiti Putra Malaysia, Serdang, Selangor, Malaysia
e-mail: abdillahfarid@upm.edu.my

M. N. F. Norrahim
Research Center for Chemical Defence, Universiti Pertahanan Nasional Malaysia, Kem Sungai Besi, Kuala Lumpur, Malaysia
e-mail: faiz@upnm.edu.my

J. Lease · Y. Andou (✉)
Department of Biological Functions Engineering, Graduate School of Life Science and Systems Engineering, Kyushu Institute of Technology, Kitakyushu, Fukuoka, Japan
e-mail: yando@life.kyutech.ac.jp

J. Lease
e-mail: lease.jacqueline708@mail.kyutech.jp

Y. Andou
Collaborative Research Centre for Green Materials on Environmental Technology, Kyushu Institute of Technology, Kitakyushu, Fukuoka, Japan

A. Anouzla et al. (eds.), *Handbook of the Effects of Landfill Emissions on Sustainable Development*, https://doi.org/10.1007/978-3-032-01951-6_8

implementation of these approaches is made. This article provides valuable information for researchers, practitioners, and policymakers interested in sustainable waste management and climate change mitigation.

Keywords Methane emissions · Greenhouse gas mitigation · Sustainable waste management · Landfill methane recovery

8.1 Introduction

Methane emissions from various sources, including natural gas production, agriculture, and landfills, significantly contribute to greenhouse gas (GHG) emissions and climate change. Efficient methane capture is crucial for reducing GHG emissions and mitigating climate change. The energy sector is a major source of methane emissions, responsible for approximately 38.1% (302.4 million tonnes CO_2 eq.) of total methane emissions in the United States (US EPA, 2021b). A 17.3% reduction in methane emissions from this sector would have a climate benefit equivalent to closing the entire world's operating coal mines (53.3 million tonnes CO_2 eq.) (Tate, 2022). Livestock production, particularly enteric fermentation in ruminants and manure management, accounts for about 14.5% of global GHG emissions, with methane (44%) being a significant contributor (FAO, 2013). Efficient methane capture from livestock manure not only reduces GHG emissions but also generates renewable energy and mitigates environmental impacts associated with manure management (Tauseef et al., 2013).

Landfills are another significant source of methane emissions, which are produced during the anaerobic decomposition of organic materials, such as food waste, paper, and yard trimmings, that are deposited in landfill (Bailey, 2022). Methane emissions from landfills contribute to climate change by trapping heat in the atmosphere, which can cause global temperatures to rise and result in severe environmental and economic consequences. According to US EPA (2021a), landfills are the third-largest source of human-related methane emissions in the country, accounting for 15% of all methane emissions. Methane has a global warming potential that is 28 times higher than carbon dioxide over a 100-year time horizon, making it a potent greenhouse gas (GHG) (European Commission, 2022). According to Maasakkers et al. (2022) and Dickie (2022), landfills are responsible for around 11% of global methane emissions and are expected to increase by about 70% by 2050 as the global population continues to climb. It is important to note that even closed landfills can continue to emit methane for several decades after closure, making the management of closed landfills a critical issue. One case study conducted by Attero, one of the largest producers of biogas in the Netherlands, on Schinnen landfill, estimates that the total amount of fugitive methane emissions from closed landfills worldwide continues for another 65 years (Castro, 2015). Therefore, proper closure and post-closure management of landfills are crucial in reducing methane emissions and mitigating their impact on climate change.

Landfill gas (LFG) collection systems are commonly used to capture methane gas generated from decomposing organic waste in landfills (Lee et al., 2017). Typically, these systems consist of vertical or horizontal gas wells that penetrate the waste mass and are connected by piping to a gas treatment system that removes moisture, impurities, and other contaminants from the gas before it is used or flared. Although traditional LFG collection systems offer benefits, they have several limitations that affect their efficiency and effectiveness. A study by Ayodele et al. (2020) states that poor waste compaction, improper landfill design, and limited access to the waste mass can result in reduced efficiency and effectiveness of LFG collection systems. The study also notes that operational difficulties, such as clogging of gas wells and damage to gas piping, can result in decreased gas collection.

The low-quality gas produced by traditional LFG collection systems also limits their effectiveness (Duan et al., 2022). According to the United States Environmental Protection Agency (EPA), the composition of landfill gas can vary widely, but it typically contains 50–55% methane, 45–50% CO_2, and 2–5% trace amounts of other gases (US EPA, 2017). The gas must be treated to remove impurities such as moisture, CO_2, and H_2S before it can be used or sold (US EPA, 2021c). This can increase costs and limit the economic feasibility of using landfill gas as a renewable energy source. Regarding flaring, the EPA reports that approximately 50 m^3 per h (438,000 m^3 per year) of collected landfill gas can be flared. This can contribute to local air pollution and greenhouse gas emissions and is also a waste of a potential energy source (US EPA, 2011).

To reduce methane emissions from landfills, several management practices have been implemented. The most common practice is landfill gas (LFG) recovery, which involves capturing methane emissions and converting them into energy, such as electricity or heat, or using them as a fuel source for industrial processes (Siddiqui et al., 2022). Traditional LFG collection systems can capture a significant amount of methane emissions from landfills, with an estimated 60% of landfill gas being collected (Witynski, 2019). However, incomplete capture of gas generated by the waste is a significant limitation of these systems. According to a study by the IPPC, traditional LFG collection systems only capture as low as 20% of the methane produced in landfills, with the remaining gas escaping into the atmosphere (Sullivan, 2010). Nevertheless, the latest studies have shown that landfill methane emissions can be reduced through the implementation of effective management practices and the use of advanced LFG collection systems (Yang et al., 2013). For example, a study found that the implementation of best management practices at landfills could result in an 80% reduction in methane emissions by 2030 (Ocko, 2021). Another study found that the use of advanced LFG collection systems at landfills in China resulted in a 50% reduction in methane emissions (Cai et al., 2018).

To overcome these limitations, strategies to develop a competent landfill gas collection system should be developed. These approaches utilize advanced technologies and strategies to enhance methane capture efficiency, reduce emissions, and optimize energy utilization. Examples of these innovations include real-time monitoring systems (Fay et al., 2011), vacuum-assisted gas collection (Winn, 2022; Wong & Zawadzki, 2023), and gas permeable barriers (Rajesh et al., 2014). Real-time monitoring systems enable continuous tracking and optimization of gas collection performance, allowing for timely adjustments to maximize methane capture. The vacuum-assisted gas collection utilizes suction to draw gas from the waste mass, increasing efficiency and improving overall capture rates. Gas permeable barriers provide an alternative method to collect gas by utilizing barriers with specialized membranes that allow gas to pass through while preventing leachate migration. This article aims to explore and evaluate these innovative approaches to landfill gas collection systems, focusing on their potential to improve methane capture efficiency in landfill sites. Case studies will be presented to showcase the performance and benefits of this present work.

8.2 Methane Collection System in Landfill

The LFG collection and LFG recovery processes exhibit a subtle distinction. The LFG collection system is not necessarily intended for gathering gas to be used in energy projects. Instead, it may be employed for gas migration control and can consist of wells designed for natural venting of the gas or interconnected wells connected through a gathering system to facilitate forced ventilation of the gas (US EPA, 2021a). On the other hand, a landfill recovery collection system is specifically designed to extract LFG for its utilization in energy projects. This system comprises extraction wells and a network of pipelines responsible for collecting and transporting the gas to the processing location. Essential components of the LFG recovery system include extraction wells, wellheads, collection piping systems, a condensate management system, a blower, and a flare.

8.2.1 Extraction Wells

Landfill gas collection systems can be designed using various approaches, including horizontal, vertical, and slotted wells. Each design has its benefits and limitations, and the choice depends on factors such as landfill geometry, waste composition, and environmental conditions. Therefore, a thorough comparison of different gas collection system designs provides valuable insights for landfill operators. Horizontal and vertical wells represent prominent options, each possessing distinct merits and demerits, as elaborated in Table 8.1.

8.2.1.1 Horizontal Wells

Horizontal wells are installed at shallow depths within the waste mass and offer the advantage of uniform gas extraction (see Fig. 8.1). They are particularly well-suited for large, flat landfill areas where gas generation is relatively consistent. It consists

Table 8.1 Comparison between gas collector designs

Types	Advantages	Challenges
Horizontal wells	• Installs at shallow depths within the waste mass • Allow for more even gas extraction and are suitable for large, flat landfill areas • Facilitates expedited LFG collecting • Minimize the need for expensive, specialized machinery • Enables gas to be extracted from the ground under a working dumping area	• Require a more extensive network of piping, leading to higher installation costs • A greater risk of air infiltration unless well protected • Landfill settling and floods are the most potential sources of failure
Vertical wells	• Typical layout • Drill deep into the landfill, and target specific gas-rich zones, making them potentially cost-effective • Inspection is easily accessible • Less impediment to landfill operations is located in a closed area	• Successful implementation relies heavily on accurate gas generation data for optimal well placement, which may be challenging to obtain in some cases • Need for expensive, specialized machinery • The collection of gases is delayed if the equipment is placed after a site or cell has been closed • Costs of operation and maintenance will be high if the system is put into use near a working landfill
Slotted wells	• Can be designed for horizontal or vertical wells • Design with perforated sections to target specific gas-rich zones and are particularly useful for heterogeneous landfill sites	• Require a considerable network of piping, necessitating careful evaluation of installation costs • Leachate influx

of a trench excavated within the waste, which is then backfilled with gravel or other materials like tire chips or broken glass. In the center of the trench, a perforated pipe with a diameter of 10–20 cm is usually placed. The spacing of these wells varies depending on the radius of influence they need to cover (Chetri & Reddy, 2021). The extraction helps to prevent localized gas accumulation and facilitates a more balanced distribution of gas capture points.

Current findings from a study by Wong and Zawadski (2023) demonstrate the effectiveness of horizontal wells in capturing methane from expansive landfill sites. The measurement showed that, at the time, the methane collection efficiency for the study area was approximately 37%. Nevertheless, an increase to 74% methane recovery could be obtained by doubling the number of horizontal collectors on a platform, installing a layer of permeable demolition waste on that level, and increasing in collection system vacuum. The calibrated gas flow model is demonstrated in Fig. 8.2a, while in Fig. 8.2b, the measurements show relatively steady methane generation rates and collection rates between 2015 and 2020, which is consistent with the relatively steady waste disposal rates during this time. These relatively consistent methane

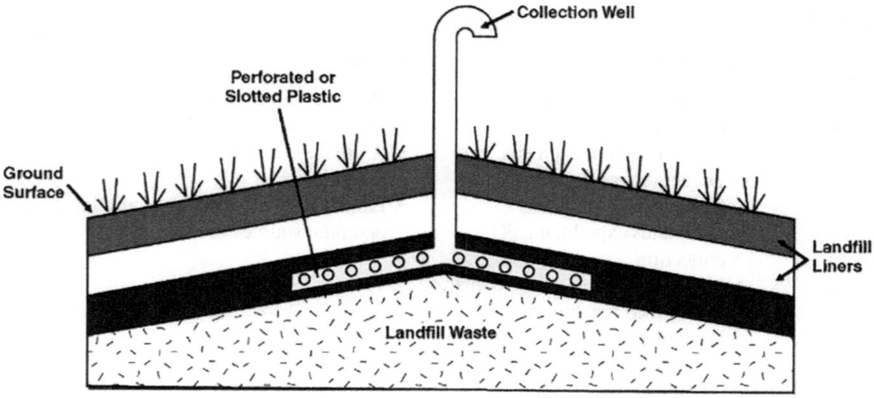

Fig. 8.1 Horizontal well installation for landfill gas recovery. Adopted from ref. (US ATSDR, 2001a)

generation rates over five years portray the reliable efficiency of using horizontal wells in the landfill. However, it is essential to note that the extensive network of piping required for horizontal wells can increase installation costs, making it less economically viable for smaller landfill operations.

8.2.1.2 Vertical Wells

Vertical wells are drilled deep into the landfill to target specific gas-rich zones (see Fig. 8.3). Generally, vertical wells are recommended for landfills with a depth exceeding 12 m. A typical vertical well configuration involves a borehole with a diameter of 30–90 cm, extending to a depth of 3–5 m above the landfill's bottom. Inside the borehole, a pipe with a diameter of 5–20 cm is inserted (Chetri & Reddy, 2021). This design is cost-effective when methane generation is concentrated in areas. It allows operators to focus on extracting gas from zones with high gas production rates of 90%, potentially leading to improved capture efficiency (Zheng et al., 2018). A recent study by Jung et al. (2011) highlights the successful implementation of vertical wells in landfills with heterogeneous waste distribution. The research found that targeted extraction from gas-rich areas using vertical wells resulted in enhanced methane capture rates, surpassing those achieved with horizontal wells in such contexts.

The waste was modeled in a landfill as five separate layers, each with different properties. These layers were meant to account for the compression of waste and the fact that deeper parts of the landfill are usually wetter. The results have shown that the gas permeability of these layers decreased exponentially from the top layer to the bottom layer. The gas permeability determines how easily gas can move through the waste. Considering this variability, based on the simulation carried out, the gas permeability can change by a factor of 250 within the 12.4 m deep landfill. This means that the ability of gas to move horizontally can be 80 times greater than its

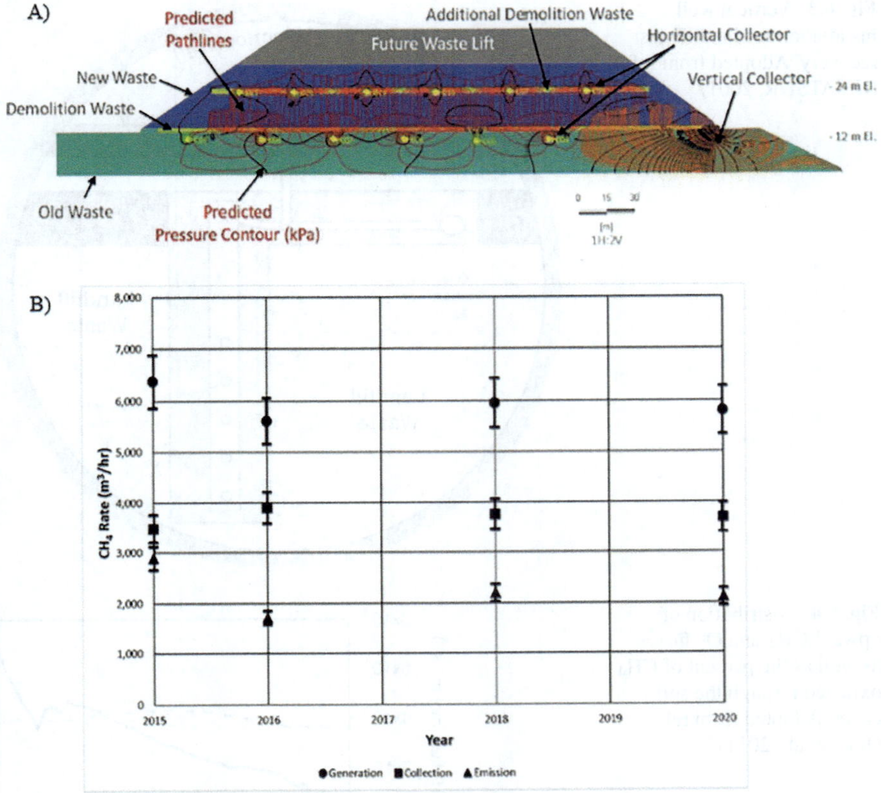

Fig. 8.2 **a** Predicted gas flow pattern using the horizontal wells in an expansive landfill and **b** methane collection, emission, and generation profiles over 6 years span. Adopted from ref. (Wong & Zawadzki, 2023)

ability to move vertically in these areas with varying waste properties. Moreover, as the distance from vertical gas recovery wells increased from 5 to 20 m, the average CH_4 flux measured at the landfill surface almost doubled (see Fig. 8.4). This insight is important for understanding how gases move through landfills and can help improve gas collection strategies.

In another scenario, a gas collection system with different vacuum strengths is investigated (Yu et al., 2009). When the vacuum is weak (0 kPa), the best distance between gas extraction wells is 20 m. As the vacuum gets stronger (going from 0 to − 2 kPa), the optimal distance increases to 60 m. This means that a stronger vacuum helps to collect more gas from the wells, but it also brings in more unwanted air. Interestingly, when the vacuum becomes even stronger (going from − 2 to − 4.5 kPa), the optimal distance between the wells only slightly increases from 60 to 80 m. This suggests that a very high vacuum does not result in much more gas recovery, but it still brings in more unwanted air. Therefore, finding the right balance between

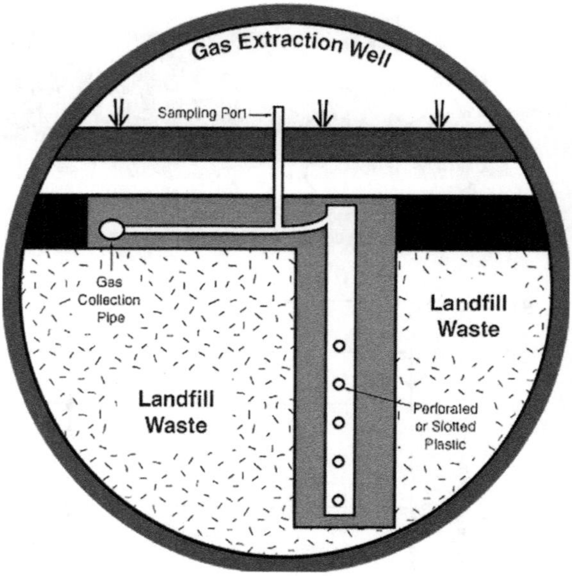

Fig. 8.3 Vertical well installation for landfill gas recovery. Adopted from ref. (US ATSDR, 2001)

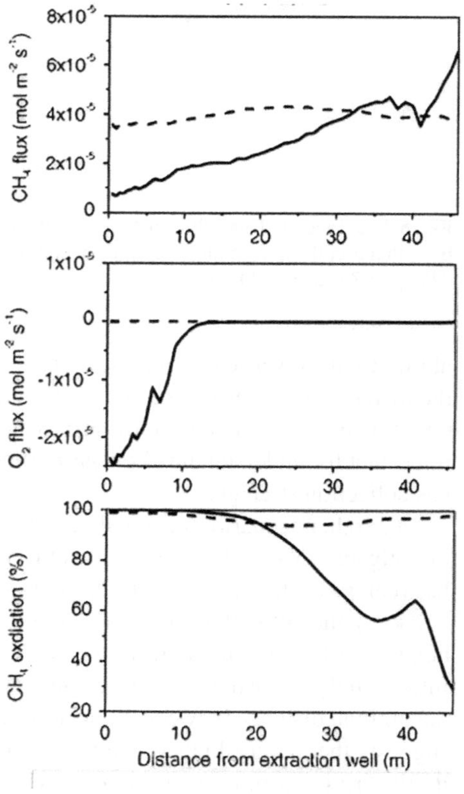

Fig. 8.4 Distribution of upward CH_4 and O_2 fluxes, as well as the percent of CH_4 oxidized through the soil cover. Adopted from ref. (Jung et al., 2011)

vacuum strength and distance between wells is crucial for efficient gas collection while minimizing undesirable air influx.

8.2.2 Wellheads

A wellhead is installed at the top of each gas recovery well to permit flow regulation and LFG sampling. For technicians to monitor the LFG's temperature, pressure, and composition, vacuum control valves and gas measuring ports are installed in the wellheads. In the event of a malfunction, the inside of the wells may be inspected thanks to the wellhead's detachable cover. Since the accumulation of leachate in the gas collecting well might impede gas flow into the wells, this feature permits detection and, if necessary, removal of leachate.

8.2.3 Collection Piping System

The LFG is transported from the collecting points to the processing facility through a network of lateral and header pipes. A network of lateral pipes extends from the header pipes. The collecting pipe must be large enough to hold the required amount of LFG while also minimizing vacuum loss and maintaining a consistent vacuum in each well. The pipe material must be physically robust to bear external stress and strains due to settling of the waste, and chemically resistant to the LFG, condensate, and leachate. Depending on the desired throughput, the diameter of a header pipe may range from 6 to 24 inches of PVC or HDPE (Chetri & Reddy, 2021). Both above-ground and underground in the cover system are acceptable locations for the collecting pipe serving the landfill. Pipes that are buried are often preferable because they are protected from vandalism, operation and maintenance equipment, and the potential for blockage caused by frozen condensate. The collecting piping's inclinations are calculated to prevent the accumulation of condensate on their interiors. Header piping systems often have low points installed for collecting condensate. Even if there is a settlement, a minimum slope of 3–5% is suggested to enable the drainage of condensate and prevent pipe blockage (US EPA, 2012). The slope is given on the lateral and header pipes that transport LFG from the gas collecting wells to the blowers and flare systems to prevent the buildup of LFG, the loss of vacuum, and the lack of consistent vacuum throughout the system (US EPA, 2012). Standard practice calls for a header pipe slope of at least 0.5% in the direction of LFG flow and 2% in the opposite direction. Similarly, laterals in the direction of LFG flow are given a minimum slope of 3%, while laterals in the opposite direction are ideally given a slope of 5% (Chetri & Reddy, 2021). Figure 8.5 is a sample site layout for LFG extraction.

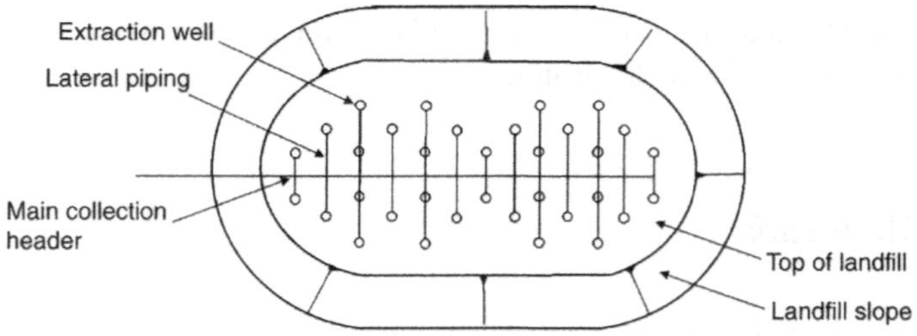

Fig. 8.5 Site layout example for LFG extraction. Adopted from (US EPA, 2021d)

8.2.4 Condensate Management System

The waste-produced LFG is elevated in moisture and temperature. During transmission from the wells to the processing unit, the heated LFG condenses into a liquid called condensate. The condensate formation is proportional to the total quantity of LFG produced or removed, the waste moisture content, the vacuum or pressure applied to the LFG, and the degree of temperature change. The LFG recovery might be hampered by condensation if the header lines get clogged. Header lines are protected by condensate knock-out tanks located at the lowest locations of the collecting pipe system. Knock-out tanks are built to isolate liquid droplets from the inflowing gas and encourage their production (US ACE, 2008). The knock-out tank for condensate is seen in Fig. 8.6. When planning the LFG collection pipe system, it is important to keep condensate control in mind. Seasonal changes will cause fluctuations in condensate generation throughout the year. During the winter, the rate of condensate generation will be maximum.

8.2.5 Blower

To move the gas from the collecting wells to the flaring unit or treatment unit, a blower creates a vacuum in the collection header. Since the flow velocity and LFG composition may fluctuate regularly, the blower must be constructed to function in a wide range of environments. Depending on the gas flow rate and the distance to flaring or other treatment operations, the size, type, and number of blowers required for the LFG collecting system will vary. Typically centrifugal blowers, rotary lobe blowers, and positive displacement blowers are employed in LFG collecting systems (US ACE, 2008).

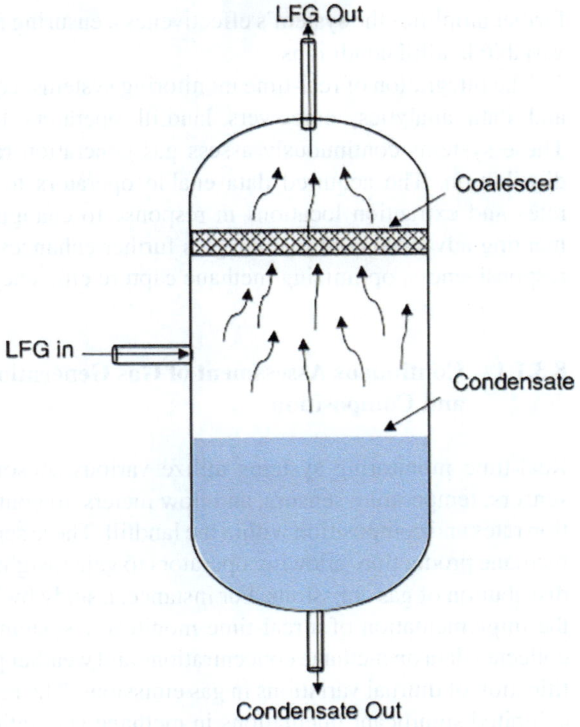

Fig. 8.6 Typical knock-out system for condensate. Adopted from ref. (Chetri & Reddy, 2021)

8.3 Latest Innovative Approaches for Landfill Gas Collection Systems

The LFG collection systems have become critical in addressing environmental concerns associated with landfill sites. This section provides the latest innovative approaches for LFG collection. Emphasis is placed on real-time monitoring systems and control strategies to optimize methane capture efficiency. Additionally, the utilization of vacuum-assisted gas collection systems and gas permeable barriers is discussed, highlighting the potential benefits and drawbacks of each approach.

8.3.1 Real-Time Monitoring Systems and Control Strategies

The integration of real-time monitoring systems with data analytics offers a novel approach to optimizing methane capture efficiency. By continuously assessing gas generation rates and composition, operators can dynamically adjust gas flow rates and extraction locations. This dynamic approach enhances LFG collection and energy recovery while minimizing emissions. The utilization of advanced control strategies further amplifies the system's effectiveness, ensuring real-time adjustments based on variable landfill conditions.

The integration of real-time monitoring systems, equipped with advanced sensors and data analytics, empowers landfill operators to make informed decisions. These systems continuously assess gas generation rates, composition, and spatial distribution. The acquired data enable operators to dynamically adjust gas flow rates and extraction locations in response to changing landfill conditions. Implementing advanced control strategies further enhances the system's adaptability and responsiveness, optimizing methane capture efficiency.

8.3.1.1 Continuous Assessment of Gas Generation Rates and Composition

Real-time monitoring systems utilize various sensors, such as gas concentration sensors, temperature sensors, and flow meters, to continuously measure gas generation rates and composition within the landfill. These sensors provide real-time data on methane production, allowing operators to gain insights into the temporal and spatial distribution of gas emissions. For instance, a study by Xia et al. (2023) demonstrated the implementation of a real-time monitoring system at a landfill site. The system collected data on methane concentrations and weather parameters, enabling the quantification of diurnal variations in gas emissions. The results indicated that the landfill exhibited significant fluctuations in methane generation, necessitating adaptive gas collection strategies to optimize capture efficiency. The findings revealed that the CH_4 levels can be different at most locations in the landfill. Therefore, it is important to determine "hotspots" with higher CH_4 concentrations were observed at perimeter locations, indicating localized releases or a broad plume, as shown by the daily maps in Fig. 8.7. The diurnal variability in CH_4 levels at landfills (see Fig. 8.8) does highlight the benefit that continuous and real-time CH_4 monitoring at landfills could provide since CH_4 levels are ever changing depending on the in-situ meteorological conditions.

8.3.1.2 Dynamic Adjustment of Gas Flow Rates and Extraction Locations

Based on the real-time data obtained from monitoring systems, operators can dynamically adjust gas flow rates and extraction locations within the gas collection network. This adaptability enables operators to respond swiftly to changing landfill conditions, such as fluctuations in waste deposition, moisture content, or temperature, which impact gas generation rates. A study by Parameswaran et al. (2023) demonstrated the application of a dynamic gas collection model via a heuristic method, driven by real-time data, to optimize methane recovery from a landfill. The work's adaptive approach considers multiple attributes (heterogeneity) of a landfill to calculate the optimal radius of influence (ROI) for gas collection by employing stochastic

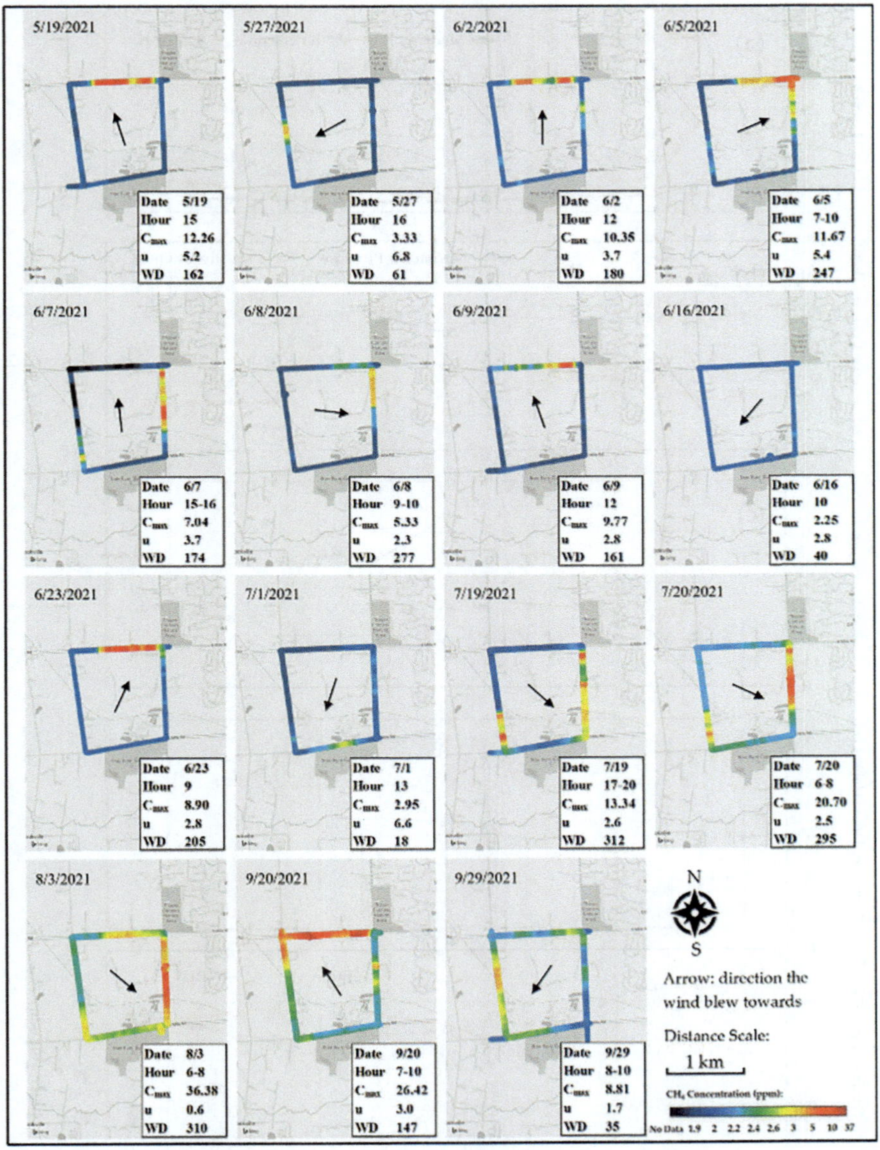

Fig. 8.7 Daily maps showing real-time monitoring of CH_4 concentrations at the landfill. Adopted from ref. (Xia et al., 2023)

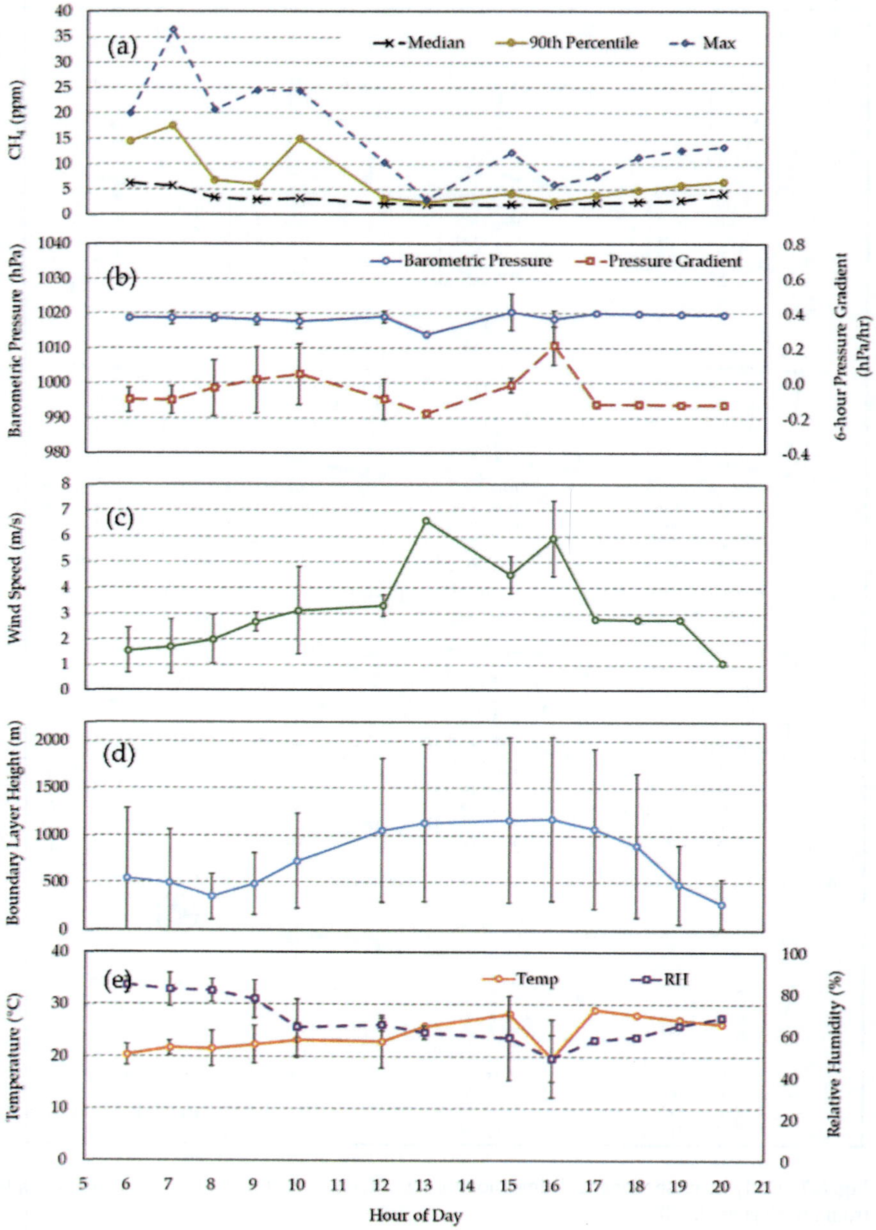

Fig. 8.8 Diurnal variation of CH_4 concentrations and hourly average of meteorological variables. Adopted from ref. (Xia et al., 2023)

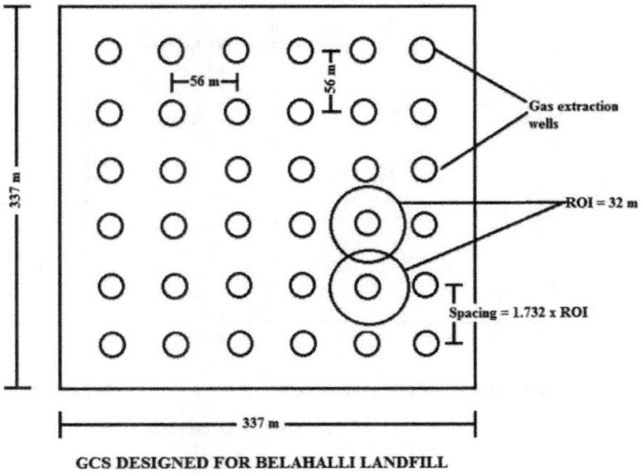

Fig. 8.9 Proposed design for gas extraction wells (GEWs) placement in the landfill. Adopted from ref. (Parameswaran et al., 2023)

modeling through three-dimensional gas flow dynamics. The stochastic approach takes into account the variability in gas generation rate and geotechnical properties that govern gas migration, including permeability, porosity, and degree of saturation. By adjusting extraction locations and flow rates, the researchers achieved a 32% increase in methane capture efficiency compared to a static control scenario.

In Fig. 8.9, we can see how the gas extraction wells (GEWs) are positioned in the landfill. The distance between the wells is 1.732 × ROI, and there is a slight overlap between the ROI of neighboring wells. This arrangement is designed to ensure efficient methane capture and prevent any gaps where methane could escape without being collected. On average, the proposed system may recover about 83% of the methane produced, with a small variation around this value. When we look at the distribution of methane recovery efficiency (Fig. 8.10b), we can see that most of the time, it falls between 75 and 92%, which means it is consistently efficient. Interestingly, Fig. 8.10a, which indicates the ROI of the gas collection system, shows a similar trend. This suggests that the variability in waste properties, both in terms of their spatial distribution and how they change over time (temporal), has a similar effect on methane recovery efficiency and the radius of influence of the gas collection system. This finding highlights the substantial benefits of dynamically adapting gas collection strategies to varying landfill conditions.

8.3.1.3 Subsurface Monitoring of LFG Migration

Real-time monitoring systems facilitate the identification of high gas generation areas and periods, leading to targeted collection efforts. By focusing on these areas during peak gas production, operators can enhance methane capture rates, thereby maximizing LFG collection efficiency. A field study conducted by Fay et al. (2011) examined the implementation of a real-time monitoring system at a landfill site to investigate subsurface migration of CH_4 and CO_2 concentration levels through

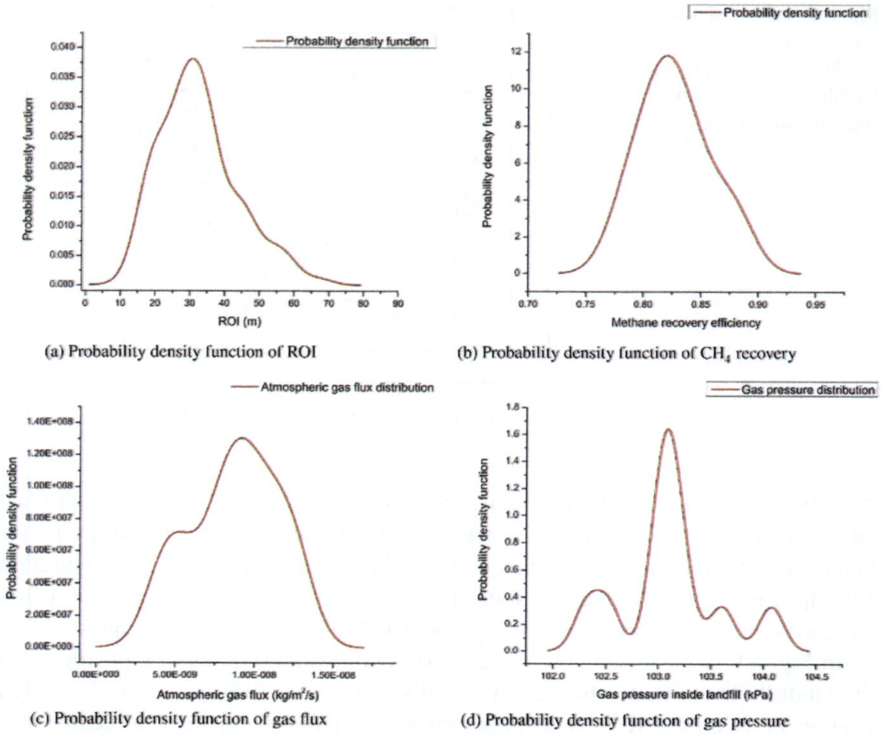

Fig. 8.10 **a** ROI and **b** CH_4 recovery efficiencies. Adopted from ref. (Parameswaran et al., 2023)

internet scale sensing (ISS) and expanded it to suit a realized sensing model for real-time landfill migration monitoring via a live website (see Fig. 8.11). This system was tested in real-world conditions for 16 months, which had collected a total of 2,445,300 measurements of gas concentrations. This extensive testing helped validate the system's effectiveness and reliability in capturing accurate data about gas levels in the environment, as shown in Fig. 8.12. The results showed that targeted gas collection, guided by real-time data, led to a 30% increase in methane recovery compared to conventional fixed extraction systems. This underscores the potential of real-time monitoring in significantly improving the ability to harvest chemical information from the landfill, which later provides the means to enforce a better LFG collection system to enhance energy recovery.

8.3.1.4 Utilization of Advanced Control Strategies

Apart from real-time monitoring, advanced control strategies further augment the effectiveness of methane capture in landfill gas collection systems. These strategies

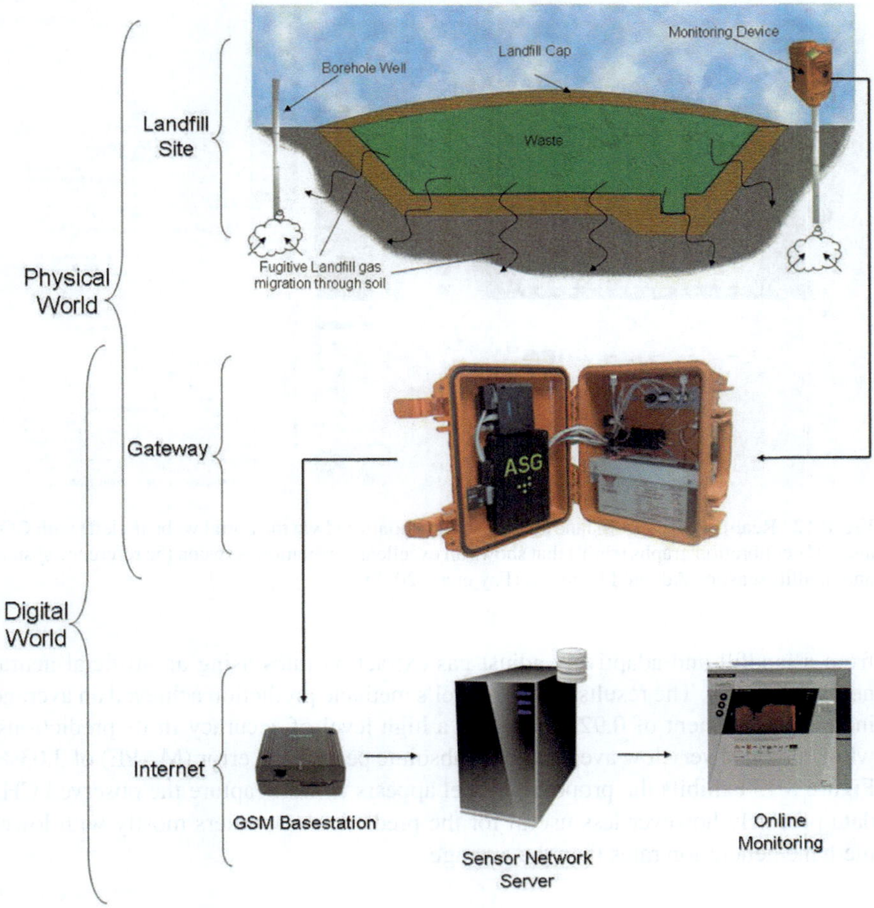

Fig. 8.11 Illustrates the landfill gas sensing model, depicting the entire process from the physical device placed in the field to the web-based user interface for visualization. The model demonstrates how the chemical data sensed in the landfill environment is transferred from the physical world to the digital world through a gateway platform. Adopted from ref. (Fay et al., 2011)

rely on data analytics, machine learning algorithms, and predictive modeling to optimize gas collection in response to dynamic landfill conditions. LFG generation is often affected by meteorological factors and landfill design. Machine learning algorithms use influencing factors such as temperature and rainfall as input features to model the process of LFG production and obtain high accuracy. An accurate and reliable prediction is needed for proper design and operation and helps eliminate environmental impacts and maximize methane harvesting.

An article by Fallah et al. (2020) presented a control system that combined real-time monitoring with predictive models to optimize gas collection at a landfill site. The control system utilized historical data to forecast the daily CH_4 generation rate

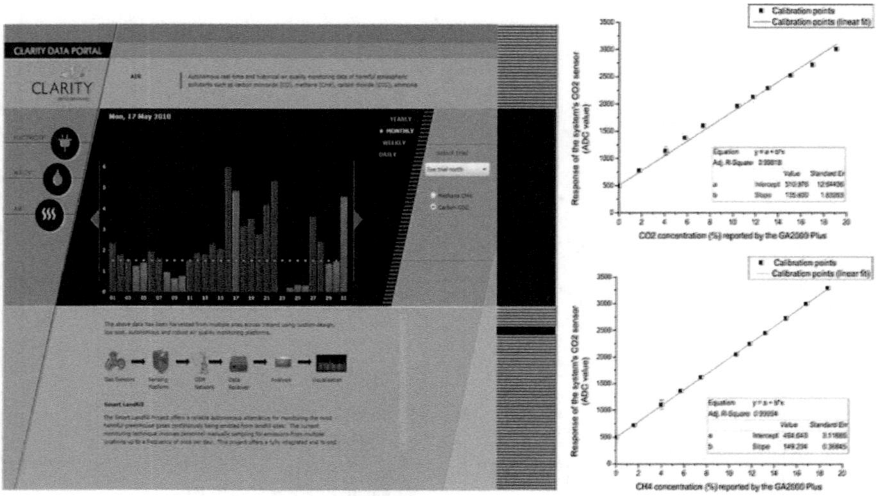

Fig. 8.12 Real-time data from landfill sites can be monitored via the portal website (left) with CO_2 and CH_4 calibration graphs (right) that shows an excellent correlation between the reference system and landfill sensors. Adopted from ref. (Fay et al., 2011)

from a landfill and adaptively adjust gas extraction rates using an artificial neural network (ANN). The results of this model's methane prediction achieved an average index of agreement of 0.92, indicating a high level of accuracy in its predictions, while having a very low average mean absolute percentage error (MAPE) of 3.03%. Figure 8.13 exhibits the proposed model appears to well capture the observed CH_4 data precisely however less useful for the prediction of outliers mostly with lower methane generation rates than the average.

8.3.2 Vacuum-Assisted Gas Collection Systems

Vacuum-assisted gas collection systems introduce negative pressure within the collection network using blowers or vacuum pumps. This suction effect accelerates gas flow toward the extraction points, significantly boosting methane capture efficiency. Vacuum-assisted systems excel in areas with low-permeability soil covers, where traditional systems may struggle to extract gases effectively. However, the additional energy consumption associated with vacuum pumps must be considered in the overall environmental impact assessment. Recent studies have demonstrated the effectiveness of vacuum-assisted gas collection systems in improving methane recovery rates. A research paper by Wong and Zawadski (2023) conducted field trials comparing the performance of vacuum-assisted systems against conventional systems in landfills with low-permeability soil covers. The calibrated gas flow model with the implementation of vacuum-assisted horizontal collectors is demonstrated in

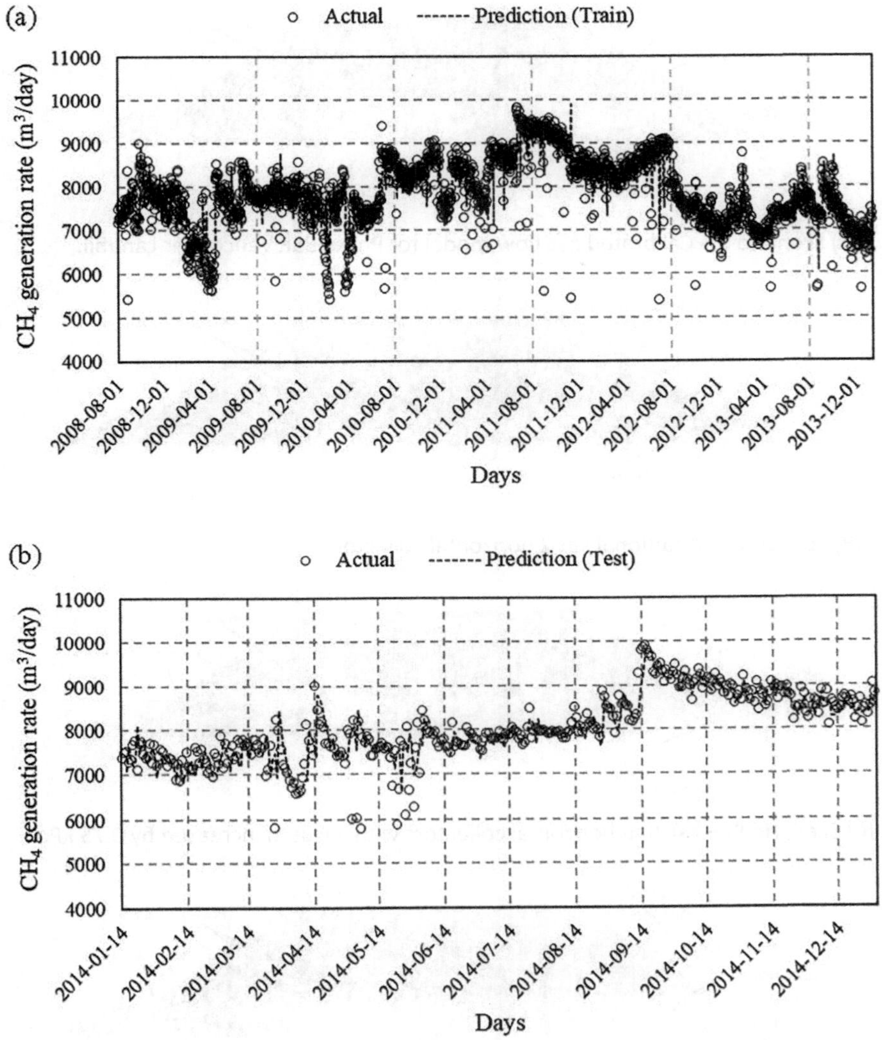

Fig. 8.13 Prediction performance of methane generation of the model proposed. Adopted from ref. (Fallah et al., 2020)

Fig. 8.14. The study found that vacuum-assisted horizontal gas collection systems showed an increase in higher methane capture efficiency from 37 to 48 and 74% with an increase of vacuum pressure from 0 to 0.75 and 1 kPa, respectively. The ability to accelerate gas flow in low-permeability conditions makes vacuum-assisted systems particularly advantageous in such scenarios.

While vacuum-assisted gas collection systems offer notable benefits in terms of improved methane capture efficiency, the additional energy consumption associated with vacuum pumps must be carefully evaluated in the overall environmental impact

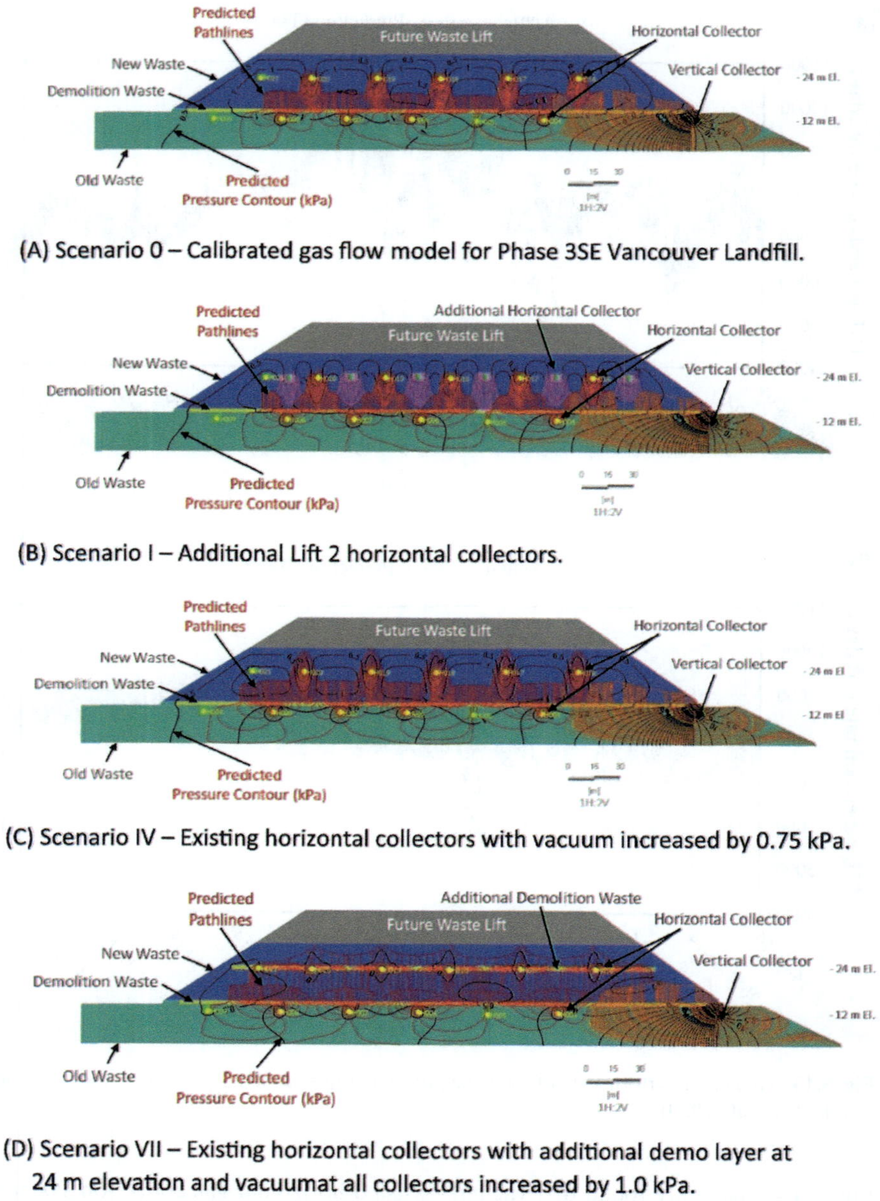

Fig. 8.14 Predicted gas flow patterns for vacuum-assisted horizontal collectors at 0.75 kPa. Adopted from ref. (Wong & Zawadzki, 2023)

assessment. Vacuum pumps require electricity to operate, and their usage contributes to an increase in energy consumption and associated greenhouse gas emissions. Landfill operators need to consider the energy source and the overall environmental footprint when deciding on the implementation of vacuum-assisted systems.

8.3.3 Gas Permeable Barriers

Gas permeable barriers, placed beneath the landfill cover, facilitate enhanced gas migration while restricting the infiltration of rainwater and leachate. By improving gas movement towards collection points, these barriers reduce the need for maintenance activities, streamlining gas capture processes. Successful implementation depends on selecting suitable barrier materials and ensuring proper installation. However, selecting appropriate barrier materials and ensuring proper installation is crucial for success, and their cost-effectiveness depends on site-specific conditions. For instance, the superiority of vacuum-assisted systems in low-permeability soil covers can be attributed to their ability to overcome the resistance posed by less permeable soils. In landfills with impermeable or less permeable soil covers, traditional gas collection systems may struggle to extract gases effectively due to limited gas movement through the soil. Vacuum-assisted systems mitigate this limitation by creating a stronger suction force, drawing the gas more efficiently toward the extraction wells.

Figure 8.15 presents data on LFG permeability and diffusion coefficients in various types of soils, obtained from both laboratory and field tests (Wang et al., 2022). The findings highlight the strong relationship between the permeability and diffusion coefficient of LFG and specific soil characteristics. These characteristics include the type of material, moisture content, soil compaction, air-filled porosity, vegetation coverage, and soil modification. It can be observed that the gas permeability of cover materials ranges from 10^{-17} to 10^{-8} m^2. The moisture content and degree of compaction are two crucial factors influencing the transport of LFG. Typically, as the moisture content increases, the gas permeability decreases (Zhan & tong, 2016). This is because higher moisture levels reduce the available airflow pathways in the soil, as the pores get filled with water. This indicates that the distribution of pore sizes in the soil has a significant impact on gas permeability. These data shown provided valuable insights into the factors affecting LFG transport through different types of cover soils as gas-permeable barriers, helping researchers and landfill operators make informed decisions for managing gas migration and collection systems.

In another study conducted by Spokas et al. (2006), methane recovery was evaluated in different landfill cells with various designs, cover materials, and gas management strategies. The research included sites with diverse cell configurations, such as those with temporary clay covers, final clay covers, geosynthetic clay liners, and geomembrane composite covers. The results showed a wide range of methane emission rates, varying from − 2.2 to over 10,000 CH_4 per m^2 in a day. The methane

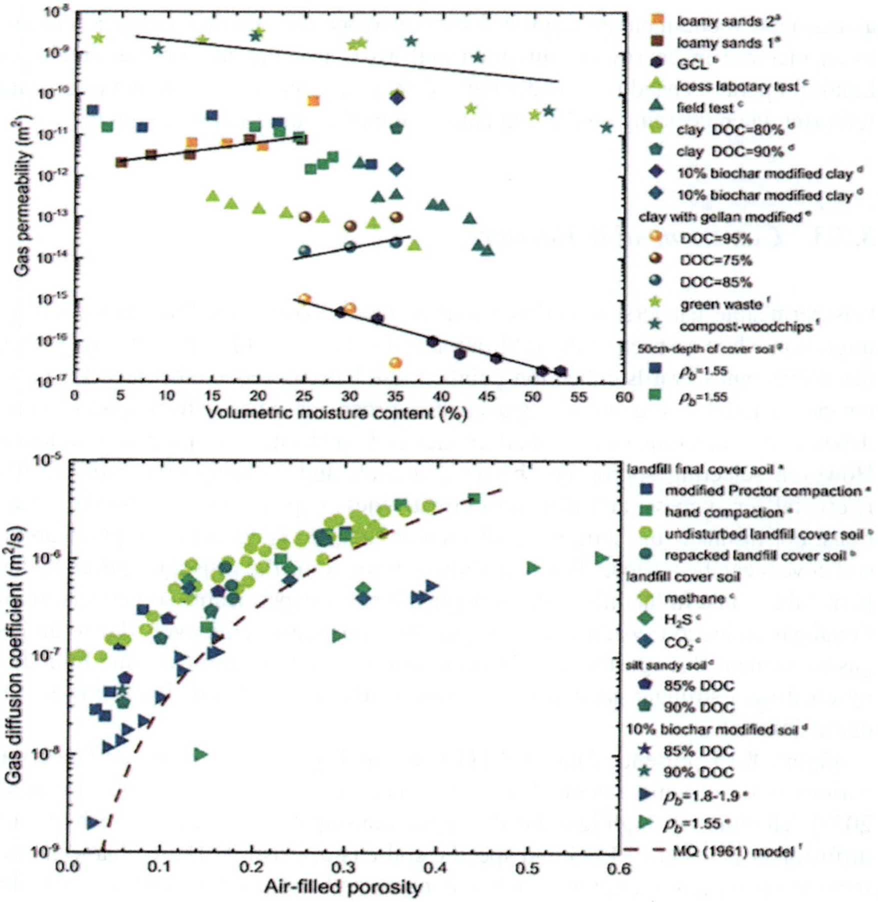

Fig. 8.15 Permeability and diffusion coefficient of cover soil from literature. Adopted from ref. (Wang et al., 2022)

recovery rates also differed based on the cell type and gas management strategy. For an operating cell with an active LFG recovery system, the methane recovery was about 35%. Temporarily covered cells with active LFG recovery achieved 65% methane recovery. Cells with clay final covers and active LFG recovery achieved an 85% methane recovery rate, while those with geomembrane final covers and active LFG recovery had the highest recovery rate at 90%.

8.4 Methane Capture from Landfill

Since LFG contains a large percentage of CH_4 (about 50%), it may be processed and utilized as a renewable energy source. Due to its greater molar adsorption coefficient for infrared radiation and longer residence duration in the atmosphere (Scheutz et al., 2009), CH_4 is a very potent GHG with a global warming potential of 25 over a period of 100 years. When considering the sources of human-caused CH_4 emissions in the United States, landfills rank third (US EPA, 2021c). Nearly half of LFG is carbon dioxide, and there are also trace amounts of non-methane organic compounds (NMOCs). While LFG emissions contribute to global warming, they may be mitigated by capturing the gas and using it as an alternative energy source. Energy recovered by LFG may be sold to offset costs and ease demands on finite resources. The recovery of CH_4 and its applications are shown in Fig. 8.16, along with the collection and processing of LFG. The depth of LFG treatment may vary depending on the nature of the application. Compared to the 1000 Btu per scf found in pipeline-grade gas (US ACE, 2008), the energy level of LFG that is collected from landfills is typically between 450 and 600 Btu (Zietsman et al., 2008). However, the energy content of the LFG varies substantially depending on the effectiveness of the LFG collecting system and the waste's state of decomposition.

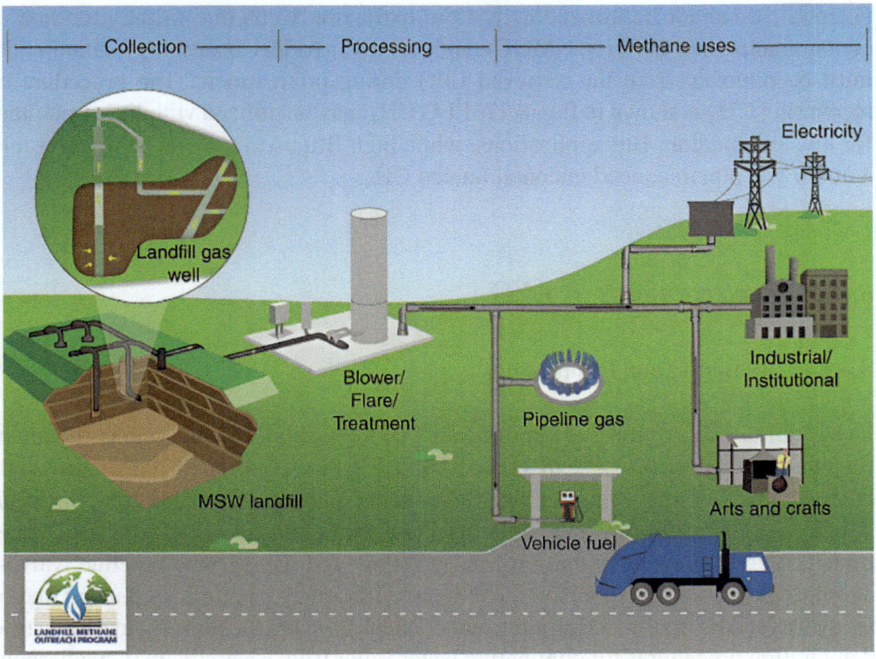

Fig. 8.16 Illustration of the landfill gas collection, processing, and utilization process. Adopted from ref. (US EPA, 2019)

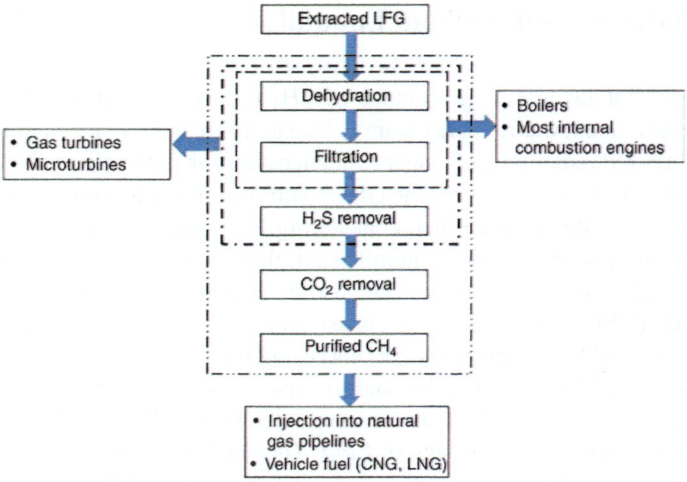

Fig. 8.17 Several steps in the process of recovering methane. Adopted from ref. (Chetri & Reddy, 2021)

Recovering CH_4 from LFG necessitates a multi-step purification procedure that accounts for the destination of the recovered gas. Although LFG has great energy potential, it cannot be utilized as is from extraction wells due to the presence of contaminants such as water, NMOCs, and carbon dioxide. Concerning contaminants must be removed from the retrieved LFG during pretreatment. The procedure of recovering CH_4 is shown in Fig. 8.17. LFG CH_4 may be utilized with little treatment for low and medium Btu applications, while high Btu uses, such as pumping it into a utility gas pipeline, need uncontaminated CH_4.

8.4.1 Pretreatments of LFG

8.4.1.1 Water Vapor Elimination

Dehydration, absorption, adsorption, and condensation are all viable treatment options for recovering CH_4 in waste. The LFG that is pumped out of the wells is often quite wet. In every gas flow, water vapor is the worst possible contamination (Chetri & Reddy, 2021). Water vapor in LFG is very damaging to vegetation. During transit, it creates hydrates that obstruct gas flow and damage the pipeline, valves, and fittings. The water vapors in the extracted LFG may be eliminated using one of many ways. Absorption, adsorption, and condensation are the most typical industrial dehydration procedures for eliminating water vapor from a gas stream (Ghanbarpour Mamaghani et al., 2023).

Water vapor in the gas stream may be removed by adsorption using a liquid desiccant. Glycol dehydration is the most prevalent absorption mechanism (Netušil & Ditl,

2012). Water and glycol have a strong attraction to one another. The most prevalent glycols used in industry for dehydration are diethylene glycol and triethylene glycol. A counter-current column made up of four to ten bubble-cap trays is used in the glycol absorption process. Glycol is supplied into the column at the top, where it flows downhill and absorbs moisture from the LFG as it passes through. The gas is released from the contactor's top.

The vaporized water bonds to the desiccant material during the adsorption process. Silica gel and beads, activated alumina, activated bauxite, and molecular sieves are some of the most frequent desiccants used in gas dehydration procedures. Water vapor is absorbed by the desiccants when the gas flows past them. In addition to water vapor, these adsorbents may also take up organic sulfur and organic halogen chemicals. Overloaded adsorbents may be regenerated by applying heat or decreasing pressure (Chetri & Reddy, 2021; Netušil & Ditl, 2012).

Mechanical dehumidification of LFG is achieved by condensation by means of cooling and compression. The water vapor in the gas is condensed when the gas is brought into contact with a surface that is colder than the dew point. The condensate is drained from the knockout tank by pump (Chetri & Reddy, 2021).

8.4.1.2 Other Contaminants Removal

Particles in the LFG might clog valves and fittings and cause harm to downstream machinery. Filter pads (usually constructed of stainless-steel wire) or cyclone separators are used to remove the particles from the LFG. Filter pads and cyclone filters can trap particles as tiny as 2 m and 15 m, respectively (SEPA, 2004).

Among the most corrosive gases is H_2S. If discharged into the air, it will harm not just machinery but also people. The burning of LFG also causes SO_x emissions. Landfills that exclusively accept municipal solid waste (MSW) often have relatively low H_2S concentrations and may not need treatment. It has been shown that the concentration of H_2S in landfills that receive building and demolition debris is greater. Several methods exist for removing H_2S from LFG. Activated carbon beds, zinc oxide, iron oxide, carbon impregnated with potassium hydroxide, and potassium carbonate are often used as adsorbents in this popular approach (Chetri & Reddy, 2021). SulfaTreat is another popular method for neutralizing H_2S at dump sites using dry, easily-dispersed granules of synthetic iron oxide blend, producing iron pyrite (Kasulla et al., 2021). Amine scrubbing is another common technique for getting rid of H_2S. A tower holding an amine solution, which has a strong affinity for H_2S, is used in this process (Moioli et al., 2013). Monoethanolamine and diethanolamine are the two most common amine solutions (Al-Baghli et al., 2001).

Nearly half of the volume of LFG is CO_2. Pipeline-grade gas has a maximum allowable CO_2 content of 2–4% by volume. Its acidic composition makes it corrosive. Absorption, adsorption, membrane separation, and cryogenic separation are only a few of the methods that may be used to get rid of CO_2 (Chetri & Reddy, 2021). Regarding absorption, gases are selectively dissolved in solvents. Solvents with a higher affinity for CO_2 are utilized to dissolve CO_2 present physically or chemically

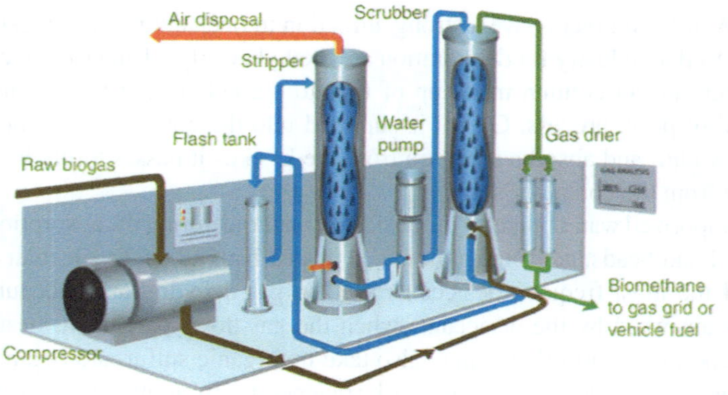

Fig. 8.18 A water scrubbing plant's schematic. Adopted from ref. (US ACE, 2008)

in the LFG. Commonly available physical solvents include Fluor solvent, Selexol, and others. On the other hand, chemical absorption involves the chemical dissolution of CO_2 in the solvent, leading to the formation of carbonates. Major chemical solvents used for this purpose are amine solutions and salts of strong alkalis with weak acids (Kikkawa et al., 2022). Water scrubbing is a widely used technique for CO_2 removal from LFG. In this process (shown in Fig. 8.18), LFG is introduced from the bottom of the scrubber, while water is supplied from the top. As water contacts the LFG, CO_2 and certain impurities like H_2S are removed by dissolving in the water. The gases dissolved in the water are subsequently stripped in a separate vessel, and the water is recirculated in the scrubber. Although the water scrubbing process effectively eliminates CO_2, it is unable to remove impurities like oxygen and nitrogen.

A bed of solid materials with a high affinity for CO_2 is used in the adsorption process, which entails passing the gas across the bed. Adsorption occurs when CO_2 is bound to the solid surface by either physical forces or chemical processes. Silica, alumina, activated carbon, silicates, and molecular sieves are all examples of common adsorbents (Chetri & Reddy, 2021; Pardakhti et al., 2019). Recent research has shown that the byproduct of the steelmaking process, steel slag, may be used as a sorbent to efficiently remove CO_2 and H_2S from biogas (Sarperi et al., 2014). Using steel slag to remove CO_2 from LFG was also shown to be effective by Chetri et al. (2020). Alkaline metal oxides, such as calcium oxide (CaO), found in steel slag combine with carbon dioxide (CO_2) in the liquid ferrous gas (LFG) to generate stable carbonates. Sulfides are produced when iron oxides in steel slag react with hydrogen sulfide. Steel slag's use in LFG treatment is shown to be both cost-effective and ecologically benign.

Polymeric membranes are used in this procedure because of their gas-selective permeability. The membranes do not function as filters, but rather as solution-diffusion systems (Bernardo et al., 2009). The pressure gradient between the upstream and the downstream parts of the system is the primary factor behind this process (Kim & Lee, 2013). The membrane separation method allows unimpeded passage of CO_2 while leaving CH_4. Separating CO_2 and CH_4 in a single step is inefficient. For thorough separation, many stages of separation are required. The membrane separation business makes extensive use of polymers, particularly rubbery polymers and glassy polymers (Bernardo et al., 2009). Poly(dimethylsiloxane), ethylene oxide/propylene oxide—amide copolymers, and silicone rubber are all examples of useful rubbery and glassy polymers. Poly(phenylene oxide), poly(carbonate), poly(imide), and poly(perfluorodioxoles) are examples of glassy polymers. Polymer membranes have been the subject of many studies to improve their performance by fixing issues like poor selectivity and permeability.

Gaseous mixtures are fractionally condensed and distilled at low temperatures in cryogenic procedures (Pellegrini et al., 2018). In landfills, the cryogenic process is used in two phases: first, CO_2 is liquefied, and then CH_4 is liquefied. Cryogenic treatment is reapplied to the CH_4 stream after the CO_2 has been stripped away to eliminate any lingering nitrogen. Each landfill's LFG to energy system is customized to meet the site's specific needs in terms of end usage and financial viability. Figure 8.19 depicts the LFG purification process used at the landfill to convert recovered LFG to LNG (Chetri & Reddy, 2021). CO_2, N_2, H_2S, moisture, and reactive organic compounds are eliminated from the LFG prior to cooling to a cryogenic fluid's typical temperature of around – 260 °F.

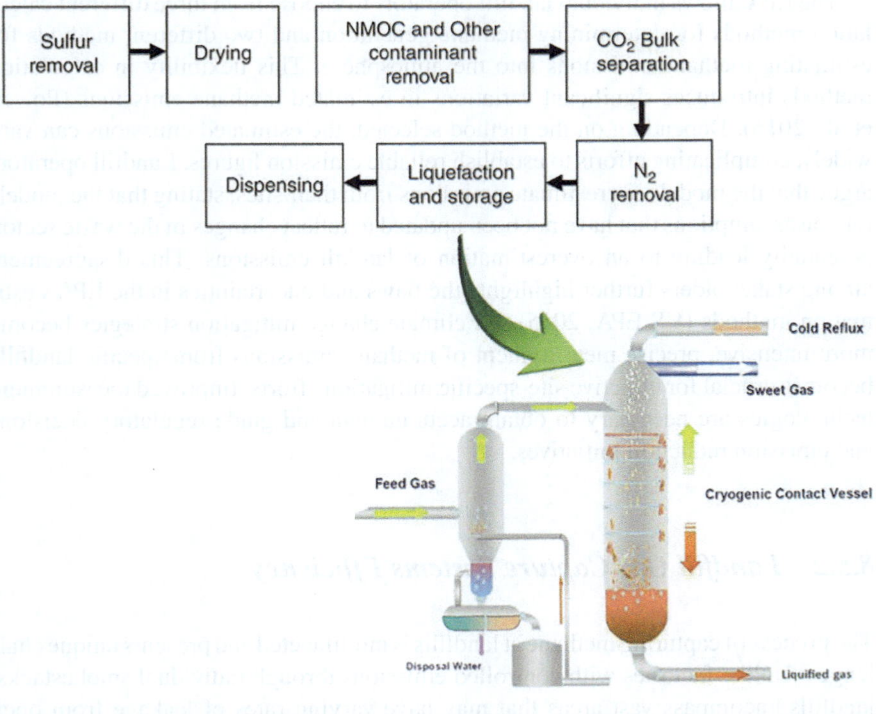

Fig. 8.19 The LFG purification and methane liquefaction system. Adopted with modest modification from ref. (Chetri & Reddy, 2021; Hassan, et al., 2022)

8.5 Challenges to Implementing Landfill Methane Capture System

8.5.1 Outdated Methane Emissions Models

The accuracy of estimating landfill methane emissions is a critical factor in developing effective mitigation strategies and addressing climate change concerns. However, the National Academy of Sciences 2018 report questioned the reliability of the Environmental Protection Agency (EPA) estimates due to uncertainties and outdated methodologies (Bruggers, 2021). This article examines the flaws in the EPA's approach to estimating landfill methane emissions and the need for improved measurement methods to guide mitigation strategies. The National Academy of Sciences report expressed "low confidence" in the EPA's estimates for landfill methane emissions (National Academic of Sciences, 2018). The report highlighted uncertainties and insufficient measurements as key factors undermining the accuracy of the EPA's estimates. It concluded that the agency's method for estimating methane emissions from landfills was "outdated" and lacked field validation, raising concerns about the reliability of the data.

The EPA allows individual landfill operators to choose from three different calculation methods for determining methane generation and two different methods for estimating methane emissions into the atmosphere. This flexibility in calculation methods introduces significant variations in estimated methane emissions (Powell et al., 2016). Depending on the method selected, the estimated emissions can vary widely, complicating efforts to establish reliable emission figures. Landfill operators argue that the models overestimate emissions from their sites, stating that the models rely on assumptions that have not been updated to reflect changes in the waste sector, potentially leading to an overestimation of landfill emissions. This disagreement among stakeholders further highlights the flaws and uncertainties in the EPA's estimation methods (US EPA, 2015). As climate change mitigation strategies become more intensive, precise measurement of methane emissions from specific landfills becomes crucial for effective site-specific mitigation efforts. Improved measurement technologies are necessary to obtain accurate data and guide regulatory decisions and emission reduction initiatives.

8.5.2 Landfill Gas Capture Systems Efficiency

The process of capturing methane at landfills is multifaceted and presents unique challenges. Unlike factories with controlled emissions through individual smokestacks, landfills encompass vast areas that may have varying rates of leakage from open sections, temporarily covered areas, or permanently closed and capped sections. Landfill plumbing is highly susceptible to disruption, emphasizing the need for meticulous attention (Bruggers, 2021). The intricate plumbing system requires ongoing maintenance and monitoring to ensure its efficient operation. New landfill sections, known as cells, have a five-year timeframe for implementing methane capture measures. However, methane pollution begins much earlier than that. The efficiency of methane collection largely relies on gas capture wells and the effectiveness of the collection system. Landfill operators are mandated to cover waste disposal areas daily, using materials such as soil, mulch, or plastic. The choice of cover material impacts methane release rates, as certain materials are more porous and result in higher methane emissions.

Weather conditions, particularly rainfall, play a significant role in methane emissions at landfills (Macias et al., 2014). Rainfall can contribute to increased methane production while simultaneously impeding gas collection systems, reducing their effectiveness. Flooded gas collection systems hinder the capture of methane, leading to elevated emissions beyond the estimates provided by the Environmental Protection Agency (EPA) (Ayandele, et al., 2022). Despite these challenges, a well-designed and well-maintained landfill gas collection system can mitigate methane emissions effectively (Ocko, 2021). A robust system can increase methane capture through wells while also leveraging methane-digesting microbes present in the landfill's soil cover. These microbes help neutralize methane, minimizing its release into the atmosphere.

Capturing methane at landfills is a complex process that requires diligent attention to plumbing systems, waste disposal practices, and the influence of weather conditions. Understanding the vulnerabilities and factors impacting methane emissions is crucial for optimizing capture efficiency. Implementing well-maintained gas collection systems and utilizing methane-digesting microbes can contribute to reducing methane emissions from landfills. Continued research and technological advancements are essential to improve the effectiveness of methane capture and mitigate its contribution to greenhouse gas emissions.

8.5.3 Modest Steps and Lack of Finances to Curb Landfill Emissions

Implementing innovative landfill gas collection systems necessitates compliance with stringent environmental regulations and permitting processes. These regulations govern aspects such as air quality, emissions standards, and waste management (US EPA, 2001). Though complex, adherence to these regulations is crucial for legal and environmental compliance. In 2016, the US EPA enacted a rule that expands methane collection requirements to an additional 93 landfills, aimed at reducing methane emissions. This rule lowers the emissions threshold that triggers the installation of gas collection systems, resulting in an estimated 7% nationwide reduction in landfill methane emissions (Bruggers, 2021). Additionally, the US EPA has notified around 40 states, overseeing approximately 1,600 landfills without approved landfill gas capture plans, to develop such plans or face enforcement actions.

Implementing innovative gas collection systems entails substantial upfront investment, encompassing the acquisition and installation of equipment, infrastructure modifications, and operational expenses. Securing sufficient funding can be a challenge, particularly for smaller landfill operators with limited financial resources. For instance, in rapidly growing regions, managing methane emissions becomes particularly challenging due to the increasing population and waste volume. The Brevard County solid waste management department (Central Florida, US) acknowledges its consistent ranking among the top 10 emitters in the US but emphasizes the difficulty in addressing the issue. Expediting the installation of a gas collection system in the landfill expansion could be a viable solution, though it depends on funding availability and proper timing (Bruggers, 2021).

These highlight the need for regulatory frameworks that offer opportunities by providing incentives and support for the adoption of innovative gas collection systems. Governments may extend grants, tax incentives, or subsidies to encourage landfill operators to embrace these technologies. Such support offsets implementation costs, facilitating the adoption of advanced systems. These could lead to more aggressive efforts being made by landfill operators to address methane emissions. While challenges remain, operators are exploring various strategies and utilizing alternative reporting options to demonstrate their commitment to reducing emissions from their facilities.

8.6 Conclusion

Waste management would not be possible without landfills. Even though sending garbage to a landfill is the least ideal option in the waste management hierarchy, enormous volumes of garbage end up there every year. In landfills, the waste decomposes via a series of stages, releasing gases including methane (CH_4). In the US, landfills account for 3% of all CH_4 emissions. While LFGs are often seen as a burden to the environment, their energy worth may be captured and recovered to transform this liability into an asset. Not only will it reduce environmental concerns, but it will also generate revenue and reduce reliance on non-renewable energy sources. This article presents new techniques for LFG harvesting. The effectiveness of methane capture may be improved using real-time monitoring systems and modern strategic control. Gas collecting system designs such as horizontal wells, vertical wells, and slotted wells are contrasted. The use of gas-permeable barriers and vacuum-assisted gas-collecting systems is also covered. There are technological, financial, and societal obstacles to recovering and using CH_4 from landfills. Unpredictability in LFG generation, limited CH_4 output, and contaminants are the primary technical hurdles. The main economic barrier is the high upfront investment to build an LFG processing complex for final applications. Lack of public understanding regarding the advantages of LFG may also present social difficulties.

References

Al-Baghli, N. A., et al. (2001). A rate-based model for the design of gas absorbers for the removal of CO_2 and H_2S using aqueous solutions of MEA and DEA. *Fluid Phase Equilibria, 185*, 31–43.

Ayandele, E., et al. (2022). Top strategies to cut dangerous methane emissions from landfills. *RMI*. [Online]. Available: https://rmi.org/top-strategies-to-cut-dangerous-methane-emissions-from-landfills/. Accessed: May 16, 2023.

Ayodele, T. R., et al. (2020). Effect of collection efficiency and oxidation factor on greenhouse gas emission and life cycle cost of landfill distributed energy generation. *Sustainable Cities Society,* 52.

Bailey, M. (2022). Backgrounder: Methane emissions from waste. *Anaergia*. [Online]. Available: https://www.anaergia.com/backgrounder-methane-emissions-from-waste/. Accessed: May 18, 2023.

Bernardo, P., et al. (2009). Membrane gas separation: A review/state of the art. *Industrial and Engineering Chemistry Research, 48*, 4638–4663.

Bruggers, J. (2021). Cutting landfill methane to fight climate change. *NPR*. [Online]. Available: https://www.npr.org/2021/07/13/1012218119/epa-struggles-to-track-methane-from-landfills-heres-why-it-matters-for-the-clima. Accessed: May 16, 2023

Cai, B., et al. (2018). CH4 mitigation potentials from China landfills and related environmental co-benefits. *Science Advances, 4*, 1–9.

Castro, A. (2015). Extending renewable electricity generation for decades at closed landfill sites. *Renewable Energy World*. [Online]. Available: https://www.renewableenergyworld.com/baseload/extending-renewable-electricity-generation-for-decades-at-closed-landfill-sites/. Accessed: May 18, 2023.

Chetri, J. K., et al. (2020). Carbon-dioxide and hydrogen-sulfide removal from simulated landfill gas using steel slag. *Journal of Environment Engineering, 146*.

Chetri, J. K., & Reddy, K. R. (2021) Methane recovery from landfills. In W. Guo, et al., (Eds.), *Sustainable resource management: Technologies for recovery and reuse of energy and waste materials*, 1st edn (pp. 699–722). WILEY-VCH GmbH.

Dickie, G. (2022) Landfills around the world release a lot of methane. *Reuters*. [Online]. Available: https://www.reuters.com/business/environment/landfills-around-world-release-lot-methane-study-2022-08-10/. Accessed: May 18, 2023.

Duan, Z., et al. (2022). Efficiency of gas collection systems at Danish landfills and implications for regulations. *Waste Management, 139*, 269–278.

European Commission. (2022). Methane emissions. *European Commision*. [Online]. Available: https://energy.ec.europa.eu/topics/oil-gas-and-coal/methane-emissions_en. Accessed: May 18, 2023.

Fallah, B., et al. (2020). Application of a multi-stage neural network approach for time-series landfill gas modeling with missing data imputation. *Waste Management, 116*, 66–78.

FAO. (2013). News Article: Key facts and findings. *Food and agriculture organization of the United Nations*. [Online]. Available: https://www.fao.org/news/story/en/item/197623/icode/. Accessed: May 18, 2023.

Fay, C., et al. (2011). Remote real-time monitoring of subsurface landfill gas migration. *Sensors, 11*, 6603–6628.

Ghanbarpour Mamaghani, Z., et al. (2023). Adsorption of CO_2 using biochar - Review of the impact of gas mixtures and water on adsorption. *Journal of Environmental Chemical Engineering, 11*, 109643.

Hassan, T.N.A.T., et al. (2022). Insights on cryogenic distillation technology for simultaneous CO2 and H2S removal for sour gas fields. *Molecules,* 27.

Jung, Y., et al. (2011). Mitigating methane emissions and air intrusion in heterogeneous landfills with a high permeability layer. *Waste Management, 31*, 1049–1058.

Kasulla, S., et al. (2021). A retrospection of hydrogen sulphide removal technologies in biogas purification. *International Journal of Trend in Scientific Research and Development, 5*, 857–863.

Kikkawa, S., et al. (2022). Direct air capture of CO_2 using a liquid amine-solid carbamic acid phase-separation system using diamines bearing an aminocyclohexyl group. *ACS Environmental Au, 2*, 354–362.

Kim, S., & Lee, Y. M. (2013). High performance polymer membranes for CO2 separation. *Current Opinion in Chemical Engineering, 2*, 238–244.

Lee, U., et al. (2017). Evaluation of landfill gas emissions from municipal solid waste landfills for the life-cycle analysis of waste-to-energy pathways. *Journal of Cleaner Production, 166*, 335–342.

Maasakkers, J. D., et al. (2022). Using satellites to uncover large methane emissions from landfills. *Science Advances, 8*.

Macias, D., et al. (2014). Impact of changes in barometric pressure on landfill methane emission. *Global Biogeochemical Cycles, 28*, 679–695.

Moioli, S., et al. (2013). Improved rate-based modeling of H_2S and CO2 removal by methyldiethanolamine scrubbing. *Industrial and Engineering Chemistry Research, 52*, 2056–2065.

National Academic of Sciences. (2018). Addressing uncertainties in anthropogenic methane emissions. In *Improving characterization of anthropogenic methane emissions in the United States* (pp. 139–170). National Academies Press

Netušil, M., & Ditl, P. (2012). Natural gas dehydration. In *Natural gas-extraction to end use* (pp. 3–22).

Ocko, I. B., et al. (2021). Acting rapidly to deploy readily available methane mitigation measures by sector can immediately slow global warming. *Environmental Research Letters*, 16.

Parameswaran, T. G., et al. (2023). Gas collection system design for a landfill via three-dimensional stochastic waste heterogeneity models and kriging. *Journal of Environmental Chemical Engineering, 11*, 110563.

Pardakhti, M., et al. (2019). Trends in solid adsorbent materials development for CO_2 capture. *ACS Applied Materials & Interfaces, 11*, 34533–34559.

Pellegrini, L. A., et al. (2018). Biogas to liquefied biomethane via cryogenic upgrading technologies. *Renewable Energy, 124*, 75–83.

Powell, J. T., et al. (2016). Estimates of solid waste disposal rates and reduction targets for landfill gas emissions. *Nature Clinical Practice Endocrinology & Metabolism, 6*, 162–165.

Rajesh, S., et al. (2014). Evaluation of gas permeability and mechanical behaviour of soil barriers of landfill cap covers through laboratory tests. *Applied Clay Science, 97–98*, 200–214.

Sarperi, L., et al. (2014). The use of an industrial by-product as a sorbent to remove CO2 and H2S from biogas. *Journal of Environmental Chemical Engineering, 2*, 1207–1213.

Scheutz, C., et al. (2009). Microbial methane oxidation processes and technologies for mitigation of landfill gas emissions. *Waste Manag. Res., 27*, 409–455.

SEPA. (2004). *Guidance on gas treatment technologies for landfill gas engines*.

Siddiqui, F. Z., et al. (2022). Pilot demonstration of clean technology for landfill gas recovery in India—A case study. *Cleaner Chemical Engineering, 2*, 100024.

Spokas, K., et al. (2006). Methane mass balance at three landfill sites: What is the efficiency of capture by gas collection systems? *Waste Management, 26*, 516–525.

Sullivan, P. (2010). *The importance of landfill gas capture and utilization in the U.S.*

Tate, R. D. (2022). Bigger than oil or gas? Sizing up coal mine methane. *Global Energy Monitor*. [Online]. Available: https://globalenergymonitor.org/wp-content/uploads/2022/03/GEM_CCM2022_final.pdf

Tauseef, S. M., et al. (2013). Methane capture from livestock manure. *Journal of Environmental Management, 117*, 187–207.

US ACE. (2008). *Engineering and design: Landfill off-gas collection and treatment systems*.

US ATSDR. (2001a). Monitoring of landfill gas. In Agency for Toxic Substances and Disease Registry (ATSDR) (Eds.), *Landfill gas primer—An overview for environmental health professionals*. United States Department of Health and Human Services (DHHS).

US ATSDR. (2001b). *Landfill gas control measures U.S. Agency for Toxic Substances and Disease Registry (ATSDR).*

US EPA. (2001). *Best practices for landfill gas collection system design and installation.*

US EPA. (2011). Available and emerging technologies for reducing greenhouse gas emissions from municipal solid waste landfills. *United States Environmental Protection Agency.*

US EPA. (2012). *International best practices guide for LFG energy projects.*

US EPA. (2017). *Landfill gas energy basics.*

US EPA. (2019). Basic information about biosolids. *U.S. Environmental Protection Agency*. [Online]. Available: https://www.epa.gov/biosolids/basic-information-about-biosolids. Accessed: May 09, 2023.

US EPA. (2021a). Overview of greenhouse gases. *United States Environmental Protection Agency*. [Online]. Available: https://www.epa.gov/ghgemissions/overview-greenhouse-gases#methane. [Accessed: May 18, 2023.

US EPA. (2021b). Greenhouse gas inventory data explorer. *United States Environmental Protection Agency*. [Online]. Available: https://cfpub.epa.gov/ghgdata/inventoryexplorer/#iallsectors/allsectors/methane/inventsect/current. Accessed: May 18, 2023.

US EPA. (2021c). Basic information about landfill gas. *United States Environmental Protection Agency (EPA).* [Online]. Available: https://www.epa.gov/lmop/basic-information-about-landfi ll-gas. Accessed: May 18, 2023.

US EPA. (2021d). *Landfill gas energy project development handbook.*

USA EPA. (2015). EPA may underestimate landfill methane. *Scientific American*. [Online]. Available: https://www.scientificamerican.com/article/epa-may-underestimate-landfill-met hane/. Accessed: May 16, 2023

Wang, Q., et al. (2022). Gas transport in landfill cover system: A critical appraisal. *Journal of Environmental Management, 321*, 116020.

Winn, Z. (2022). Reducing methane emissions at landfills. *Massachusetts Institute of Technology*. [Online]. Available: https://news.mit.edu/2022/loci-methane-emissions-landfills-0202. Accessed: May 19, 2023.

Witynski, M. (2019). Disputed ground: The future of landfill gas-to-energy. *Waste Dive*. [Online]. Available: https://www.wastedive.com/news/disputed-ground-the-future-of-landfill-gas-to-ene rgy/557706/. [Accessed: May 18, 2023.

Wong, C. L. Y., & Zawadzki, W. (2023). Emissions rate measurement with flow modelling to optimize landfill gas collection from horizontal collectors. *Waste Management, 157*, 199–209.

Xia, T., et al. (2023). Mobile measurements of atmospheric methane at eight large landfills: An assessment of temporal and spatial variability. *Atmosphere (Basel)., 14*, 906.

Yang, N., et al. (2013). Greenhouse gas emissions during MSW landfilling in China: Influence of waste characteristics and LFG treatment measures. *Journal of Environmental Management, 129*, 510–521.

Yu, L., et al. (2009). Gas flow to a vertical gas extraction well in deformable MSW landfills. *Journal of Hazardous Materials, 168*, 1404–1416.

Zhan, L. T., et al. (2016). Field measurement of gas permeability of compacted loess used as an earthen final cover for a municipal solid waste landfill. *Journal of Zhejiang University-SCIENCE A*, 17, 541–552.

Zheng, Q. T., et al. (2018). Design of vertical landfill gas collection wells considering non-homogeneity with depth. *Waste Management, 82*, 26–36.

Zietsman, J., et al. (2008). Feasibility of landfill gas as a liquefied natural gas fuel source for refuse trucks. *Journal of the Air and Waste Management Association, 58*, 613–619.

Chapter 9
Air Pollution Control Strategies for Sustainable Landfill Management

Mohammed Abdillah Ahmad Farid, Mohd Nor Faiz Norrahim, Jacqueline Lease, and Yoshito Andou

Abstract The integration of waste management and air pollution control strategies is essential for sustainable landfill management. Landfill sites are a significant source of air pollution, and current waste management practices and air pollution control technologies have limitations in reducing emissions. This article provides an overview of current practices and challenges in landfill management and discusses principles of sustainable landfill management. The article identifies key opportunities for integrating waste management and air pollution control strategies, including innovative approaches to landfill design and operation and advanced air pollution control technologies. The benefits of integrated strategies are discussed, including reduction of air pollution and associated health risks, energy recovery, and greenhouse gas emissions reduction, as well as economic and social benefits. Finally, the article presents challenges to the implementation of integrated strategies and identifies future research directions, emerging technologies, and trends. This article provides valuable information for researchers, policymakers, and decision-makers interested in sustainable waste management and air pollution control.

M. A. Ahmad Farid
Institute of Tropical Forestry and Forest Products (INTROP), Universiti Putra Malaysia, Serdang, Selangor, Malaysia
e-mail: abdillahfarid@upm.edu.my

M. N. F. Norrahim
Research Center for Chemical Defence, Universiti Pertahanan Nasional Malaysia, Kem Sungai Besi, Kuala Lumpur, Malaysia
e-mail: faiz@upnm.edu.my

J. Lease · Y. Andou (✉)
Department of Biological Functions Engineering, Graduate School of Life Science and Systems Engineering, Kyushu Institute of Technology, Kitakyushu, Fukuoka, Japan
e-mail: yando@life.kyutech.ac.jp

J. Lease
e-mail: lease.jacqueline708@mail.kyutech.jp

Y. Andou
Collaborative Research Centre for Green Materials on Environmental Technology, Kyushu Institute of Technology, Kitakyushu, Fukuoka, Japan

A. Anouzla et al. (eds.), *Handbook of the Effects of Landfill Emissions on Sustainable Development*, https://doi.org/10.1007/978-3-032-01951-6_9

9.1 Introduction

Landfill operations involve the deposition of waste in engineered cells, followed by the application of cover materials. These processes can result in emissions of a range of pollutants, including volatile organic compounds (VOCs), methane, and particulate matter (PM) (Karthikeyan et al., 2011). Current air pollution control technologies, such as gas collection systems and flares, are effective at reducing some of these emissions, but they are not always sufficient or efficient (Winn, 2022). Moreover, their waste management practices often prioritize cost over sustainability, resulting in suboptimal waste diversion and energy recovery.

The accumulation of waste at landfills is a global environmental issue that has been linked to various environmental concerns, including air pollution. According to the United States Environmental Protection Agency (EPA), landfills are the third-largest source of human-related methane emissions in the US, accounting for approximately 15% of total methane emissions in 2021 (US EPA, 2021a). The World Bank estimates that methane emissions from landfills represent around 5% of global greenhouse gas emissions (World Bank, 2018). In addition to methane, landfills also emit other volatile organic compounds, such as benzene and toluene, which can have negative health effects (Moolla et al., 2013). A study conducted in Italy found that people living within 2 km of a landfill had a higher risk of hospitalization for respiratory diseases (Fazzo et al., 2023). The World Health Organization (WHO) estimates that outdoor air pollution is responsible for around 4.2 million premature deaths worldwide each year (WHO, 2022). While landfills are just one source of outdoor air pollution, they can contribute to this overall figure. In terms of specific regions, a study conducted in India found that landfill emissions accounted for 4% of the nation's total greenhouse gas emissions (ICLEI, 2018).

As the volume of waste continues to increase, so does the amount of emissions generated by landfills. According to the World Bank, global waste generation is expected to rise from 2.01 billion tonnes in 2016 to 3.40 billion tonnes by 2050 (World Bank, 2018). This increase in waste production will likely lead to a corresponding increase in emissions generated by landfills. The emissions from landfills can also contribute to climate change by releasing methane, a potent greenhouse gas, into the atmosphere. Methane, one of the gases produced by decomposing waste in landfills, is a potent greenhouse gas with a global warming potential 28 times higher than carbon dioxide over a 100-year period (European Commission, 2022). Landfills are the third-largest source of anthropogenic methane emissions in the United States, accounting for approximately 15% of total emissions (US EPA, 2021b).

Moreover, landfill sites are often located in low-income or marginalized communities, which are disproportionately affected by the adverse health effects of air pollution. This raises concerns about environmental injustice and the need for equitable distribution of environmental risks and benefits. A study conducted in Nigeria found that residents living near a landfill had higher levels of respiratory and skin diseases compared to those living further away from the landfill (Abiola et al., 2021). Similarly, a study in Malaysia found that residents living within 3 km of a landfill had higher rates of respiratory illness compared to those living further away (Esphylin et al., 2018). Furthermore, landfill sites are often located in low-income or marginalized communities. A study in the United States found that zip codes with a higher percentage of minority and low-income residents had a greater density of landfills and waste facilities compared to zip codes with higher-income and predominantly white residents (Taylor et al., 2007). This disproportionate siting of landfills in marginalized communities can exacerbate environmental injustice and result in increased exposure to harmful pollutants.

The objective of this article is to explore the principles, challenges, and benefits of integrating waste management and air pollution control strategies, and to provide case studies and best practices that demonstrate the potential of this approach. By considering waste management and air pollution control as interrelated elements of a system, rather than as independent processes, we can develop more effective and sustainable solutions to the challenges posed by landfills.

9.2 Management of Municipal Solid Waste in a Global Scenario

9.2.1 Global Demographic Waste Generation

Figure 9.1 shows a global pattern of increasing amounts of municipal solid garbage. There are now three nations that produce the most municipal solid waste: the United States (258 million metric tonnes), China (220 million metric tonnes), and India (169 million metric tonnes) (Alves, 2023; Sebastian, 2020). Large urban populations that are promptly thriving and adopting high-consumption lifestyles comparable to those of wealthy countries have propelled China, India, Brazil, Indonesia, and Mexico into the top ten list of countries that generate the most municipal solid waste. Municipal solid waste production is directly proportional to a country's level of economic and social development; as a result, developed nations produce more garbage per person than less developed countries or those in the third world. By 2025, most developing countries are forecast to see a sharp uptick in their rates of producing municipal solid trash per capita, as shown in Fig. 9.2. This is in stark contrast to the developed nations, where these rates are forecast to climb or decline more gradually (Nanda & Berruti, 2021).

Municipal solid waste production by the world income group is shown in Fig. 9.3. Both as it is 2020 and as it is predicted to be in 2025. As of 2020, those high-income nations now create more municipal solid waste than low-income countries. Lower-middle-income countries include some of the world's most populous countries, such as India, Indonesia, Pakistan, and Nigeria, therefore, their daily generation of municipal solid wastes is significantly higher than that of upper-middle-income countries. Countries with lower middle incomes are among the top 20 producers of municipal solid trash (Hoornweg & Bhada-Tata, 2012). However, due to fast population growth,

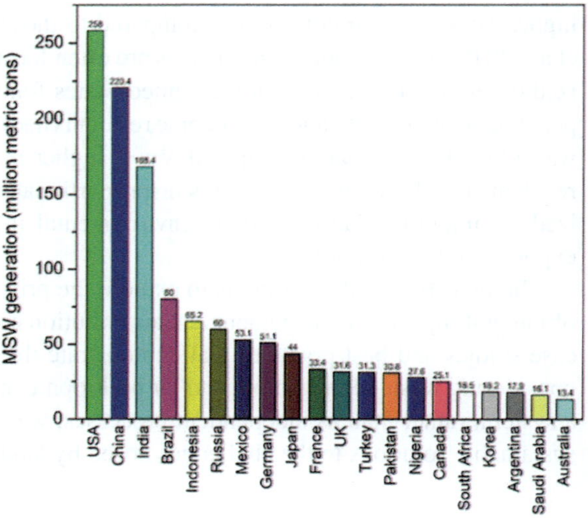

Fig. 9.1 Worldwide generation of municipal solid waste. It should be noted that the top five solid-waste-generating countries are the USA, China, India, Brazil, and Indonesia. Adopted from ref. (Nanda & Berruti, 2021)

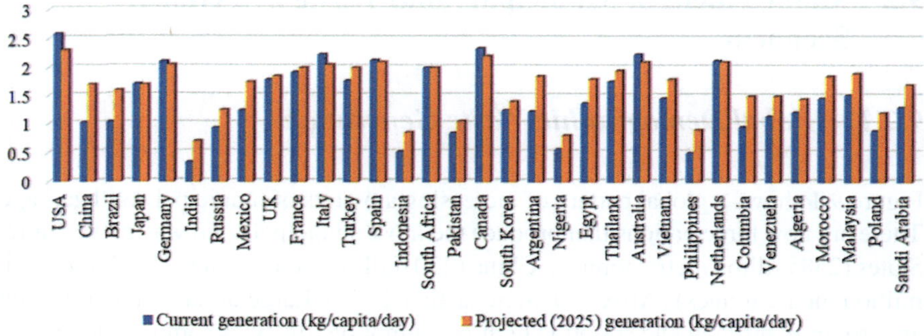

Fig. 9.2 Current and projected generation of global municipal solid waste per capita

additional waste will be produced in 2025. This is especially true in countries like India and China, which both have large and relatively poor rural populations with extreme low-income groups.

9.2.2 Typical Waste Compositions

Biodegradable and non-biodegradable fractions from organic and inorganic materials, respectively, make up a large portion of municipal solid waste, although its composition and categorization greatly vary across various municipalities throughout the globe. Table 9.1 lists the common components of municipal solid waste, which includes food scraps, yard trimmings, paper, cardboard, plastic, rubber, metal, glass, electronics, inert materials, and other random trash. The organic component of MSW

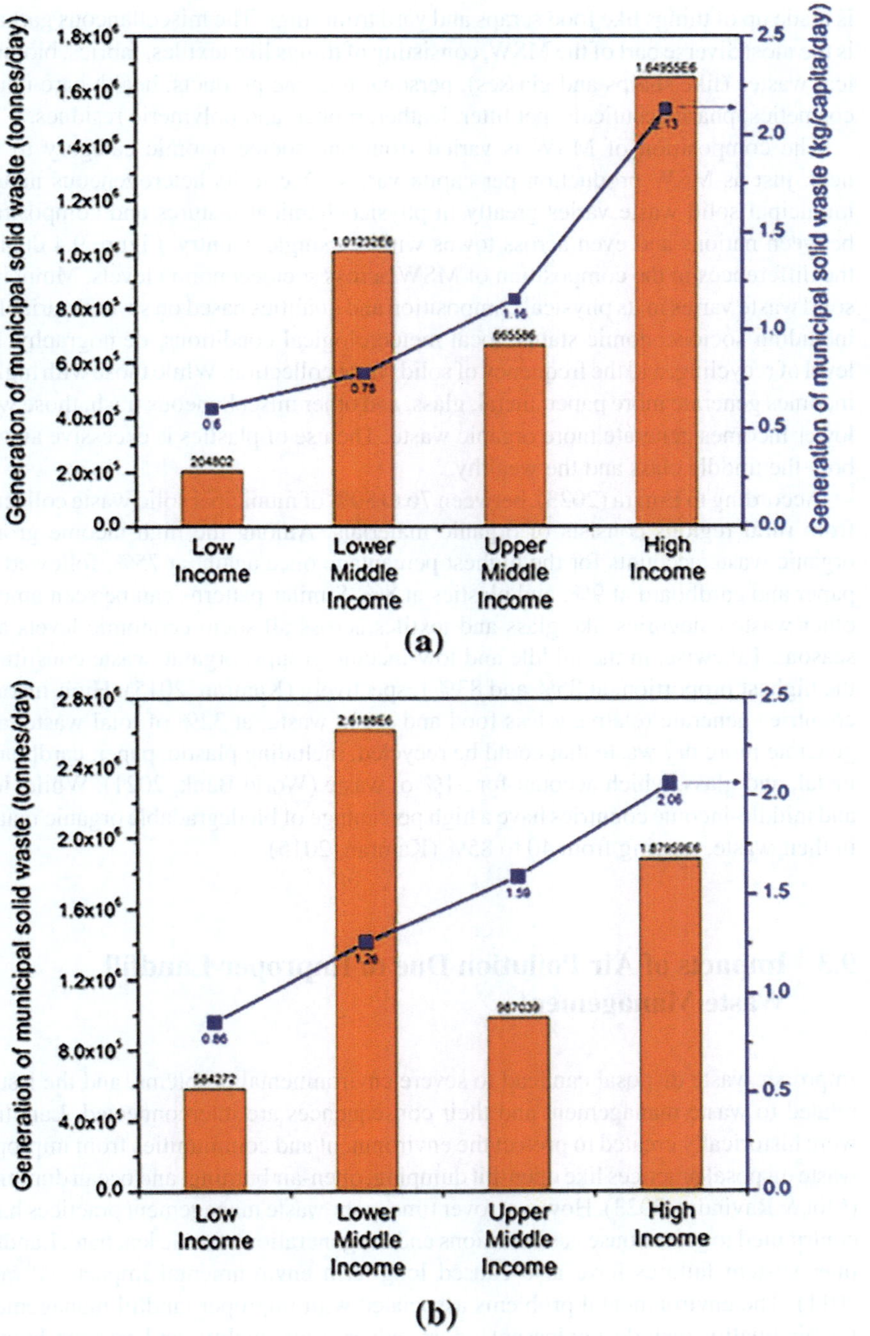

Fig. 9.3 Rates of MSW production in various countries and regions, as a function of per capita income. **a** 2020 estimates and **b** projected estimates in 2025. Adopted from ref. (Hoornweg & Bhada-Tata, 2012)

is made up of things like food scraps and yard trimmings. The miscellaneous garbage is the most diverse part of the MSW, consisting of things like textiles, fabrics, biomedical wastes (like sharps and glasses), personal hygiene products, health care items, cosmetics, pharmaceuticals, pet litter, leather, rubber, and polymeric residues.

The composition of MSW is varied from one socioeconomic category to the next, just as MSW production per capita varies. Due to its heterogeneous nature, municipal solid waste varies greatly in physicochemical features and composition between nations and even across towns within a single country. Figure 9.4 depicts the differences in the composition of MSW across socioeconomic levels. Municipal solid waste varies in its physical composition and qualities based on several variables, including socioeconomic status, local meteorological conditions, demography, the level of recycling, and the frequency of solid waste collection. While those with higher incomes generate more paper, metal, glass, and other miscellaneous trash, those with lower incomes generate more organic waste. The use of plastics is excessive among both the middle class and the wealthy.

According to Emara (2023), between 76 to 80% of municipal solid waste collected from rural regions consists of organic materials. Among the high-income group, organic waste accounts for the highest percentage once again, at 75%, followed by paper and cardboard at 9%, and plastics at 8%. Similar patterns can be seen among other waste categories like glass and textiles across all socio-economic levels and seasons. Likewise, in the middle and low-income groups, organic waste constitutes the highest proportion, at 85% and 83% respectively (Kamran, 2015). High-income countries generate relatively less food and green waste, at 32% of total waste, and generate more dry waste that could be recycled, including plastic, paper, cardboard, metal, and glass, which account for 51% of waste (World Bank, 2021). While, low and middle-income countries have a high percentage of biodegradable organic matter in their waste, ranging from 40 to 85% (Kamran, 2015).

9.3 Impacts of Air Pollution Due to Improper Landfill Waste Management

Improper waste disposal can lead to severe environmental problems, and the issues related to waste management and their consequences are interconnected. Landfills were historically created to protect the environment and communities from improper waste disposal practices like open-pit dumping, open-air burning, and ocean dumping (Mor & Ravindra, 2023). However, over time, poor waste management practices have contributed to greenhouse gas emissions and the generation of toxic leachate. Landfill liner system failures have also caused long-term environmental impacts (Pivato, 2011). The environmental problems associated with improper landfill management for air quality include unpleasant odors, adverse air quality, and various human health risks. A diagram illustrating the environmental issues connected with improper landfill site management is shown in Fig. 9.5.

Table 9.1 Typical composition of municipal solid waste by component type

Waste components	Compositions
Kitchen waste	• Food waste (e.g., spoiled meat, fish, bones, eggshells, crab, mussel and lobster shells, and expired solid food) • Agro-food residues (e.g., vegetable refuse, fruit shells, corncob, husk, kernels, peels, rind, seeds) • Coffee residues and waste tea leaves
Yard waste	Leaves, grass, tree trimmings, twigs, bark, straw, stalks, charred wood
Paper and cardboard	Newsprint, advertisement flyers, magazines, books, tissue paper, parchment paper, thermal paper, copy paper, multiuse paper, shredded office paper, folders, paper bags, cardboard boxes, packaging boxes, corrugated fiberboard
Plastic and rubber	• Low-density polyethylene (e.g., shampoo bottles, detergent bottles, edible oil containers, and plastic cans) • High-density polyethylene (e.g., three-dimensional printer filament, bottle caps, coax cable insulation, electrical plumbing boxes, food storage containers, shoe last, plastic lumber, piping for water and sewer, plastic surgery and storage sheds) • Polypropylene (e.g., bottles, chairs, and desks) • Polyethylene terephthalate (e.g., potable water bottles and beverage bottles) • Polyethylene (e.g., low-grade polythene bags, plastic packaging) • Polycarbonate (water and milk packaging bottles/cans, compact disks, digital versatile disks, and safety goggles) • Polyvinyl chloride (e.g., vinyl records, wire rope, construction flooring, ceiling tiles, home playground, foam, and toys) • Polystyrene (e.g., flexible plastics, plastic cups) • Clear warp; ziplock bags; rubber bands
Metal	Food packaging cans, cans, aluminum foil, cookware, metal utensils and culinary equipment, knives, wires, fences, bottles, lids
Glass	Food storage containers, glass bottles, casserole dishes, utensils, light bulbs, home décor, decorative and framed mirrors
Electronic waste	• Dead batteries • Electronic devices (e.g., thrashed computers, monitors, laptops, tablets, mobile phones, sound systems, clocks, and watches)
Miscellaneous	• Ceramics • Discarded clothes • Rags (e.g., torn bags, tote, and luggage) • Biomedical wastes (e.g., needles, syringes, and broken household health devices) • Pharmaceuticals (e.g., medicine, pills, capsules, creams, and cosmetics products) • Diapers; sanitary napkins; contraception • Pet litter • Leather and textiles • Carpet
Inert materials	• Broken and discarded furniture • Stones, soil, silt, concrete, ash, dust, drywall, and other inorganic materials • Construction, demolition, and renovation wastes

Adopted from ref. (Nanda & Berruti, 2021)

Fig. 9.4 Composition of municipal solid waste according to income level. Adopted from ref. (Kumar & Samadder, 2017)

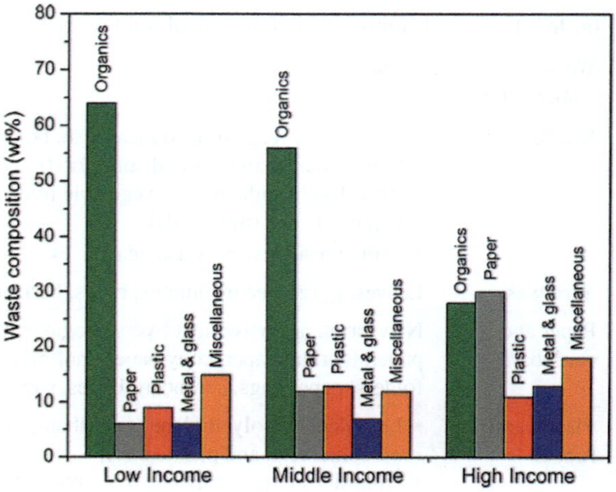

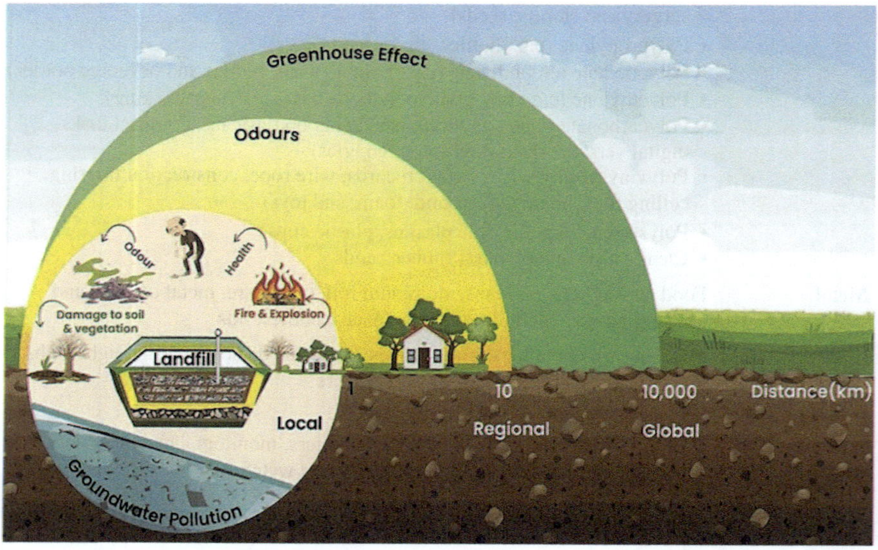

Fig. 9.5 Distance-based analysis of the environmental effects of landfills. Adopted from ref. (Mor & Ravindra, 2023)

9.3.1 Odor Issues

The main odorous substances emitted from landfill sites are styrene, toluene, xylene, acetone, methanol, n-butanone, n-butyl aldehyde, acetic acid, dimethyl sulfide, dimethyl disulfide, and others. However, H_2S and NH_4 are responsible for most of the odors at landfills (Fang et al., 2012). These odors are spread by strong winds, impacting nearby settlements, and affecting inhabitants' well-being. Extended exposure to landfill odours can lead to psychological discomfort, including anxiety and unease, as well as physical symptoms such as eye irritation, headaches, nausea, and respiratory issues (ASTDR, 2009). To address these issues, odor management has become essential in regulations to protect the environment and human health. Studies utilize weather data and air dispersion modeling to assess pollution and odor exposures around landfill sites. Alternative daily cover (ADC) materials and odor emission factors (OEFs) are employed to mitigate odorous emissions and control odor levels (Siddiqua et al., 2022).

9.3.2 Climate Change

Landfilling of waste produces significant gaseous emissions, including greenhouse gases (GHGs), through the anaerobic decomposition of organic materials present in the solid waste (US EPA, 2021b). The waste sector, including landfills, contributes around 3.2% of GHGs emissions globally (European Commission, 2020). Landfills contribute about 15% to global anthropogenic CH_4 emissions, which is a potent GHG. Open burning of MSW is also an important contributor to particulate matter and air pollutant emissions (Gómez-Sanabria et al., 2022). In the United States, waste activities generated emissions of 163.7 million metric tons (MMT) of CO_2 equivalent in 2019, accounting for 2.5% of total U.S. GHG emissions (US EPA, 2021c).

CH_4 is a major component of LFG, contributing to approximately 45–60% of LFG composition, along with 40–60% of CO_2. Other minor constituents include N_2, O_2, $SO_4{}^{2-}$, NH_4, H_2, and CO (ATSDR, 2001). Global GHG emissions are predicted to increase by up to 76% by 2050, partly due to uncontrolled waste disposal (Monni et al., 2006). Volatile organic compounds (VOCs) are emitted from landfills and can significantly impact tropospheric chemistry (Nair et al., 2019). Inadequate landfill waste management release trace gases like N_2, sulfur, and particulate matter (PM) into the air (Siddiqua et al., 2022). These particles can include dust, ash, and other fine particulate matter, which can contribute to air pollution and have adverse health effects when inhaled.

Polybrominated diphenyl ethers (PBDEs) are found in waste disposal sites and may leach from waste due to various reactions. PBDEs leach from waste because they have weak chemical bonds and are susceptible to physical, chemical, and biological factors during waste disposal (Zhang et al., 2021). E-waste recycling sites have been identified as a source of PBDE emissions into the ambient air during reclamation processes (Ghimire et al., 2019). Landfills also create environmental microbiological concerns, leading to the formation of persistent organic pollutants (POPs) like dioxins and polycyclic aromatic hydrocarbons (PAHs) (Weber et al., 2008).

9.3.3 Health Risks

LFG contains VOCs like BTEX compounds (benzene, toluene, ethyl-benzene, and xylene isomers) with carcinogenic and teratogenic properties. Some VOCs can also harm the central nervous and reproductive systems and affect the lungs, liver, and kidneys (Lakhouit et al., 2016; Minnesota Pollution Control Agency, 2023). Landfills are also associated with hazardous air pollutants (HAP) like dioxins and furans, which are carcinogenic (Davoli et al., 2010). Landfill plumes can lead to health issues such as diarrhea, dysentery, vomiting, abdominal pain, and hepatitis (Mor & Ravindra, 2023).

9.3.4 Water Resources

Poor waste management practices, including inadequate landfill waste management, can lead to the contamination of drinking water sources. This contamination can have indirect effects on air quality as pollutants from the landfill can leach into the groundwater and potentially affect the surrounding environment. LFG generation can cause water pollution due to CO_2 dissolution in water, leading to carbonic acid formation, lowering pH, and affecting water hardness and mineral content (Nolasco et al., 2008). CH_4 is less dense than air, and CO_2 is heavier, causing CO_2 to migrate toward the landfill's bottom, reaching groundwater. LFG also emits hazardous gases like volatile hydrocarbons and vinyl chloride, posing a threat to air and groundwater resources (Mor & Ravindra, 2023). These substances interact with groundwater, forming an aqueous-gaseous phase that contaminates groundwater supplies.

9.3.5 Soil and Vegetation

Gases, including LFG, are one of the major causes of poor vegetation growth in landfills, but other soil characteristics can also inhibit growth (Flower et al., 1981). Landfills can render surrounding soil and land unusable, and inadequate management of landfills can lead to issues concerning leachate collection and LFG generation, which give rise to groundwater contamination and air pollution (Ozbay et al., 2021). Improper landfill gas management can lead to gas migration and expose plant roots to high CO_2 and CH_4 concentrations, causing damage due to oxygen deprivation in the root zone (see Fig. 9.6). Methanotrophs, CH_4-consuming bacteria, contribute to oxygen deprivation by oxidizing CH_4 near the surface (Mor & Ravindra, 2023). Soil temperature rises due to CH_4 oxidation's release of heat, which may lead to a lack of oxygen for plants (Attalage et al., 2023; Iravanian & Ravari, 2020). CO_2 generated by CH_4 oxidation and landfill gas may be detrimental to plant growth, particularly when present in high concentrations (30–45%) (Mor & Ravindra, 2023).

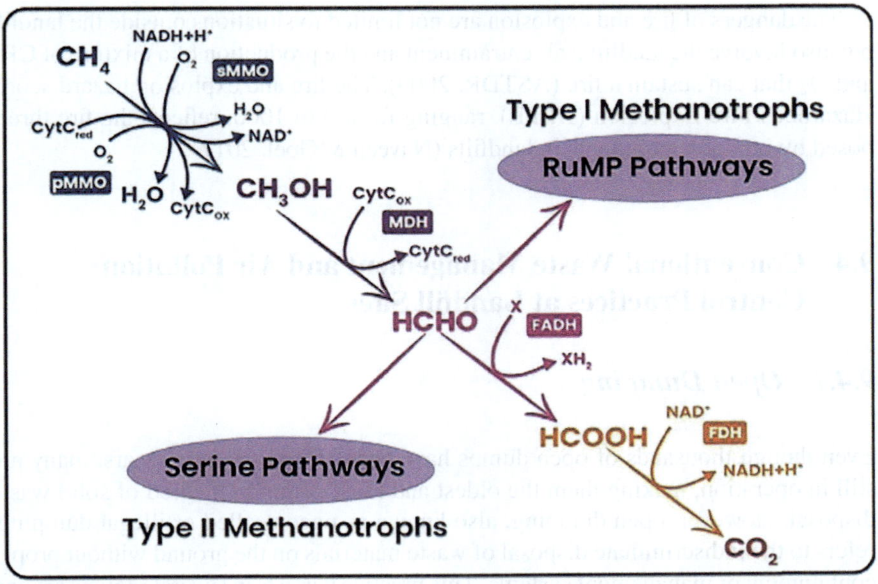

Fig. 9.6 The steps used by methanotrophic bacteria to oxidize methane. Adopted from ref. (Mor & Ravindra, 2023)

9.3.6 *Fire and Explosion Hazards*

Landfill gas (LFG) can be trapped in landfill sacks, slowly leak out, or build up pressure, leading to the risk of explosion or uncontrollable fire. The average calorific value of landfill gas is 20 MJ m^{-3}, making it suitable for conversion into electricity. However, due to its density, which is 1.04 relative to air, landfill gas has the potential to explode (Mor & Ravindra, 2023). Burning municipal solid waste (MSW) at landfill sites is a common practice to recover valuable materials and reduce waste volume.

Unsorted debris in landfills may contain combustible items that cause spontaneous fires. Landfill fires release different volatile organic compounds (VOCs), and the accumulation of CH_4 gas poses an explosion risk. CH_4 is highly flammable and can form explosive mixtures with air, with a flammability range between 5 to 15$\%_{vol}$ in enclosed spaces where people live or work (Milosevic et al., 2018). The combination of CH_4 with other components also affects its flammability or explosion potential.

When the O_2 level falls below 12.8% and the diluent is CO_2 or N_2, no combustible mixtures exist (Mor & Ravindra, 2023). However, landfill gas can interact with air during migration, leading to its depletion in CH_4 and bringing it within the flammability range. Landfill gas can migrate through pathways and enter structures or underground infrastructure near the landfill, creating gas pockets and posing a risk of explosion. Previous studies have reported fires and explosions related to gas migration from landfills, highlighting the importance of controlling landfill gases to manage environmental impacts.

The dangers of fire and explosion are not limited to situations outside the landfill but also involve the landfill's air entrainment and the production of a mixture of CH_4 and O_2 that can sustain a fire (ASTDR, 2009). The fire and explosion hazard score, Hazardous Fire Explosion (HRFE), ranging from 0 to 1000, reflects the fire threat posed by ignitable pollutants in landfills (Naveen & Goel, 2017).

9.4 Conventional Waste Management and Air Pollution Control Practices at Landfill Sites

9.4.1 Open Dumping

Even though thousands of open dumps have been closed in recent years, many are still in operation, making them the oldest and most common method of solid waste disposal. However, open dumping, also known as uncontrolled or illegal dumping, refers to the indiscriminate disposal of waste materials on the ground without proper containment or management systems. This practice persists in several regions around the world due to limited waste management infrastructure and ineffective enforcement of regulations. Unfortunately, many third-world nations have a widespread problem with unlawful and unregulated landfills, often known as open dumpsites (Ferronato & Torretta, 2019). The widespread prevalence of open dumping presents a serious threat to environmental quality, natural ecosystems, and public well-being.

Open dumping exposes waste to the elements, leading to the leaching of hazardous substances into the soil. These contaminants can infiltrate groundwater, leading to long-term pollution of water sources and the subsequent spread of pollutants to adjacent areas (Wijekoon et al., 2022). The uncontrolled waste decomposition process in open dumps generates leachate, a toxic liquid effluent containing high concentrations of heavy metals, organic compounds, and other harmful substances. When leachate percolates into nearby water bodies, it significantly deteriorates water quality and threatens aquatic ecosystems.

Open dumping contributes to the generation of GHGs, such as methane and carbon dioxide, during the decomposition of organic waste (Karthikeyan et al., 2011). These gases have potent global warming potential and are known to exacerbate climate change. Open dumps attract scavengers, including birds and animals, which disturb the waste and inadvertently scatter it across the surrounding environment (Asim et al., 2012). Such activities lead to habitat destruction and can disturb the delicate balance of local ecosystems.

Open dumping was also cited as a source of soil contamination, which negatively impacted agricultural output (Lastname et al., 2021). Infectious infections proliferate due to open waste dumps as well (Paul et al., 2019). In addition, they noted that waste mineralization happens with landfill age, which improves waste leaching qualities.

Different MSW landfilling scenarios were examined in the work by Yadav and Samadder (2018), including open dumping and unsanitary landfilling without energy recovery. It was discovered that there was a range of environmental impacts across various situations, and open dumping has contributed to the most harmful substance to the marine environment. The effect of open dumping on the water was studied, whereby most water parameters, including conductivity, total dissolved solids (TDS), hardness, and alkalinity, were found to have greatly risen owing to waste disposal (see Table 9.2). Water quality concerns such as bad odor, bacterial contamination, and discoloration were also reported. Consequently, nitrogen and phosphorus have leached from the dumpsite, leading to a rise in heavy metal pollution such as mercury, lead, etc., and leading to excessive algal blooms in rivers and streams.

Chromium (Cr) and lead (Pb) concentrations in the groundwater were three times higher than the permissible limits, indicating a significant increase compared to the past decade (Mohan & Joseph, 2021). The presence of heavy metals was attributed to hazardous waste fractions such as metal parts, chemical wastes, paint wastes, batteries, and e-waste, accumulating at the dumpsite over time. A total of 19 chemicals were found in the air sample taken from the open dumping yard (CEM & GAIA, 2018). The detected compounds include Carbon Disulphide, Carbonyl Sulphide, Propene, Chloro-methane, 1–3 Butadiene, Ethanol, Acetonitrile, Acrolein, Acetone, n-Hexane, Tetrahydrofuran, Benzene, n-Heptane, Toluene, n-Octane, Ethylbenzene, Styrene, n-Nonane and D-Limonene. Two of the compounds, 1–3 butadiene and benzene, and one of the substances, chloro-methane, were found to be at or over the US Environmental Protection Agency's lifetime acceptable exposure standards. Both benzene and 1–3 butadiene were found at concentrations more than eight times higher than the US EPA's recommended screening threshold. The reference concentration (RfC) screening limit is the maximum safe level of exposure to a chemical throughout an individual's lifetime. Of these 19 chemicals, 1 affects the blood, heart, and bone marrow; 2 affects the cardiovascular and peripheral nervous systems; 5 affect the kidneys and reproductive system; 6 affect the liver; 13 affect the skin and eyes; 15 affect the lungs and central nervous system; and 16 affect the skin and eyes.

An analysis is conducted on the soil found at the landfill to assess how the practice of open disposal has affected its characteristics. The soil is found to have elevated levels of various heavy metals, such as Pb, Ni, Cr, Cu, and Zn, making it inappropriate for any kind of activity (Ali et al., 2014). Additionally, the dumpsite is noted to

Table 9.2 Water quality parameters comparison between open dumping site with standards

Parameter	Unit	Open dumping site	Standards
pH	–	7.5–7.8	6.5–8.5
Conductivity	mS/cm	220–274	–
TDS	mg/l	771–1440	500
Total hardness	mg/l	256–306	300
Alkalinity	mg/l	215–270	200
Chloride	mg/l	183–243	250
Residual chlorine	mg/l	0.26	0.2

Adopted from ref. (Mohan & Joseph, 2021)

trigger swift acidification of the soil. Fluctuations are observed in the concentrations of essential plant nutrients like calcium, potassium, and magnesium.

9.4.2 LFG Flaring

Landfills have long been recognized as significant sources of methane emissions, a potent greenhouse gas. Flaring all the collected LFG is an option for landfills that must or voluntarily choose to keep collecting and destroying LFG but have no plans to use energy recovery (US EPA, 2023a). This method often results in the lowest CH_4 emissions and the least offensive odors. Landfills whose gas output is on the decline may need to modify or replace their flare system. To keep the collecting system functioning as intended and able to tolerate falling gas flows, changes, and maintenance will be necessary. There is also the possibility of capping and decommissioning wells in unused parts of the landfill to stop them from adding contaminating air to the LFG collecting system.

In recent years, landfill gas flaring has emerged as a viable solution for managing these emissions while generating energy. Flaring is the process of burning off potentially explosive gases. LFG is primarily composed of methane (CH_4) and carbon dioxide (CO_2), along with trace amounts of various volatile organic compounds (VOCs) and hazardous air pollutants (HAPs) (Nanda & Berruti, 2021). The anaerobic decomposition of organic matter within landfills is the main driver of LFG generation (Spokas et al., 2006). This process creates an opportunity to harness the potential energy contained within landfill gas while mitigating its adverse environmental impacts.

When the LFG is flared, the methane is burned with oxygen to produce carbon dioxide. Combustion of LFG may produce particularly toxic gases (SO_x, NO_x, CO), depending on its composition of H_2S and N_2. When flaring, the temperature must be at least 950 °C (Ennovo, 2023). The generation of poisonous chemicals like dioxins is a problem at lower temperatures. Flaring occurs often in landfills that discard LFG instead of capturing it for reuse. Flaring LFG from anaerobic or semi-aerobic landfills (with low CH_4 content) necessitates the addition of CH_4 to the fuel mix. This treatment strategy is thus directed against anaerobic LFG.

Landfill gas flaring involves controlled combustion of the extracted LFG, converting CH_4 into CO_2 and water vapor, which are less potent GHG. The combustion process follows the principles of stoichiometry to ensure efficient conversion and appropriate flaring systems are employed to optimize combustion efficiency. Advanced flaring technologies, such as air-assisted flares and staged flaring, have shown improved performance and reduced emissions (Sorrels et al., 2019).

Landfill gas flaring plays a crucial role in reducing CH_4 emissions, thus contributing to climate change mitigation (US EPA, 2023a). Methane has a global warming potential many times greater than carbon dioxide, making its conversion through flaring a significant tool in curbing overall greenhouse gas emissions. Furthermore, LFG flaring assists in reducing odorous and toxic emissions, enhancing local air quality. Several factors influence the efficiency of landfill gas flaring systems. These include landfill gas composition, flow rate, and temperature, as well as the design and operation of the flaring equipment. Optimal efficiency can be achieved through continuous monitoring, control, and maintenance of the flaring systems.

Flares are utilized in the waste management sector, but they go by a variety of names, some of which are descriptive or historical in character and hence may be difficult to keep straight (SEPA, 2002). Many other words for flares are used, including "open," "candle," "elevated," "diffusion," "pre-aerated," "conventional," "enclosed," "shrouded," "muffled," and "ground." To distinguish between the two primary types of flare, open and enclosed are used in this chapter, and their differences are detailed in Table 9.3. Flares often eliminate over 99% of CH_4, and, by assuming no maintenance or faults, a flare will always be active even after the closure of the landfill.

Despite its potential benefits, landfill gas flaring faces challenges that need to be addressed for broader adoption. One significant challenge is the variability in LFG composition, affecting the flaring process's stability and efficiency. Additionally, the ongoing cost to operate the gas collection system and flare may hinder the implementation of LFG flaring in certain regions. Ongoing expenses for running and maintaining infrastructure, such as USD 2600 per well and USD 5100 for each flare, on average, each year. For a typical landfill using 800 cfm of LFG, the cost of running the electric blower system would be USD 74,000 per year (US EPA, 2023b). Landfills with leachate recirculation or large amounts of gas condensate incur maintenance costs for pumps to remove moisture from the system. Sites with inadequate gas quality may need to spend more on propane as a supplementary fuel to keep combustion going. Therefore, mitigating these challenges requires strategic planning and policy support.

Table 9.3 Types of flares used in landfill

Type	Description
Open flare	• Open flames are used to burn waste gas; however, a windscreen is usually installed • Combustion control, if present, is minimal • Also known as an elevated flare
Enclosed flare	• In a vertical, cylindrical, or rectilinear enclosure, landfill gas is burned with the help of a combustion control system • The enclosure is usually insulated to prevent heat loss and allow for greater temperatures of operation • Also known as ground flares

Adopted from ref. (SEPA, 2002)

9.4.3 Conventional Landfill Cover

The management of solid waste is a pressing concern in modern society, and landfills remain a primary means of disposal for non-recyclable and non-compostable materials. The landfill covers typically consist of layers that aim to prevent the infiltration of precipitation and the escape of landfill gases. The key components of such covers include the soil layer, geomembrane liner, and vegetation.

Conventional cover systems typically have one or more barrier layers to restrict infiltration and gas migration. The primary barrier is the soil layer, which serves as the principal medium for retaining moisture and impeding the migration of gases. The selection of appropriate soil types and their compaction significantly influence the overall effectiveness of the landfill cover (Chetri & Reddy, 2021a). The clay layer in the cover system of a landfill plays a crucial role in providing an effective barrier to prevent the migration of leachate and landfill gases into the surrounding environment (Wang et al., 2022). In certain cases, a geomembrane liner is employed as an additional protective measure. Although this liner enhances the barrier properties, its long-term durability and potential for degradation need to be carefully evaluated. Incorporating vegetation atop the landfill cover aids in reinforcing soil stability and promoting evapotranspiration. However, maintaining vegetation requires continuous management and can be challenging in arid climates.

While primitive covers may reduce the immediate environmental impact, they may not be entirely reliable in preventing leachate generation and the subsequent contamination of groundwater. Conventional cover systems such as high construction and maintenance costs since it is susceptible to damage due to desiccation cracking and freezing (Chetri & Reddy, 2021a). Other than that, damages can be done due to physical and biological processes which can further lead to increased infiltration. The potential for leachate escape through cracks and fissures warrants ongoing monitoring and evaluation. The structural integrity of primitive covers can degrade over time due to settlement, erosion, or weathering. The conventional covers may be less effective in managing LFG emissions compared to modern engineered covers, consisting of CH_4, CO_2, and trace amounts of non-methane organic compounds (NMOCs). The increasing concerns regarding fugitive CH_4 gas emissions from landfills have led to extensive research on alternative cover materials which can mitigate the CH_4 emissions (Wang et al., 2022). Increased methane generation can lead to safety concerns and contribute to greenhouse gas emissions (Fig. 9.7).

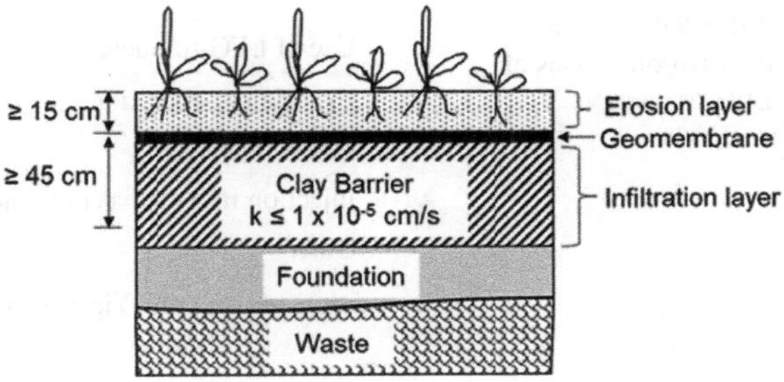

Fig. 9.7 Cross section for a conventional cover system in the landfill. Adopted from ref. (Chetri & Reddy, 2021a)

9.5 Advanced Practices of Waste Management and Air Pollution Control in Landfills

9.5.1 Landfill Gas Collection and Utilization

Landfill gas collection is demonstrated to be a cost-effective way to reduce methane emissions from landfills. Almost often, landfill gas is extracted by active collection from gas-wells in waste by applying suction pressure with a blower. The gas is then sent to either an equipment burner or a flare for removal. To be put to use, landfill gas is either burned directly in furnaces or kilns or else it is combusted in a gas engine and transformed into power. Using landfill gas is crucial because it helps to defray the expense of collecting and regulating the gas. Methane is transformed into carbon dioxide (CO_2) when burned or flared. With a stronger global warming potential than carbon dioxide, methane has a direct impact on reducing greenhouse gas emissions. This direct impact is about equivalent to 10 kg CO_2eq per m^3 of landfill gas. When used as fuel, landfill gas eliminates the need for fossil fuels. This additional decrease in emissions amounts to around 1 kg CO_2eq per m^3 (Oonk, 2012).

Today's LFG collection technology is an established technique that has been refined since the 1970s. There are different ways to use LFG to get energy from it. The most cost-effective way for landfills located within two to three km is to use the gas directly for things like heating and as fuel for boilers. Efficiency levels as high as 90% may be achieved when using LFG as direct fuel. Considering the LFG has to be extracted around the clock, the efficiency drops to 45%. However, this method is not used as much, which constituents only 20% of the energy recovered at landfills, because there is not enough demand willing to buy the gas for these purposes (see Table 9.4).

Instead, more than 70% of LFG energy recovery plants use reciprocating engines. These engines help convert the gas into power, which can be used to generate electricity. A gas/petrol engine is the most common type of engine used for this purpose (Yip & Chua, 2008). Even if the overall investment costs are much more than for direct consumption and the efficiency rate is lower (30%), it is important to take into account that energy generation may be carried out throughout the year. If the cost of producing electricity from LFG is the same as the cost of paying the utility

Table 9.4 Typical distribution in uses of LFG-to-energy

Use of LFG-to-energy	Distribution (%)
Electricity generation	72
Direct heating	21
Injection into the gas pipeline	4
Other	3

Adopted from ref (Yip & Chua, 2008)

company, then it is the most cost-effective option. While the price of power supplied to the utility company may be higher in other countries, in Germany it is only worth roughly half as much as its retail cost. Electricity generation via internal combustion engines is a common practice in many parts of the globe.

9.5.1.1 LFG-to-Energy

Landfill gas predominantly consists of CH_4 and carbon dioxide CO_2, with trace amounts of other gases like nitrogen, oxygen, and volatile organic compounds. The formation of LFG occurs through a series of complex biochemical processes, involving anaerobic decomposition of organic waste by microorganisms. As organic materials degrade in an oxygen-depleted environment, methane and carbon dioxide are generated as byproducts. Effective gas collection and extraction systems are pivotal for the success of LFG-to-energy projects. Vertical and horizontal wells are installed strategically throughout the landfill to capture the gas released during the waste decomposition process, as illustrated in Fig. 9.8. Vacuum pumps or blowers are then employed to extract the gas from these wells and direct it toward the treatment and utilization facilities.

Prior to utilization, LFG undergoes treatment to remove impurities and enhance its energy content. The treatment typically involves processes like compression, removal of moisture, hydrogen sulfide, and volatile organic compounds. These steps improve the gas quality and reduce the likelihood of equipment corrosion and environmental harm. Figure 9.9 portrays the treatment stages for LFG before utilization. A knockout pot, filter, and blower remove moisture from crude LFG during the first treatment. The secondary treatment includes moisture removal using an after-cooler, siloxane/sulfur removal, and compression. After the secondary treatment, LFG may be used to generate electricity or medium-Btu fuel. The advanced treatment eliminates CO_2, N_2, O_2, and VOCs and compresses LFG into high-Btu gas for car fuel or pipeline injection, while tail gas is flared or thermally oxidized.

Various energy conversion technologies can be employed to utilize LFG effectively. The two primary methods are combustion and anaerobic digestion. In combustion, LFG is burned directly in a boiler or internal combustion engine to produce heat, which can be used for district heating or industrial processes. Additionally, electricity can be generated through steam turbines connected to generators. In anaerobic digestion, LFG is fed into a biogas plant where it undergoes a controlled decomposition

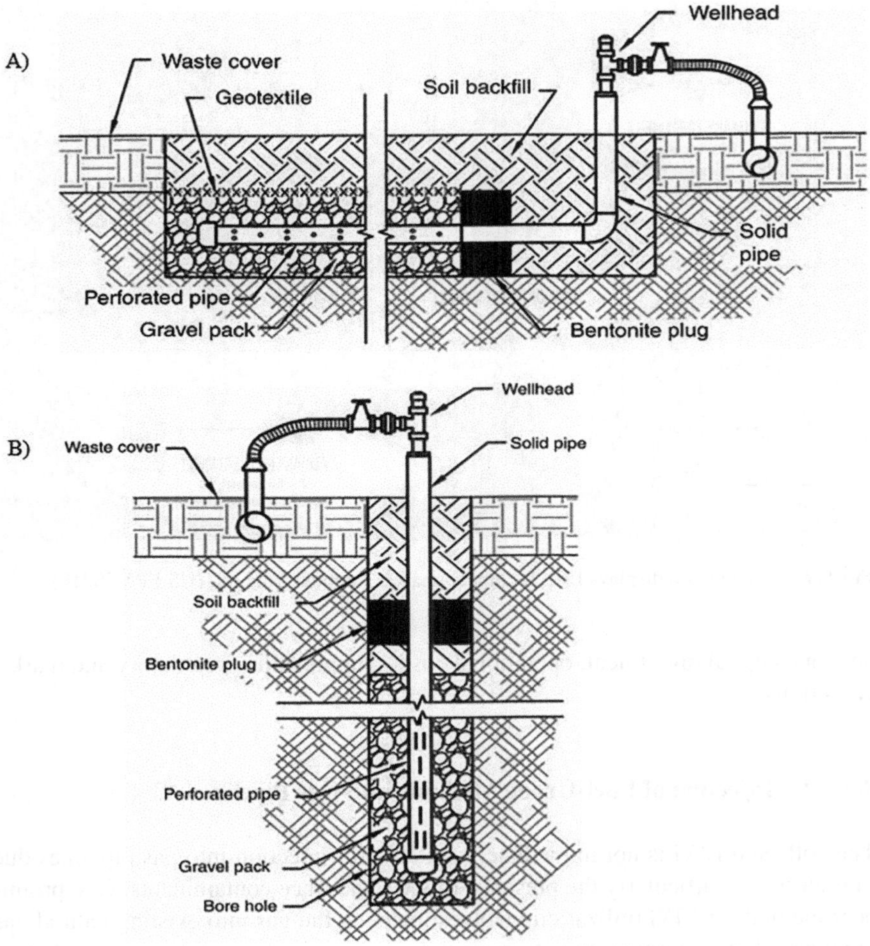

Fig. 9.8 Typical **a** horizontal and **b** vertical well configurations. Adopted from ref. (Chetri & Reddy, 2021b)

process to produce biogas. The biogas, which has a higher methane content than raw LFG, can be used for electricity generation or upgraded to biomethane for injection into the natural gas grid.

The conversion of LFG into energy offers several environmental benefits. By capturing and utilizing LFG, methane emissions are significantly reduced, mitigating the landfill's contribution to climate change. The utilization of LFG also displaces the need for fossil fuels, further lowering greenhouse gas emissions and promoting a cleaner energy mix. LFG-to-Energy projects can be financially attractive due to revenue streams from electricity sales and potential carbon credits. However, economic feasibility depends on factors such as the scale of the landfill, energy

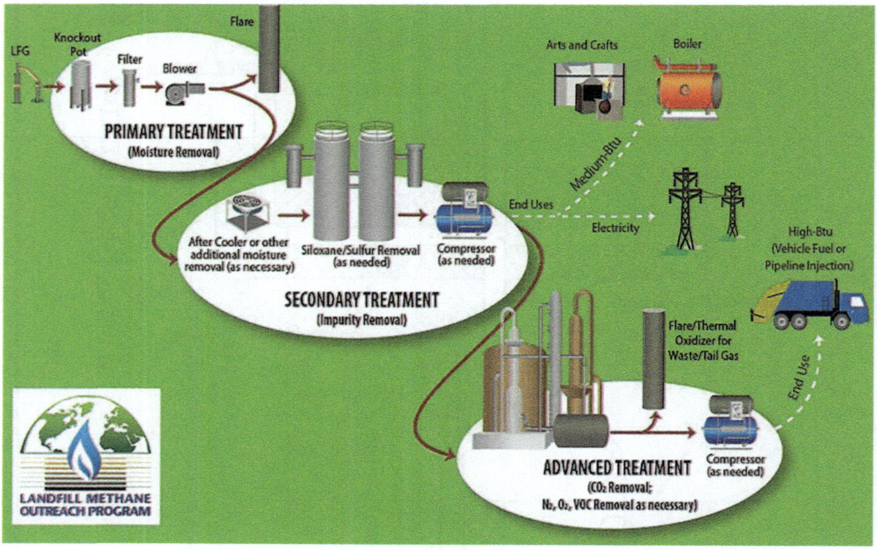

Fig. 9.9 This diagram displays LFG treatment phases. Adopted from ref. (US EPA 2021b)

demand, capital investment, operational costs, and prevailing regulatory and market conditions.

9.5.1.2 Injection of Fuel-Grade Methane into Gas Pipelines

The collected LFG is not immediately suitable for injection into gas pipelines due to impurities, particularly the presence of CO_2 and trace contaminants. One prominent method for LFG utilization involves injecting the gas into existing natural gas pipelines after upgrading them to meet the required pipeline-quality standards (see Table 9.5). This process, known as biogas upgrading, represents a promising avenue for renewable energy integration. Various biogas upgrading technologies exist, with the most common methods being pressure swing adsorption (PSA), water scrubbing, and amine scrubbing. Each technique aims to remove impurities, such as CO_2 and trace contaminants, to elevate the CH_4 content in the biogas to pipeline-quality levels. While PSA offers high-purity separation, water scrubbing, and amine scrubbing are cost-effective and operationally efficient (see Fig. 9.10).

PSA relies on selective adsorption of CO_2 and other impurities by a solid adsorbent material. The process alternates between adsorption and desorption phases under varying pressure conditions, effectively separating CO_2 from methane. At high pressures, CO_2 is adsorbed onto the adsorbents from biogas, while the subsequent pressure drop allows the gas to escape. The alternate pressurization-depressurization step ultimately yields a high-purity CH_4 (96–98%) as the product. The process relies on the molecular properties of the CO_2 and the adsorbent material's affinity for the gas,

Table 9.5 Natural gas pipeline quality standards

Empty Cell	EU	USA
Min. CH_4 (%, mol)	70–98	93.5–95.5
Max. CO_2 (%, mol)	1–8	2–3
Max. O_2 (%, mol)	0.01–1	0.2–3
Max. N_2 (%, mol)	2–10	–
Max. H_2 (%, mol)	0.1	0.1
Max. H_2S	2–15 mg/m^3	6 - 88 mg/m^3
Max. total sulfur	10–150 mg/m^3	265 mg/m^3
Water dew point	– 5 °C/8.4 mPa – 0 °C/0.4 mPa	–
Water content	0.05–1 g/m^3	65 mg/m^3

Adopted from ref. (Sun et al., 2015)

which is commonly silica-gel zeolites and activated carbon. Adsorption of CO_2 onto an adsorbent surface occurs due to van der Waals forces (Iglesias et al., 2021).

Water scrubbing involves passing the LFG through a water column, where CO_2 dissolves, leaving methane-enriched gas. The CO_2-laden water is then separated, and the methane is further purified. This upgrading method is by far the most popular option, and it is offered by a broad variety of commercial suppliers. Water is used as a selective absorbent, and its usage is common at an industrial scale since it can tolerate contaminants better and is cheaper than other absorbents. Since CO_2 is more soluble than CH_4 according to Henry's Law, it becomes more dispersible as pressure rises. With CO_2 being 26 times more soluble in water at 25 °C than CH_4, hence the method takes advantage of this simple fact (Iglesias et al., 2021).

Amine scrubbing is a chemical absorption process in which CO_2 is absorbed by an amine solvent. The absorbed CO_2 is later separated from the solvent, yielding methane-enriched gas. Amines including diethanolamine (DEA), monoethanolamine (MEA), and methyl diethanolamine (MDEA) are often used to neutralize CO_2 and H_2S. Because of its greater CO_2 absorption capacity compared to MDEA alone, a blend of MDEA and piperazine (PZ) is now the most preferred solvent combination (Iglesias et al., 2021).

Using membranes for separating gases has been a commercially viable option for many years. Membranes are selectively permeable barriers, allowing substances to flow through based on their permeability and the concentration, temperature, or pressure gradients across the membrane. Membrane separation makes use of two different process models: pore-flow and solution-diffusion. Gas separation via pore-flow using polymeric membranes is widely employed. CO_2 is allowed to pass through the membrane during the biogas upgrading process, while CH_4 is kept in the input. However, traditional membranes also retain N_2 while permeating H_2O, O_2, H_2S, and CO_2 throughout the biogas upgrading process (Munoz et al., 2015).

Alternatives to water for CO_2 absorption include organic solvents like methanol (CH_3OH), N-methyl pyrrolidone (NMP), and polyethylene glycol ethers (PEG). To

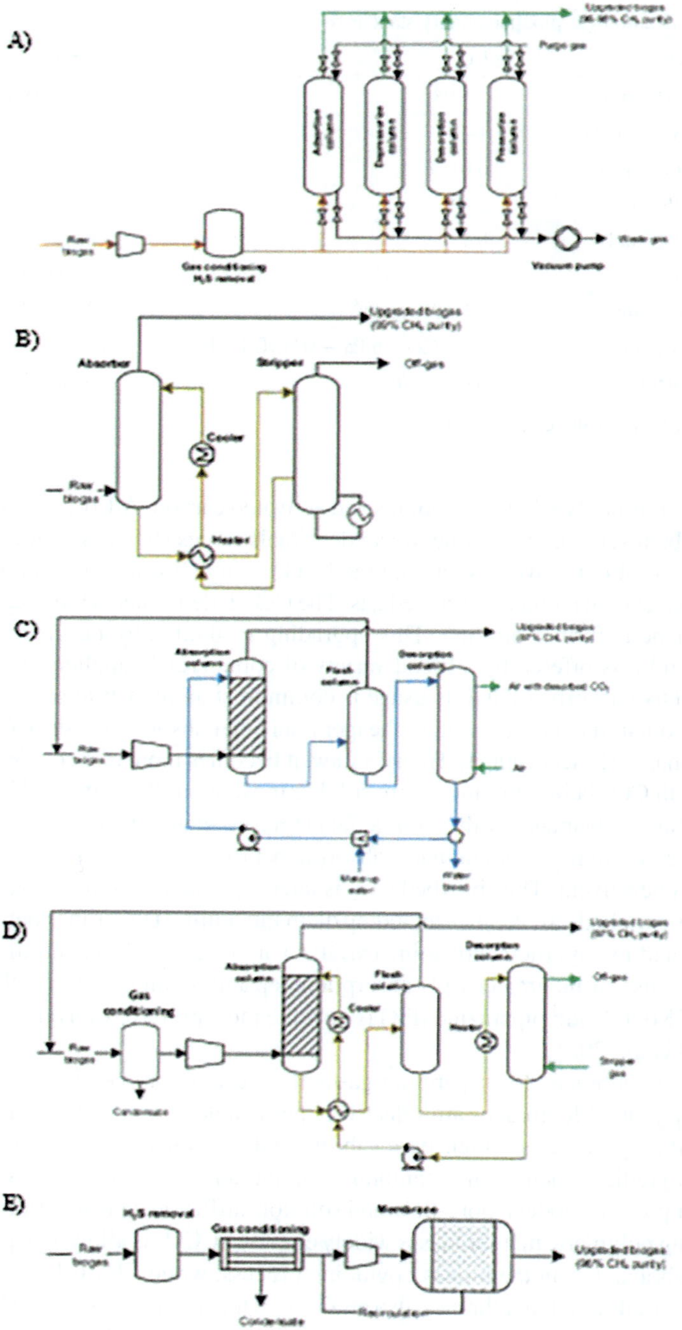

Fig. 9.10 Biogas upgrading technologies with **a** pressure swing adsorption (PSA), **b** amine scrubber, **c** water scrubber, **d** organic solvent scrubber, and **e** membrane. Adopted from ref. (Munoz et al., 2015)

keep conditions favorable for solvent absorption, it is advised that H_2S be removed beforehand in this procedure. This method is more costly to run than water scrubbing because it uses more energy and more expensive organic solvents to achieve the same CO_2 separation and anticorrosion characteristics. The removal of water from biogas and the introduction of heat are both necessary for the desorption of CO_2 at 40 °C when using an organic solvent. In addition, the organic solvent must be regenerated, which often involves heating it to very high temperatures. Using organic solvents has a similar effect to water scrubbing, with losses of less than 2% CH_4 (Adnan et al., 2019).

Landfill gas is considered a renewable energy source, as its production stems from the anaerobic decomposition of organic waste. By integrating LFG into the natural gas grid, renewable energy penetration can increase, diversifying the energy mix and promoting sustainable development. Biogas upgrading and injection projects can offer economic benefits through revenue generation from selling the upgraded biogas to gas utilities or through emission reduction credits. Additionally, the utilization of LFG can offset the costs associated with traditional landfill gas management, making it an economically viable solution.

However, to inject LFG into existing natural gas pipelines, the gas must meet stringent quality standards, including methane concentration, calorific value, and trace contaminant limits. The gas upgrading process must ensure consistent compliance with these requirements. Adapting existing gas pipeline infrastructure to accommodate LFG injection necessitates careful planning and coordination to ensure safe and reliable operations. The compatibility of materials, pressure ratings, and monitoring systems must be evaluated to prevent potential issues. Landfill gas production is subject to variability based on waste composition, landfill age, and environmental factors. This fluctuation can pose challenges to maintaining a stable supply of injected gas.

9.5.1.3 Biological Upgrading for CO_2 and H_2S Mitigation

The integration of biogas upgrading in landfill sites presents a feasible and innovative approach to maximize energy recovery, mitigate greenhouse gas emissions, and contribute to sustainable waste management. The conversion of the CO_2 fraction in crude LFG finds a crucial partner in synergizing energy recovery potential from landfill within the circular economy. The crux of this technology hinges on employing bioreactors housing autotrophic microorganisms adept at utilizing the CO_2 fraction as a carbon source and H_2 as the electron donor to produce high-quality biomethane (as depicted in Eq. 9.1) (Awe et al., 2017).

$$4H_2 + CO_2 \rightarrow CH_4 + 2H_2O(\Delta G_0 = -131 \text{ KJ}) \tag{9.1}$$

Broadly, from an industrial standpoint, two autotrophic metabolism pathways come to the fore; (1) the photoautotrophic pathway, facilitated by microalgae that harness solar radiation for CO_2 fixation, and (2) the chemoautotrophic pathway, wherein bacteria reduce CO_2 to carbon monoxide, methane, or alternative organic compounds. Luo et al. (2012) worked shown that biogas upgrading may produce CH_4 at 95% purity under various conditions.

Hydrogenotrophic Biogas Upgrading

Hydrogenotrophic biogas upgrading is a method that refines biogas composition, particularly its methane content, rendering it more suitable for energy applications and reducing its carbon footprint. This process involves the utilization of microorganisms, specifically hydrogenotrophic methanogens, which are capable of catalyzing the conversion of CO_2 and H_2 into CH_4. This reaction contributes to the enrichment of CH_4 content, thereby elevating the energy content of the biogas.

The central biochemical pathway underpinning hydrogenotrophic biogas upgrading involves the acetogenesis pathway, the hydrogenotrophic methanogenesis pathway, and the syntrophic acetate oxidation pathway. In the acetogenesis pathway, acetogenic bacteria convert organic acids into acetate and H_2. Subsequently, hydrogenotrophic methanogens utilize the produced H_2 and CO_2 to synthesize CH_4. Concurrently, the syntrophic acetate oxidation pathway involves a symbiotic relationship between acetate-oxidizing bacteria and hydrogenotrophic methanogens, where acetate oxidation generates H_2 that is utilized for CH_4 production. These intricate interdependencies demonstrate the collaborative nature of the hydrogenotrophic biogas upgrading process (Fig. 9.11).

The hydrogenotrophic methanogenic upgrading demonstrates robust impurity tolerance and moderate operational conditions, hence high reliability. More than 99% conversion efficiencies and biomethane productivities of up to 38–40 $m^3\ mR^{-3}\ d^{-1}$ are possible using plug-flow bioreactors (Savvas et al., 2017) and biotrickling filters (Dupnock & Deshusses, 2019). This setup allows different types of cultures that do not need oxygen to adjust easily to the conditions that enhance biogas quality (Porté et al., 2019). In these conditions, certain types of microorganisms that use hydrogen and acetate become more common, e.g., hydrogenotrophic and acetotrophic archaea.

The process of upgrading biogas involves different types of microorganisms working together. Specifically, about 70% of the methane produced in this process is a result of the relationship between methane-producing microorganisms called acetoclastic methanogens (*Methanosarcinals*) and other bacteria that help break down acetate (*Firmicutes*). The remaining 30% of methane comes from another group of microorganisms (methanogenic hydrogenotrophs) that use H_2 to turn CO_2 into CH_4 (Bassani et al., 2015).

Photosynthetic Biogas Upgrading

Photosynthetic biogas upgrading leverages the metabolic prowess of photosynthetic microorganisms to enhance the quality of biogas. By capitalizing on the synergies between microbial photosynthesis and anaerobic digestion, this approach offers a

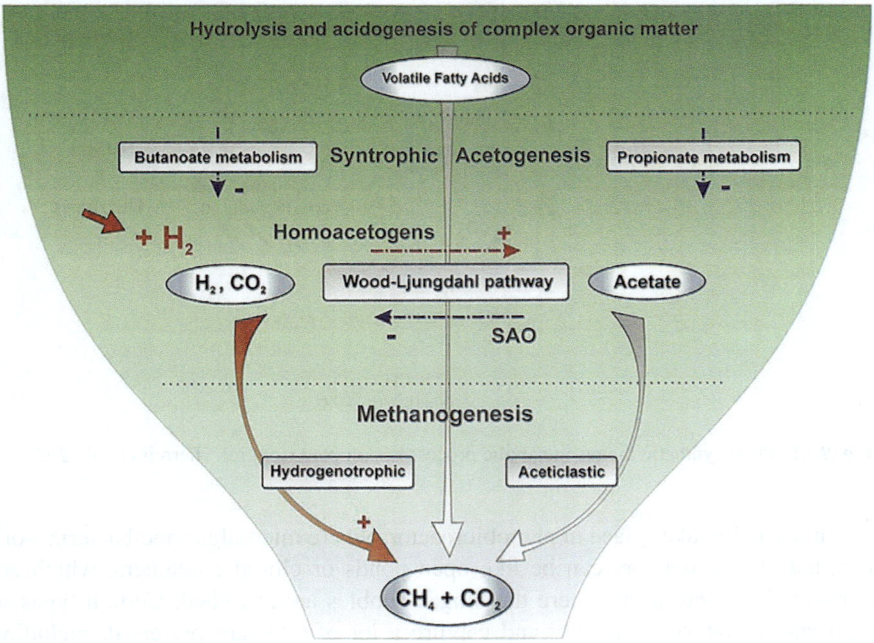

Fig. 9.11 Hydrogen-assisted methanogenesis metabolic processes. Adopted from ref. (Angelidaki et al., 2018)

sustainable means of increasing the methane content of biogas while concurrently mitigating CO_2 emissions. This also means we can get rid of unwanted gases like H_2S from the biogas. Microalgae are tiny plants that can soak up CO_2 while they grow, and bacteria can also help by turning H_2S into something harmless.

At the heart of photosynthetic biogas upgrading lies the intricate interaction between photosynthetic microorganisms and methanogens. During photosynthesis, microorganisms like algae and cyanobacteria utilize solar energy to fix CO_2 and produce organic compounds. When integrated into anaerobic digestion systems, these photosynthates serve as substrates for methanogenic archaea, thereby promoting the conversion of carbon dioxide into methane. This dynamic interplay not only enriches the methane content of the biogas but also reduces the carbon footprint of the process (Iglesias et al., 2021).

The success of photosynthetic biogas upgrading hinges on the formation of synergistic microbial consortia. The photosynthetic microorganisms contribute organic matter, while methanogens metabolize these compounds to yield methane. Furthermore, other microorganisms, such as sulfate-reducing bacteria, play a crucial role in preventing hydrogen sulfide accumulation, which can be detrimental to the overall process. The intricate network of interactions within these consortia underscores the complexity of photosynthetic biogas upgrading (Leong & Chang, 2022) (Fig. 9.12).

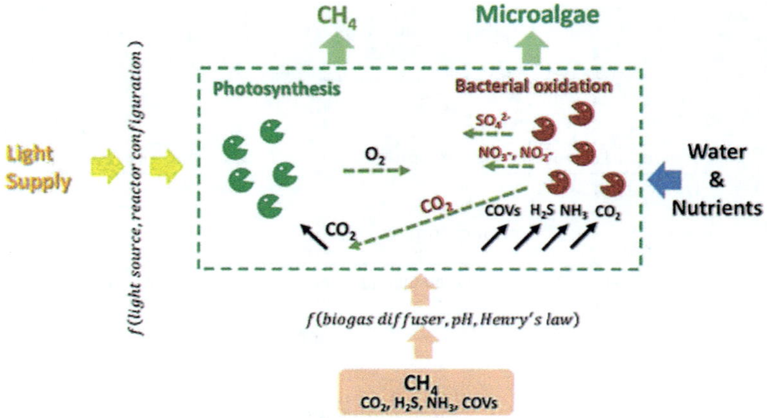

Fig. 9.12 Photosynthetic biogas metabolic processes. Adopted from ref. (Ferreira et al., 2022)

This usually takes place in photobioreactors where microalgae and bacteria work together. These reactors can be like open ponds or closed containers, which are connected to other parts where the biogas bubbles are absorbed. Certain types of microalgae that grow quickly and capture a lot of CO_2 are preferred, including *Nannochloropsis*, *Chlorella*, and *Scenedesmus*. However, these microalgae may change as a result of seasonal variations, such as temperature, sunlight, and the stuff in the digestate. The bacteria consume digestate, which comes from the liquid part left over after the biogas, which contains H_2S (as an energy source) and CO_2 (as a carbon source). This is like recycling nutrients from the biogas process to help the microorganisms grow (Bahr et al., 2014).

Microaerobic H_2S Removal

Biogas, which is produced from organic waste, often contains a gas called hydrogen sulfide (H_2S) that needs to be removed because it is harmful and smells bad. Some methods involve using physical or chemical treatments to remove the H_2S, but these methods are more expensive to run. On the other hand, there is a method called biological biogas desulfurization, which is more cost-effective. This method uses living organisms to get rid of the H_2S. It can be done by adding biotrickling filters to the biogas system or anaerobic digesters, i.e., by applying microaerobic conditions during the digestion process. One effective way to remove H_2S is by introducing a tiny amount of air or oxygen into the anaerobic digester (0.005–5 L O_2 L^{-1} day^{-1}). This is known as microaeration. It is simple and efficient, and it can remove almost all the H_2S (> 90%) (Khoshnevisan et al., 2017). Interestingly, the technology only requires minimal modification to the existing anaerobic digester system with low operating costs (0.0019–0.0039 € m^{-3}) (Díaz et al., 2015) (Fig. 9.13).

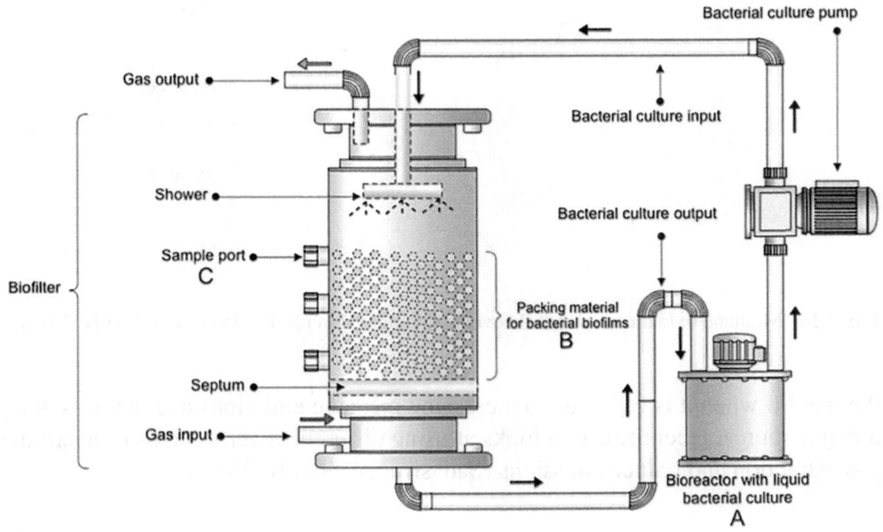

Fig. 9.13 Biotrickling filter setup. Adopted from ref. (Valdebenito-Rolack et al., 2021)

9.5.1.4 Methane Oxidation Soil Cover

Methane oxidation in soil cover is a process that occurs in landfills, where a considerable fraction of the methane produced is oxidized by the covering soil before it can reach the atmosphere. The mechanism of methane oxidation in landfill soil covers is illustrated in Fig. 9.14. The process involves the diffusion of methane from the landfill into the soil cover, where it is oxidized by methanotrophic bacteria, which use methane as their energy source. The composition and characteristics of the soil, as well as the presence of aged refuse, can enhance the methane oxidation potential. Methane oxidation rates observed in topsoil covering a retired landfill were reported to be the highest (45 g per m^2 daily) for any environment (Whalen et al., 1990). It encourages the microbial oxidation of CH_4 as LFG seeps out from within the landfill. The oxidation of CH_4 in the cover soil is mainly due to methanotrophic bacteria, which employ CH_4 monooxygenase (MMO) to assimilate CH_4 as their primary source of carbon and energy for cell propagation (Sadasivam & Reddy, 2014).

The efficiency of LFG extraction systems is highly dependable on the cover type, ranging from 50 to 95% (Chanton et al., 2009). As mentioned in the previous section, these systems often work with flaring, which turns CH_4 into less harmful CO_2, though it can have its issues e.g., releasing harmful byproducts that pose health concerns. Installing extraction systems can be tough, especially if landfill sites are abandoned, and gas generation rates are low. Real cases show these systems are operational less than half the time due to setbacks. Gas can also escape through vents, increasing emissions. Therefore, landfill covers play a big role in controlling emissions.

Landfill cover systems vary based on landfill needs: daily covers after each workday, intermediate covers for areas not used for a year, and final covers to seal

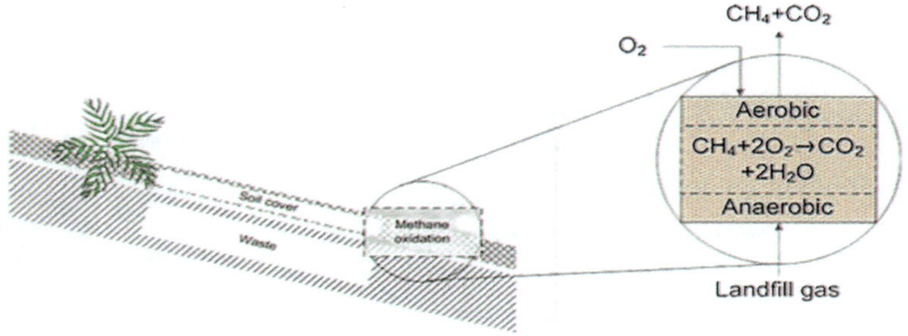

Fig. 9.14 Methane oxidation in landfill covers. Adopted from ref. (Sadasivam & Reddy, 2014)

the landfill when it is full. Due to increasing methane emissions and landfills being a major source, recent research looks at using different cover systems to maximize gas extraction and reduce emissions (Sadasivam & Reddy, 2014).

Biocovers

In recent times, there has been a global shift towards using biocovers instead of soil covers, driven by innovative efforts aimed at further reducing CH_4 emissions. Unlike soil covers, biocovers are engineered systems comprising an upper layer designed to enhance methane oxidation, along with a basal gas distribution layer (GDL) that evenly spreads landfill gas to achieve effective oxidation. Typically, the oxidation layer consists of organic-rich materials like composts, sewage sludge, peat, etc., either on their own or as amendments to the cover soil (Sadasivam & Reddy, 2014). These materials actively support and facilitate the growth of methanotrophic bacteria.

There are two pathways for landfill gas to enter the gas distribution layer: directly from the waste material beneath, through the compost materials, or into the biocover directly (Gebert et al., 2022). Similar to soil cover, the methanotrophic bacteria in the compost materials oxidize the methane in the landfill gas, which is subsequently used by the bacteria as a source of energy. The oxidation rate of methane may be affected by the thickness of the biocover, with thicker biocovers often having greater oxidation rates (Yusnan, 2020). It has been established via research that biocovers may significantly cut methane emissions from landfills by 98% (Pecorini & Iannelli, 2020). Recent research also indicates that adding aged waste to landfill cover soil might increase its methane oxidation potential (He et al., 2022) (Fig. 9.15).

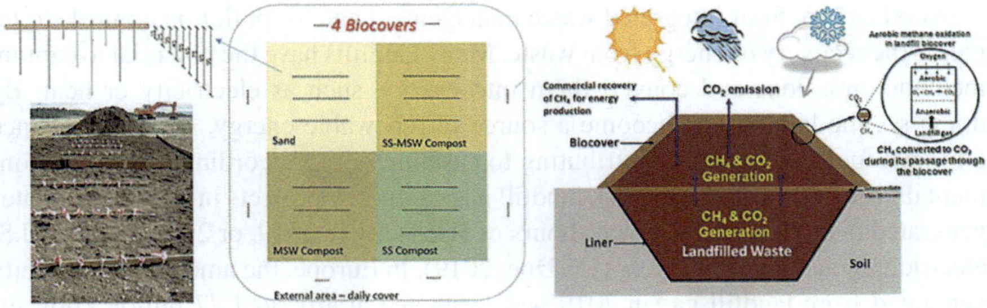

Fig. 9.15 Biocovers landfill schematic with different soil cover used. Adopted from ref. (Bajwa, 2012; Pecorini & Iannelli, 2020)

9.6 Potential of Integrated Air Pollution Controls in Landfills

To address these issues, waste management and air pollution control strategies must be integrated to achieve sustainable landfill management. By doing so, air pollution from landfills can be minimized, reducing the associated health risks, and contributing to a more sustainable future. One of the main benefits of integrating waste management and air pollution control strategies is the reduction of air pollution emissions. Effective waste management practices, such as waste reduction, recycling, and composting, can reduce the amount of waste that is sent to landfills. This, in turn, reduces the amount of emissions generated by landfills, including greenhouse gases and volatile organic compounds.

For instance, a study conducted by the Environmental Protection Agency (EPA) in the United States found that effective waste management practices, such as waste reduction and recycling, can significantly reduce the amount of greenhouse gas emissions from landfills. The study reported that the amount of carbon dioxide equivalent (CO_2e) emissions from landfills was reduced by 348 million metric tons in 2018 due to these practices (US EPA, 2022a). Similarly, a study conducted in the European Union (EU) found that waste reduction, recycling, and composting can reduce the emissions of greenhouse gases (GHG) by up to 87% compared to landfilling (Vahk, 2020). The study also reported that landfilling was responsible for approximately 3% of total greenhouse gas emissions in the European Union, the fourth largest GHG emission sector in the EU. In addition to reducing greenhouse gas emissions, integrating waste management and air pollution control strategies can also reduce the emissions of volatile organic compounds (VOCs) from landfills. VOCs are a significant contributor to air pollution and can have negative health impacts on humans, including respiratory issues and cancer (US EPA, 2022b). A study found that implementing a waste reduction program (garbage sorting policy and monitoring odor pollution) resulted in a 60% reduction in VOC emissions from landfills (Liu et al., 2020).

Another benefit of integrated waste management and air pollution control strategies is the recovery of energy from waste. Many landfills have the potential to capture methane emissions and convert them into energy, such as electricity or heat. By doing so, the landfill can become a source of renewable energy, reducing reliance on fossil fuels and further contributing to sustainability. According to the Environmental Protection Agency (EPA), landfill gas-to-energy projects in the United States generated over 79 billion kilowatt-hours of electricity in 2019, or 2% of current U.S. electricity generation for a year (US Doe, 2019). In Europe, the amount of electricity generated from landfill gas in 2019 was estimated at around 1.77 billion kilowatt-hours (Fernandes, 2023), enough to power around 58.9 million homes at a 30 kWh a day consumption rate. The recovery of energy from landfill gas not only reduces greenhouse gas emissions but also provides economic benefits. The US EPA estimates that landfill gas-to-energy projects can generate revenue and create jobs in local communities, while also reducing emissions (USA EPA, 2023a). It is assumed that 1.2 billion tons of post-recycling MSW are generated annually in the world, yet only 16.6% of it is treated for energy recovery, estimated to generate 13 gigawatts (GW) of electricity on a global scale (Dadario et al., 2023). Additionally, a study conducted by the European Commission estimated that by 2020, energy recovery from renewable resources contributed up to 21.8% of the European Union's energy consumption (Fernandez, 2023) with waste-to-energy capacity covering 3% amounted to 5.1 GW (Aizarani, 2022), helping to meet renewable energy targets and reduce greenhouse gas emissions.

9.7 Challenges and Future Research Directions

9.7.1 Challenges to Implementation of Integrated Strategies

The implementation of integrated strategies for air pollution control in landfills can be hindered by various barriers, including technical and financial constraints, as well as regulatory and institutional barriers, as summarized in Table 9.6. The lack of adequate infrastructure for waste management and air pollution control is a major technical barrier to the implementation of integrated strategies for air pollution control in landfills (Williams et al., 2023). In many developing countries, the infrastructure for waste management is insufficient or outdated, which can lead to inadequate collection and disposal of waste. For example, in India, only about 60% of the waste generated is collected and only about 15% is treated, leading to widespread dumping in open landfills (Swaminanthan, 2018).

Similarly, the infrastructure for air pollution control, such as gas collection and treatment systems, may not be available or functioning properly. In some cases, the gas collection systems may be inadequate, leading to low capture rates and increased emissions of greenhouse gases and volatile organic compounds. For example, a study concludes that many South American countries are struggling with many landfills that

Table 9.6 Challenges related to implementing waste management and air pollution control technologies in landfills

Challenge	Description
Technicality	Insufficient infrastructure for waste management and air pollution control
Site-specific factors	Influence of waste composition, gas generation rates, and geographic location
Financial	High implementation costs and inadequate revenue generation
Institutional	Lack of coordination, political will, and community involvement

lack proper gas collection and treatment systems, resulting in significant emissions of methane and other gases (Abubakar et al., 2022). To put into perspective, a study in China found that there were less than 20% of methane is recovered from landfills, while Western nations recover up to 60% (Abubakar et al., 2022).

Site-specific factors have a significant impact on the effectiveness and suitability of pollution control technologies in landfill sites. Waste composition, landfill gas generation rates, and geographic location are among the key characteristics that influence the performance of these technologies (Hays & Yoon, 2021). The composition of waste varies from one landfill site to another, with variations in organic content, moisture levels, and the presence of contaminants. These variations can affect the efficiency of pollution control technologies, as different waste compositions may require specific treatment approaches. For example, a landfill with high organic content may benefit from biofiltration systems that harness microbial activity to remove pollutants (Han et al., 2020). On the other hand, a landfill with a significant presence of volatile organic compounds (VOCs) may require membrane separation technologies for selective removal (McWatters & Rowe, 2009).

Landfill gas generation rates also vary based on factors such as waste decomposition rates and the age of the landfill. This impacts the sizing and capacity requirements of pollution control technologies. A higher gas generation rate may necessitate larger gas collection and treatment systems to handle the volume of gas produced. Conversely, a lower gas generation rate may allow for the implementation of smaller-scale technologies. Geographic location is another crucial site-specific factor. Climate conditions, such as temperature and humidity, can influence the performance of pollution control technologies. For instance, biofiltration systems may be more effective in regions with moderate temperatures and adequate moisture levels, as microbial activity is sensitive to environmental conditions (Sheoran et al., 2022). In contrast, membrane separation technologies may be better suited for sites with extreme temperatures or dry climates, as they can operate efficiently under a wide range of conditions (McWatters & Rowe, 2009).

In addition, financial constraints can pose a significant challenge to the implementation of integrated strategies for air pollution control in landfills. The cost of implementing air pollution control measures can be high, and the revenue generated from waste management activities may not be sufficient to cover these costs. In some cases, the cost of implementing air pollution control measures may even exceed the revenue generated from waste management, making it economically unfeasible for some landfill operators. For example, a study conducted in India found that the cost of implementing a gas collection and treatment system in landfills was high, and the revenue generated from waste management activities was not enough to cover these costs (Talyan et al., 2008). Another study conducted in the United States found that the cost of implementing air pollution control measures in landfills was a major concern for landfill operators and that many operators were hesitant to invest in such measures due to financial constraints (Bruggers, 2021). Furthermore, financial constraints can also affect the maintenance and operation of air pollution control systems in landfills. Without proper maintenance, these systems can malfunction or become less effective, leading to increased emissions and air pollution. A study conducted in South Africa found that the lack of funding for the operation and maintenance of gas collection and treatment systems was a significant barrier to the effective control of landfill gas emissions (Mbazima et al., 2022).

Institutional barriers can hinder the implementation of integrated strategies for air pollution control in landfills. These barriers may arise due to a lack of coordination among different agencies and stakeholders responsible for waste management and air pollution control. Such coordination is necessary to ensure that the different entities work towards the same goal of reducing air pollution emissions from landfills. One institutional barrier is the lack of clear mandates and responsibilities among different agencies involved in waste management and air pollution control. This lack of clarity can result in overlap or gaps in responsibilities, leading to confusion and delays in decision-making. For example, a municipality may be responsible for waste collection and disposal, while an environmental agency is responsible for air pollution control. The lack of coordination between these agencies can lead to inadequate air pollution control measures at landfills. One study published highlighted the lack of coordination and communication among different stakeholders as a key institutional barrier to effective waste management (Liang et al., 2023). The authors noted that overlapping responsibilities and lack of clear mandates can lead to delays and inadequate implementation of waste management and control measures.

Another institutional barrier is the lack of political will and commitment to implementing integrated strategies for air pollution control. In some cases, politicians may prioritize other issues over waste management and air pollution control. This lack of political will can lead to inadequate funding and resources for air pollution control measures and may result in a lack of enforcement of regulations and standards. Another study identified the lack of political will and commitment as a key institutional barrier to effective waste management in sub-Saharan Africa (Debrah et al., 2022). The authors argued that without the necessary political will, adequate funding and resources may not be provided for air pollution control measures, and regulations and standards may not be enforced.

Also, there may be a lack of community involvement and consultation in the decision-making process. Communities living near landfills may not be adequately informed about the risks associated with air pollution emissions from landfills, and their input may not be taken into account when developing integrated strategies for air pollution control (Gallagher et al., 2008). This lack of community involvement can lead to mistrust and resistance to air pollution control measures and may result in delays or even abandonment of projects. Furthermore, a report by the World Bank on solid waste management in Latin America and the Caribbean identified the lack of community involvement and consultation as a major institutional barrier to effective waste management (Hettiarachchi et al., 2018). The report noted that community resistance and mistrust can lead to delays and even abandonment of waste management projects.

9.7.2 Research Needs and Opportunities

Air pollution is a major environmental concern, with landfills being one of the major sources of air pollution. Air pollution from landfills can cause a variety of health problems, including respiratory diseases and cancer. To mitigate the negative effects of air pollution from landfills, there is a need to integrate air pollution control measures into landfill management (Table 9.7). Research has identified several opportunities and research needs for the integration of air pollution control in landfills. One area of research is the development of cost-effective air pollution control technologies. Landfill gas (LFG) is a major contributor to air pollution from landfills, and technologies to capture and treat LFG have been developed. However, these technologies can be expensive, and there is a need for research to develop cost-effective solutions (Manasaki et al., 2021).

Another research need is the development of effective monitoring systems to track emissions from landfills. Monitoring systems can help identify sources of pollution and track the effectiveness of air pollution control measures (US ATSDR, 2001). However, current monitoring systems are often inadequate, and there is a need to develop more effective monitoring technologies. Research has also identified the need for a better understanding of the factors that influence the emissions of air pollutants from landfills. The composition of waste, the design of the landfill, and the management practices used all affect emissions of air pollutants. Understanding these factors can help identify ways to reduce emissions. Additionally, there is a need for research to understand the social and economic factors that influence the adoption of air pollution control measures in landfill management (Adzawla et al., 2019). For example, there may be cultural or political barriers to the adoption of certain technologies or practices. Understanding these factors can help identify ways to overcome barriers to adoption.

Several studies have demonstrated the effectiveness of air pollution control measures in reducing emissions from landfills. A study conducted in the United States found that the implementation of LFG recovery systems reduced emissions of methane, a potent greenhouse gas, by 60 to 90% (US EPA, 2023a). Another study conducted in Malaysia found that the implementation of a gas collection and treatment system reduced emissions of methane by 89% (Abushammala et al., 2011). However, despite the effectiveness of air pollution control measures, there are still

Table 9.7 Research needs and opportunities for integrating waste management and air pollution control in landfills

Research area	Description	Examples	Research opportunities
Cost-effective pollution control technologies	Extensive research efforts have yielded the development of efficient and environmentally friendly technologies for gas collection and landfill treatment systems	i. LFG system ii. Bioreactor landfill iii. Anaerobic digestion iv. Plasma gasification v. Biofiltration vi. Advanced gas and leachate treatment systems	These technologies can be expensive, and there is a need for research to develop cost-effective solutions, especially for low- and middle-income countries
Factors influencing emissions	Multiple factors, including waste composition, landfill design, and management practices, can influence emissions of air pollutants from landfills	i. Organic waste, for example, tends to produce higher levels of methane, a potent greenhouse gas, during decomposition compared to other waste types ii. Landfills with inadequate gas collection systems or improper sealing can lead to higher emissions of methane and other gases iii. Effective waste compaction reduces the void spaces within the landfill, limiting oxygen availability and promoting anaerobic conditions that reduce methane emissions	A comprehensive understanding of these factors can provide valuable insights and research opportunities to effectively mitigate emissions

(continued)

Table 9.7 (continued)

Research area	Description	Examples	Research opportunities
Social and economic factors influencing adoption	Additional factors, such as social and economic considerations, may exert influence on the adoption of sustainable control measures in landfill management	i. Landfill operators may face financial constraints or limited budgets, making it challenging to invest in these measures ii. A weak regulatory environment or inconsistent enforcement may create a less compelling incentive for operators to adopt these measures iii. Concerns over odor, noise, visual impact, and potential health risks associated with landfills and their emissions can influence public opinion and community support iv. Growing public concern about climate change, air quality, and environmental sustainability can create pressure on landfill operators to implement measures that mitigate air pollutant emissions	A thorough understanding of these factors can facilitate the identification of strategies to overcome barriers to adoption

(continued)

Table 9.7 (continued)

Research area	Description	Examples	Research opportunities
Effectiveness of pollution control measures	Multiple studies have provided evidence showcasing the efficacy of sustainable waste management practices and air pollution control measures in mitigating emissions from landfills	i. Composite covers with LFG recovery systems may reduce methane by 60–90% compared to traditional soil covers ii. Removal of VOCs emissions can be achieved using activated carbon adsorption iii. Treatment with membrane bioreactors and reverse osmosis, effectively remove pollutants and improve the quality of discharged leachate iv. Landfill mining offers resource recovery and minimizes greenhouse gas emissions iv Implementing integrated waste management systems that combine various strategies, such as waste reduction, recycling, composting, and energy recovery, have proven effective in minimizing the environmental impact of landfills	Environmental policymakers and technology experts have expressed scepticism regarding the effectiveness of certain measures, highlighting the need for thorough assessments of environmental impacts to inform decision-making. This emphasizes the importance of conducting comprehensive evaluations to ensure accurate and reliable assessments of the effectiveness of these measures

challenges to their widespread adoption. One of the main challenges is the high cost of implementing and operating these technologies (Penza et al., 2015). In many low and middle income countries, waste management budgets are limited, making it difficult to invest in expensive air pollution control technologies (Wijekoon et al., 2022). Additionally, there may be political or cultural barriers to the adoption of certain technologies or practices.

9.8 Conclusion

Effective air pollution control strategies are paramount for achieving sustainable landfill management. The urgency to address air quality concerns in and around landfills cannot be overstated, as the repercussions of unchecked pollution are far-reaching and can impact both the environment and public health. This article has explored various proactive approaches, ranging from advanced cover systems and gas collection technologies to regular monitoring. To ensure successful implementation, collaboration among governmental bodies, industries, and communities is essential. By adopting a multifaceted approach that combines innovative engineering solutions, rigorous monitoring, and a commitment to continuous improvement, the goals of minimizing air pollutants, curbing greenhouse gas emissions, and protecting air quality can be achieved. As we strive for more sustainable waste management practices, it is imperative that policymakers, researchers, and stakeholders work in tandem to establish robust strategies that not only mitigate air pollution but also contribute to the overall well-being of our planet and its inhabitants. In the journey towards sustainable landfill management, effective air pollution control emerges as a pivotal factor that paves the way for a cleaner, healthier, and more resilient future.

References

Abiola, A. H. O., et al. (2021). Comparison of respiratory and skin disorders between residents living close to and far from Solous landfill site in Lagos State, Nigeria. *African Journal Primary Health & Care Family Medicine, 13*, 1–7.

Abubakar, I. R., et al. (2022). *Environmental sustainability impacts of solid waste management practices in the global south* (p. 19). Public Health.

Abushammala, M. F. M., et al. (2011). Regional landfills methane emission inventory in Malaysia. *Waste Management Research, 29*, 863–873.

Adnan, A. I., et al. (2019). Technologies for biogas upgrading to biomethane: A review. *Bioengineering, 6*, 1–23.

Adzawla, W., et al. (2019). Do socioeconomic factors influence households' solid waste disposal systems? Evidence from Ghana. *Waste Mancgemaent Research, 37*, 51–57.

Aizarani, J. (2022). Europe: waste-to-energy capacity by country 2022. *Statista*. [Online]. Available: https://www.statista.com/statistics/1122082/europe-waste-to-energy-capacity-by-country/. Accessed: May 22, 2023.

Ali, S. M., et al. (2014). Open dumping of municipal solid waste and its hazardous impacts on soil and vegetation diversity at waste dumping sites of Islamabad city. *Journal King Saud University-Science, 26*, 59–65.

Alves, B. (2023). Municipal waste generation by select country. *Statista.* [Online]. Available: https://www.statista.com/statistics/916749/global-generation-of-municipal-solid-waste-by-country/. Accessed: Aug 10, 2023.

Angelidaki, I., et al. (2018). Biogas upgrading and utilization: Current status and perspectives. *Biotechnology Advances, 36*, 452–466.

Asim, M., et al. (2012). Scavengers and their role in the recycling of waste in Southwestern Lahore. *Resources, Conservation and Recycling, 58*, 152–162.

ASTDR. (2009). Landfill gas safety and health issues. *Agency for Toxic Substances and Disease Registry (ASTDR).* [Online]. Available: https://www.atsdr.cdc.gov/hac/landfill/html/ch3.html. Accessed: Aug 10, 2023.

ATSDR. (2001). Landfill gas basics. *U.S. Agency for toxic substances and disease registry.* [Online]. Available: https://www.atsdr.cdc.gov/hac/landfill/html/ch2.html. Accessed: Aug 11, 2023.

Attalage, D. S., et al. (2023). *Impact of landfill gas exposure on vegetation in engineered landfill biocover systems implemented to minimize fugitive methane emissions from landfills* (p. 20). Public Health.

Awe, O. W., et al. (2017). A review of biogas utilisation, purification and upgrading technologies. *Waste and Biomass Valorization, 8*, 267–283.

Bahr, M., et al. (2014). Microalgal-biotechnology as a platform for an integral biogas upgrading and nutrient removal from anaerobic effluents. *Environmental Science and Technology, 48*, 573–581.

Bajwa, T. M. (2012). *Experimental characterization of the thermal, hydraulic and mechanical (THM) properties of compost based landfill covers.* University of Ottawa.

Bassani, I., et al. (2015). Biogas upgrading via hydrogenotrophic methanogenesis in two-stage continuous stirred tank reactors at mesophilic and thermophilic conditions. *Environmental Science and Technology, 49*, 12585–12593.

Bruggers, J. (2021). Cutting landfill methane to fight climate change. *NPR.* [Online]. Available: https://www.npr.org/2021/07/13/1012218119/epa-struggles-to-track-methane-from-landfills-heres-why-it-matters-for-the-clima. Accessed: May 16, 2023.

CEM and GAIA. (2018). *Report of the sample taken downwind of the Kodungaiyur dumpyard*

Chanton, J. P., et al. (2009). Methane oxidation in landfill cover soils, is a 10% default value reasonable? *Journal of Environmental Quality, 38*, 654–663.

Chetri, J. K., & Reddy, K. R. (2021) Methane recovery from landfills. In W. Guo, et al., (Eds.), *Sustainable resource management: Technologies for recovery and reuse of energy and waste materials*, 1st edn (pp. 699–722). WILEY-VCH GmbH.

Chetri, J. K., & Reddy, K. R. (2021a). Advancements in municipal solid waste landfill cover system: A review. *Journal of the Indian Institute of Science, 101*, 557–588.

Dadario, N., et al. (2023). Waste-to-energy recovery from municipal solid waste: Global scenario and prospects of mass burning technology in Brazil. *Sustain, 15*, 1–20.

Davoli, E., et al. (2010). Waste management health risk assessment: A case study of a solid waste landfill in South Italy. *Waste Management, 30*, 1608–1613.

Debrah, J. K., et al. (2022). Barriers and challenges to waste management hindering the circular economy in sub-Saharan Africa. *Urban Science, 6*, 57.

Díaz, I., et al. (2015). Economic analysis of microaerobic removal of H2S from biogas in full-scale sludge digesters. *Bioresource Technology, 192*, 280–286.

Dupnock, T. L., & Deshusses, M. A. (2019). Detailed investigations of dissolved hydrogen and hydrogen mass transfer in a biotrickling filter for upgrading biogas. *Bioresource Technology, 290*, 121780.

Emara, K. (2023). Sustainable solid waste management in rural areas: A case study of Fayoum governorate, Egypt. *Energy Nexus, 9*, 100168.

Ennovo. (2023). Lo-Cal flare range. *Ennovo.* [Online]. Available: https://ennovo.com.au/equipment/lo-cal-flare-range/. Accessed: Aug 11, 2023.

Esphylin, D., et al. (2018). The association of reported respiratory symptoms among children in Malaysia with particulate matter exposure in municipal solid waste landfills. *Malaysian Journal Medicine and Health Science, 14*, 2–11.

European Commission. (2020). Greenhouse gas emissions from waste. *Eurostat*. [Online]. Available: https://ec.europa.eu/eurostat/web/products-eurostat-news/-/ddn-20200123-1. Accessed: Aug 11, 2023.

European Commission. (2022). Methane emissions. *European Commision*. [Online]. Available: https://energy.ec.europa.eu/topics/oil-gas-and-coal/methane-emissions_en. Accessed: 18 May, 2023.

Fang, J. J., et al. (2012). Odor compounds from different sources of landfill: Characterization and source identification. *Waste Management, 32*, 1401–1410.

Fazzo, L., et al. (2023). *The health impact of hazardous waste landfills and illegal dumps contaminated sites: An epidemiological study at ecological level in Italian Region* (p. 11). Public Heal.

Fernandes, L. (2023) EU: Energy production from landfill gas. *Statista*. [Online]. Available: https://www.statista.com/statistics/863329/landfill-gas-energy-production-in-the-european-union-eu/. Accessed: May 19, 2023.

Fernandez, L. (2023). EU-28: renewables in final energy consumption. *Statista*. [Online]. Available: https://www.statista.com/statistics/864900/share-of-renewable-energy-electricity-consumption-european-union-eu28/. Accessed: May 22, 2023.

Ferreira, A. F., et al. (2022). The experiences of success and failure in the pilot and real-scale photosynthetic biogas production. In *3rd Generation biofuels disruptive technology to enable commercial production*. https://doi.org/10.1016/B978-0-323-90971-6.00004-8

Ferronato, N., & Torretta, V. (2019). *Waste mismanagement in developing countries: A review of global issues* (p. 16). Public Health.

Flower, F., et al. (1981). Landfill gas, what it does to trees, and how its injurious effects may be prevented. *Arboric. Urban for., 7*, 43–52.

Gallagher, L., et al. (2008). Host community attitudes towards solid waste landfill infrastructure: Comprehension before compensation. *Journal of Environmental Planning and Management, 51*, 233–257.

Gebert, J., et al. (2022). Design of microbial methane oxidation systems for landfills. *Frontiers in Environmental Science, 10*, 1–28.

Ghimire, A., et al. (2019). Polybrominated diphenyl ethers (PBDEs) in ambient air samples at the electronic waste (e-waste) reclamation site. *Waste Disposal Sustainable Energy, 1*, 79–89.

Gómez-Sanabria, A., et al. (2022). Potential for future reductions of global GHG and air pollutants from circular waste management systems. *Nature Communications, 13*, 1–12.

Han, Y., et al. (2020). Biofilters for the co-treatment of volatile organic compounds and odors in a domestic waste landfill site. *Journal of Cleaner Production, 277*, 1–9.

Hays, L., & Yoon, H. (2021) Landfill gas generation and collection. *Geo-engineer*. [Online]. Available: https://www.geoengineer.org/education/web-class-projects/ce-176-environmental-geotechnics/assignments/landfill-gas-generation-collection. Accessed: Aug 15, 2023.

He, H., et al. (2022). Enhanced methane oxidation potential of landfill cover soil modified with aged refuse. *Atmosphere (Basel)., 13*, 1–12.

Hettiarachchi, H., et al. (2018). Municipal solid waste management in Latin America and the Caribbean: Issues and potential solutions from the governance perspective. *Recycling, 3*.

Hoornweg, D., & Bhada-Tata, P. (2012). *What a waste: A global review of solid waste managment.*

ICLEI. (2018). Waste sector. In *GHG Platform India* (pp. 1–32).

Iglesias, R., et al. (2021). Biogas from anaerobic digestion as an energy vector: Current upgrading development. *Energies, 14*, 1–30.

Iravanian, A., & Ravari, S. O. (2020). Types of contamination in landfills and effects on the environment: A review study. In *IOP Conference Series Earth Environmental Science* (Vol. 614).

Kamran, A., et al. (2015). Effects of socio-economic status and seasonal variation on municipal solid waste composition: A baseline study for future planning and development. *Environmental Science Europe, 27*.

Karthikeyan, O. P., et al. (2011). Characterization of particulate matters and volatile organic compounds in the ambient environment of open dump sites. *Universal Journal of Environmental Research & Technology, 1*, 140–150.

Khoshnevisan, B., et al. (2017). A review on prospects and challenges of biological H_2S removal from biogas with focus on biotrickling filtration and microaerobic desulfurization. *Biofuel Research Journal, 4*, 741–750.

Kumar, A., & Samadder, S. R. (2017). A review on technological options of waste to energy for effective management of municipal solid waste. *Waste Management, 69*, 407–422.

Lakhouit, A., et al. (2016). Two novel biofilters to remove volatile organic compounds emitted by landfill sites. *Water, Air, and Soil Pollution, 227*, 1–9.

Nyiramigisha, et al. (2021). Harmful impacts of heavy metal. *Reviews in Agricultural Science*, 9271–282.

Leong, Y. K., & Chang, J. S. (2022). Integrated role of algae in the closed-loop circular economy of anaerobic digestion. *Bioresource Technology, 360*, 127618.

Liang, C., et al. (2023). Identifying waste supply chain coordination barriers with Fuzzy MCDM. *Sustainability, 15*, 1–23.

Liu, Y., et al. (2020). VOCs released from municipal solid waste at the initial decomposition stage: Emission characteristics and an odor impact assessment. *Journal of Environmental Sciences (China), 98*, 143–150.

Luo, G., et al. (2012). Simultaneous hydrogen utilization and in situ biogas upgrading in an anaerobic reactor. *Biotechnology and Bioengineering, 109*, 1088–1094.

Manasaki, V., et al. (2021). Techno-economic assessment of landfill gas (LFG) to electric energy: Selection of the optimal technology through field-study and model simulation. *Chemosphere, 269*, 128688.

Mbazima, S. J., et al. (2022). Waste-to-energy in a developing country: The state of landfill gas to energy in the Republic of South Africa. *Energy Exploration & Exploitation, 40*, 1287–1312.

McWatters, R. S., & Rowe, R. K. (2009). Transport of volatile organic compounds through PVC and LLDPE geomembranes from both aqueous and vapour phases. *Geosynthetics International, 16*, 468–481.

Milosevic, L. T., et al. (2018). Identification of fire hazards due to landfill gas generation and emission. *Polish Journal Environmental Studies, 27*, 213–221.

Minnesota Pollution Control Agency. (2023). Volatile organic compounds (VOCs). *Minnesota Pollution Control Agency*. [Online]. Available: https://www.pca.state.mn.us/pollutants-and-contaminants/volatile-organic-compounds-vocs. Accessed: Aug 10, 2023.

Mohan, S., & Joseph, C. P. (2021). Potential hazards due to municipal solid waste open dumping in India. *Journal of the Indian Institute of Science, 101*, 523–536.

Monni, S., et al. (2006) *Global climate change mitigation scenarios for solid waste management*, 603

Moolla, R., et al. (2013). Occupational health risk assessment of benzene and toluene at a landfill site in Johannesburg, South Africa. *WIT Transactions on The Built Environment, 134*, 701–712.

Mor, S., & Ravindra, K. (2023). Municipal solid waste landfills in lower- and middle-income countries: Environmental impacts, challenges and sustainable management practices. *Process Safety and Environment Protection, 174*, 510–530.

Munoz, R., et al. (2015). A review on the state-of-the-art of physical/chemical and biological technologies for biogas upgrading. *Reviews Environmental Science Bio/Technology, 14*, 727–759.

Nair, A. T., et al. (2019). Emerging perspectives on VOC emissions from landfill sites: Impact on tropospheric chemistry and local air quality. *Process Safety and Environment Protection, 121*, 143–154.

Nanda, S., & Berruti, F. (2021). Municipal solid waste management and landfilling technologies: A review. *Environmental Chemistry Letters, 19*, 1433–1456.

Naveen, B. P., & Goel, A. (2017) A hazard ranking system for landfill rehabilitation in India. In *Indian Geotechnical Conference.*

Nolasco, D., et al. (2008). Non-controlled biogenic emissions to the atmosphere from Lazareto landfill, Tenerife, Canary Islands. *Environmental Science and Pollution Research, 15*, 51–60.

Oonk, H. (2012). Efficiency of landfill gas collection for methane emission reduction. *Greenhouse Gas Measurement and Management, 2*, 129–145.

Ozbay, G., et al. (2021). *(2021) Design and operation of effective landfills with minimal effects on the environment and human health.* Public Health.

Paul, S., et al. (2019). Assessing the ecological impacts of ageing on hazard potential of solid waste landfills: A green approach through vermitechnology. *Journal of Cleaner Production, 236*, 117643.

Pecorini, I., & Iannelli, R. (2020). Landfill GHG reduction through different microbial methane oxidation biocovers. *Processes, 8*, 1–12.

Penza, M., et al. (2015). A case-study of microsensors for landfill air-pollution monitoring applications. *Urban Climate, 14*, 351–369.

Pivato, A. (2011). Landfill liner failure: An open question for landfill risk analysis. *Journal of Environmental Protection, 2*, 287–297.

Porté, H., et al. (2019). Process performance and microbial community structure in thermophilic trickling biofilter reactors for biogas upgrading. *Science of the Total Environment, 655*, 529–538.

Sadasivam, B. Y., & Reddy, K. R. (2014). Landfill methane oxidation in soil and bio-based cover systems: A review. *Reviews in Environmental Science & Biotechnology, 13*, 79–107.

Savvas, S., et al. (2017). Biological methanation of CO_2 in a novel biofilm plug-flow reactor: A high rate and low parasitic energy process. *Applied Energy, 202*, 238–247.

Sebastian, A. (2020) Countries that produce the most waste. *Investopedia.* [Online]. Available: https://www.investopedia.com/articles/markets-economy/090716/5-countries-produce-most-waste.asp. Accessed: Aug 10, 2023.

SEPA. (2002). *Guidance on landfill gas flaring.*

Sheoran, K., et al. (2022). Air pollutants removal using biofiltration technique: A challenge at the frontiers of sustainable environment. *ACS Engineering Au, 2*, 378–396.

Siddiqua, A., et al. (2022). An overview of the environmental pollution and health effects associated with waste landfilling and open dumping. *Environmental Science and Pollution Research, 29*, 58514–58536.

Sorrels, J. L., et al. (2019) *VOC controls.*

Spokas, K., et al. (2006). Methane mass balance at three landfill sites: What is the efficiency of capture by gas collection systems? *Waste Management, 26*, 516–525.

Sun, Q., et al. (2015). Selection of appropriate biogas upgrading technology-a review of biogas cleaning, upgrading and utilisation. *Renewable and Sustainable Energy Reviews, 51*, 521–532.

Swaminanthan, M. (2018). How can India's waste problem see a systemic change?. *Economic and Political Weekly.* [Online]. Available: https://www.epw.in/engage/article/institutional-framework-implementing-solid-waste-management-india-macro-analysis. Accessed: May 22, 2023.

Talyan, V., et al. (2008). State of municipal solid waste management in Delhi, the capital of India. *Waste Management, 28*, 1276–1287.

Taylor, W. C., et al. (2007). Environmental justice: A framework for collaboration between the public health and parks and recreation fields to study disparities in physical activity. *Journal of Physical Activity & Health, 4*(Suppl), 1.

US ATSDR. (2001). Monitoring of landfill gas. In Agency for Toxic Substances and Disease Registry (ATSDR) (Eds.), *Landfill gas primer—An overview for environmental health professionals.* United States Department of Health and Human Services (DHHS)

US DOE. (2019). *Waste-to-energy from municipal solid wastes.*

US EPA. (2021). Waste. In *Inventory of U.S. Greenhouse gas emissions and sinks: 1990–2019* (pp. 1–64). U.S. Environmental Protection Agency.

US EPA. (2021). Overview of greenhouse gases. *United States Environmental Protection Agency*. [Online]. Available: https://www.epa.gov/ghgemissions/overview-greenhouse-gases#methane. Accessed: May 18, 2023.

US EPA. (2021). Basic information about landfill gas. *United States Environmental Protection Agency (EPA)*. [Online]. Available: https://www.epa.gov/lmop/basic-information-about-landfi ll-gas. Accessed: May 18, 2023.

US EPA. (2022). Volatile organic compounds' impact on indoor air quality. *United States Environmental Protection (EPA)*. [Online]. Available: https://www.epa.gov/indoor-air-quality-iaq/volatile-organic-compounds-impact-indoor-air-quality. Accessed: May 19, 2023.

US EPA. (2022). National overview: Facts and figures on materials, wastes and recycling. *United States Environmental Protection Agency (EPA)*. [Online]. Available: https://www.epa.gov/facts-and-figures-about-materials-waste-and-recycling/national-ove rview-facts-and-figures-materials. Accessed: May 19, 2023.

US EPA. (2023b). Continue to operate gas collection system and flare. *U.S. Environmental Protection Agency*. [Online]. Available: https://www.epa.gov/lmop/continue-operate-gas-collection-system-and-flare. Accessed: Aug 06, 2023.

US EPA. (2023). Benefits of landfill gas energy projects. *United States Environmental Protection Agency*. [Online]. Available: https://www.epa.gov/lmop/benefits-landfill-gas-energy-projects. Accessed: May 19, 2023.

Vahk, J. (2020). *Landfill emission reductions only tell half the story as GHG emissions from Waste-to-Energy incineration double.*

Valdebenito-Rolack, E., et al. (2021). Markers for the comparison of the performances of anoxic biotrickling filters in biogas desulphurisation: A critical review. *Processes, 9*, 1–22.

Wang, Q., et al. (2022). Gas transport in landfill cover system: A critical appraisal. *Journal of Environmental Management, 321*, 116020.

Weber, R., et al. (2008). Dioxin- and POP-contaminated sites -contemporary and future relevance and challenges: Overview on background, aims and scope of the series. *Environmental Science and Pollution Research, 15*, 363–393.

Whalen, S. C., et al. (1990). Rapid methane oxidation in a landfill cover soil. *Applied and Environment Microbiology, 56*, 3405–3411.

WHO. (2022). Ambient (outdoor) air pollution. *World Health Organization (WHO)*. [Online]. Available: https://www.who.int/news-room/fact-sheets/detail/ambient-(outdoor)-air-quality-and-health. Accessed: May 19, 2023.

Wijekoon, P., et al. (2022). Progress and prospects in mitigation of landfill leachate pollution: Risk, pollution potential, treatment and challenges. *Journal of Hazardous Materials, 421*, 126627.

Williams, P. A., et al. (2023). Review of barriers to effective implementation of waste and energy management policies in Ghana: Implications for the promotion of waste-to-energy technologies. *Waste, 1*, 313–332.

Winn, Z. (2022) Reducing methane emissions at landfills. *Massachusetts Institute of Technology*. [Online]. Available: https://news.mit.edu/2022/loci-methane-emissions-landfills-0202. Accessed: May 19, 2023.

World Bank. (2018). Global waste to grow by 70 percent by 2050 unless urgent action is taken. *The World Bank*. [Online]. Available: https://www.worldbank.org/en/news/press-release/2018/09/20/global-waste-to-grow-by-70-percent-by-2050-unless-urgent-action-is-taken-world-bank-report

World Bank. (2021). Trends in solid waste management. *The World Bank*. [Online]. Available: https://datatopics.worldbank.org/what-a-waste/trends_in_solid_waste_management.html. Accessed: Aug 10, 2023.

Yadav, P., & Samadder, S. R. (2018). Environmental impact assessment of municipal solid waste management options using life cycle assessment: A case study. *Environmental Science and Pollution Research, 25*, 838–854.

Yip, C. H., & Chua, K. H. (2008). An overview on the feasibility of harvesting landfill gas from MSW to recover energy. In *International Conference on Construction and Building Technology* (pp. 303–310).

Yusnan, et al. (2020). Effect of biocover thickness on methane oxidation rate of landfill gas emission from municipal solid waste landfill in tropical climate region. In *IOP Conference Series Materials Science Engineering* (Vol. 778, pp. 0–7). IOP Publishing.

Zhang, Y., et al. (2021). Release, transformation, and risk factors of polybrominated diphenayl ethers from landfills to the surrounding environments: A review. *Environment International, 157*, 106780.

Chapter 10
Emission from Engineered Landfills and Impacts on Climate

Sukhendu Dey, Priyajit Chatterjee, Megha Santra, Apurba Ratan Ghosh, and Palas Samanta

Abstract A modern landfill is an engineered method for depositing waste in specially constructed and protected cells on the land surface or in excavations into the land surface. Despite the fact that an increasing amount of waste is reused, recycled or energetically valorized, landfills still play an important role in waste management strategies. The degradation of wastes in the landfill results in the production of leachate and gases. These harmful gases are the potential threats to the human and animals' health as well as degraded the natural environment. The leachate from landfill is rich in organic, inorganic and suspended particles which may cause threat to ecosystem. Engineered method landfill gas consists mainly of methane and carbon dioxide, both important greenhouse gases. This chapter recommends an engineered landfill design helpful for landfill gas generation which replaces the fossil fuel as a compressed natural gas or liquefied natural gas. Finally, the impact of these released gaseous emissions has been analyzed completely and this review is complemented with suggestions to minimize the environmental burden of landfills and to re-introduce the buried resources to the material cycle.

Keywords Landfill · Gas emission · Waste · Climate · Environmental impact

S. Dey · M. Santra · A. R. Ghosh
Department of Environmental Science, The University of Burdwan, Burdwan, West Bengal, India

P. Chatterjee
University Science Instrumentation Centre, The University of Burdwan, Burdwan, West Bengal, India

P. Samanta (✉)
Department of Environmental Science, Sukanta Mahavidyalaya, University of North Bengal, Dhupguri, West Bengal, India
e-mail: samanta.palas2010@gmail.com

A. Anouzla et al. (eds.), *Handbook of the Effects of Landfill Emissions on Sustainable Development*, https://doi.org/10.1007/978-3-032-01951-6_10

10.1 Introduction

Each and every year near about 1.3 billion tons of waste are generated throughout world. The World Bank has declared that the amount of generated annual waste will be raised up to 1.4 billion tons by the year 2100 due to population growth, standard of living, industrialization etc. where popular waste disposal method is landfill (Simmons, 2016; Surroop & Mohee, 2011). Landfill can be denoted as the physical facilities or places where waste residuals and solid waste are dumped and it is a most common and old popular waste management choice for developing countries (Zabaleta, 2008). In recent years, there has been a growing concern about the environmental impacts of improper waste management, particularly in the context of landfill disposal. From a social standpoint, the accelerated use of landfills for waste disposal has raised significant issues. The sheer volume of waste being generated has put immense pressure on existing landfill sites, leading to their rapid exhaustion. This has necessitated the search for new areas to establish landfills, further encroaching upon natural habitats and potentially causing irreversible damage to the environment. From an environmental perspective, the improper construction and management of landfills pose substantial threats (Kaur et al., 2022). There is different multiple landfill design with their own design and characteristics having various degree of sustainability, engineered landfill is one of them. Engineered landfills sites are constructed through proper planning and engineered techniques which may control waste and avoid surface water contamination by the help of well-constructed and well-designed surface drainage. Others characteristics include excavation, compacting waste materials into smaller layers, spreading of soil materials to cover the waste, removal of leachate from wastes into lagoons or similar layers and most importantly planned isolation of landfills from surrounding geology. This landfill referred as sanitary landfills which may require protected bottom layer. The design, construction and development of these landfills require sufficient planning, location siting, construction and operational requirements are more stringent than the other types of landfills and this landfill have the impact on public health and the environment (Ozbay et al., 2021). One of the major environmental problems associated with landfills is the production of leachate. Leachate refers to the contaminated liquid that is generated as water percolates through the waste in a landfill. There is ample evidence that improper disposal of waste leading to the production of leachate may cause contamination of air, surface water, groundwater, soils, sediments, and biota (Makgae, 2011). Contamination of groundwater by landfill leachate is considered a major environmental concern, as it poses a risk to downstream surface waters and wells. Additionally, leachate can have adverse effects on the ecosystem and public health (Weng et al., 2010). The adverse environmental impacts associated with landfill management necessitate the implementation of effective waste management strategies. In order to mitigate the environmental burden of landfills, it is crucial to employ engineered landfill design that focuses on reducing and managing leachate and gas (Siddiqua et al., 2022).

One such strategy is the implementation of engineered landfill designs that prioritize gas generation. These designs aim to harness the energy potential of landfill gases, such as methane, by capturing and utilizing them as an alternative to fossil fuels. These engineered landfill designs include the installation of gas collection systems, where methane and other gases produced during waste decomposition are captured and extracted (Chetri & Reddy, 2021). The air around engineered landfills seemed to have a strange quality to it, one that the scientific community had only begun to understand. As part of their research into the environmental impact of landfill emissions, a team of researchers had gathered at a landfill site in the middle of the day (Mor & Ravindra, 2023). The sun shone brightly overhead, shedding light on the intricate network of pipes snaking through the landfill and leading up into the atmosphere. The team, equipped with an array of sensors and detectors, had been tasked with measuring the level of emissions produced by the landfill. The resulting readings showed that these emissions were far higher than expected, and contained a number of noxious compounds. It was clear that the landfill was having a serious impact on the environment, and the team had concluded that the landfill was a major contributor to climate change. The team spent the remainder of the day discussing the findings and strategizing on how to mitigate the emissions. As they packed up their equipment and prepared to leave the site, the team was struck by the gravity of their discoveries. They had, through their research, uncovered a threat to the world that had only recently become apparent. As the team drove away from the landfill, they discussed the implications of their findings. What measures could be taken to reduce emissions from landfills? Could this threat be stopped, or at least minimized? These questions lingered in the minds of the researchers, and their work had only just begun.

This chapter will focus on emissions from engineered landfills and their potential impacts on climate. It will review the different types of gases produced by landfills, as well as the technologies used to mitigate their release. Finally, this paper will discuss the overall impact of landfills on climate, as well as the potential solutions for reducing landfills' contribution to global warming. By understanding the full scope of landfill emissions, we can make informed decisions to reduce our collective carbon footprint and protect our planet for future generations.

10.2 Cumulative Impacts of the Waste Treatment Facility

Due to urbanization and industrialization huge amount of waste are generated and this huge amount of waste are needed to minimize or required to manage. Here landfill is the most common and popular technique for municipal and industrial waste throughout the world (Adesina et al., 2018; Gandhimathi et al., 2013; Hamoda, 2006; Kale et al., 2010; Krcmar et al., 2018). At the landfill sites soil and ground water got contaminated additionally gases and odour also emitted. Different gases and odour are generated due to the anaerobic decomposition of solid waste (Abdul-Wahab et al., 2017; Kumar et al., 2016; Liu et al., 2016). Naturally gases and odour are generated through the biodegradation of organic carbon which contain by waste (Chemel et al., 2012; Fang et al., 2012). From the waste management facilities different air pollutants are generated like methane (CH_4), carbon dioxide (CO_2), carbon monoxide (CO), oxide of nitrogen (NO), sulphur dioxide (SO_2), particulate matter (PM), hydrocarbon (HC) etc. Negative environmental impact and health consequences observed due to emission of landfill gases and odour. Anaerobic decomposition of wastes in landfills promotes the production of CH_4 which has a global warming potential of about 21 multiples of that of CO_2. Epidemiological studies have established that there is a high correlation between the concentration level of air contaminants and human health (Ancona et al., 2015; Ashworth et al., 2016; Hamoda, 2006). Acid gases such as nitrogen dioxide and sulphur dioxide emitted from integrated waste management facility can cause inflammation and bronchoconstriction (Kim et al., 2013a, 2013b; Kim et al., 2013a, 2013b; Spengler et al., 1990). They irritate airways, affect the immune cells in the lungs, and increase susceptibility to respiratory infections. While people suffering from Asthma are most susceptible, although a high level of NO_2 and SO_2 may also produce effects on the lung function of non-asthmatics (Porta et al., 2009).

10.3 Engineered Landfills and Landfill Emission

Landfills are an important means of disposing of human waste and other unwanted materials, and they are found both near urban areas and in remote parts of the world. The term landfilling is used to describe that the process where solid waste is dumped in a landfill. Landfills are classified in three categories such as hazardous, non-hazardous and inert waste. The major components of the landfill construction are the bottom liner, leachate collection and management system, gas management facilities, storm water management, intermediate covers and final cover (Tchobanoglous et al., 1993; Vesilind et al., 2002). An engineered landfill is a trash disposal site that is meticulously planned, built, and operated to reduce negative effects on the environment and safeguard public health. An engineered landfill is made up of a number of systems and parts that cooperate to control waste, keep it contained, and keep the surrounding area clean. Safety, long-term sustainability, and environmental preservation are the main priorities in the planning and building of designed landfills. In order to handle trash responsibly, reduce the risk of environmental contamination, and lessen the danger to public health, responsibly designed landfills are necessary. An engineered landfill typically has the following structure (Fig. 10.1):

Liner Framework: Two types of liner are generally used in engineered landfilling. The *primary liner* is often composed of impermeable materials such as clay, geomembrane made of high-density polyethylene (HDPE), or both. In order to stop waste and leachate from seeping into the groundwater and underlying soil, it is positioned at the bottom of the landfill. Secondary liner is generally utilized to add another

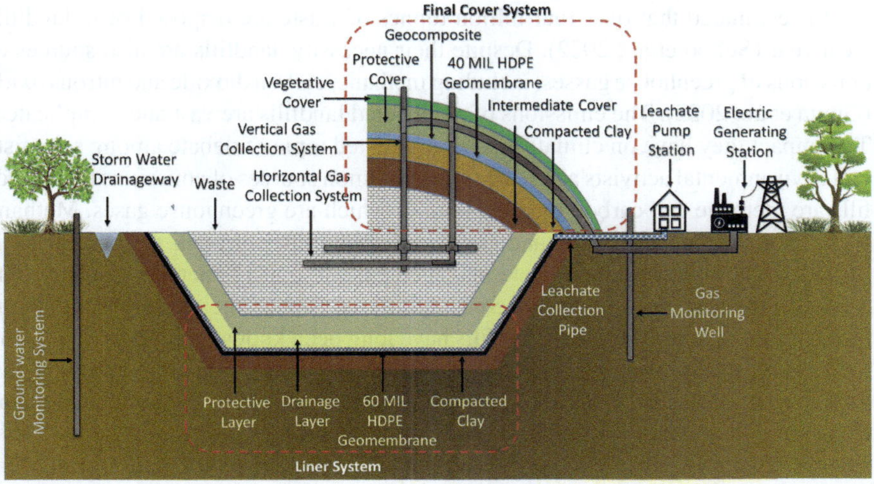

Fig. 10.1 Schematic diagram of a engineered landfill (Mondal et al., 2023; Permission taken from Elsevier, licence number 5656510065678)

degree of defense against leachate migration, certain landfills use secondary liners. It is frequently built over the main liner.

Leachate Gathering Mechanism: Above the primary liner, a system of pipes and drainage layers is put in place to collect leachate—a liquid that occurs when water seeps through the garbage. Usually, collected leachate is piped to a processing plant for analysis.

System for Collecting Gas: Methane (CH_4) and carbon dioxide (CO_2) are the main gases extracted from landfills by installing gas collection wells and pipelines. These gasses, which are produced as garbage breaks down, have the potential to be dangerous if discharged into the atmosphere. Recovered gas can be managed to reduce emissions in various ways, for as by flaring it or using it to generate electricity.

Stormwater Control System and Trash Cell: Engineered landfills are equipped with stormwater management technologies to keep runoff away from garbage. By gathering and controlling stormwater runoff, these systems make sure that it doesn't combine with leachate.

Separate trash cells or disposal sites make up the landfill. Garbage is progressively added to these cells, and the location of the garbage is carefully regulated to reduce settlement and avoid air pockets.

Last Coverage Framework: After a cell is completely filled, a last cover system is put in place to cap and seal it. Usually composed of many layers, this cover contains plants to prevent erosion and lessen water infiltration, a protective soil layer, and a geomembrane.

It is estimated that over four billion tonnes of waste are disposed of in landfills each year (Sohoo et al., 2022). Despite their necessity, landfills are also sources of emissions of greenhouse gasses, including methane, carbon dioxide and nitrous oxide (Gupta et al., 2022). The emissions of engineered landfills are vast and complicated. The impact they have on climate has been a heated topic of debate among scientists and environmental activists alike. The most common sources of emissions from landfills are methane and carbon dioxide, both of which are greenhouse gases. Methane is the more potent of the two and can be released either through the landfill itself or through activities such as waste management and management of leachate. The release of methane gas from engineered landfills has been an increasingly contentious issue in recent years. Although landfills have long been known to produce the potent greenhouse gas, it wasn't until recently that scientists fully understood the magnitude of its impacts on the environment, with some estimates suggesting that methane emissions from landfills account for over 10 percent of global greenhouse gas emissions (Chai et al., 2016). Carbon dioxide can be released from the decomposition of organic matter or from combustion of landfill gas. The extent to which these emissions contribute to climate change has been the subject of much discussion (Bogner et al., 2011). Studies have shown that landfill gas emissions are up to four times more potent than carbon dioxide as a greenhouse gas. However, some dispute this, arguing that methane's potency is overstated due to its shorter lifespan in the atmosphere. Further, there is debate over how much of these emissions are actually released from landfills, as some of the methane may be produced by other sources such as wetlands or agriculture (Johari et al., 2012; Voigt et al., 2017).

In response, landfill operators have been increasingly striving to reduce their methane emissions. To this end, many landfills are investing in advanced engineering techniques designed to capture and control the emission of methane and other pollutants (Ozbay et al., 2021). These engineering efforts include the installation of methane collection systems, which trap and process methane released from the landfill and then use it to produce energy. Additionally, landfill operators are also employing the use of gas controls, such as aeration systems, that reduce the amount of methane released into the atmosphere (Mehrdad et al., 2021; Vitkov, 2019). Finally, engineers are developing more effective methods of capturing and treating landfill leachate, the contaminated liquid that often forms pools at the bottom of landfills. By capturing this liquid and treating it off-site, operators create a system that reduces the amount of methane and other pollutants released into the atmosphere (Abdel-Shafy et al., 2023). By controlling the emission of methane and other pollutants, landfill operators are not only helping to reduce their own environmental footprint, but they are also helping to lessen the impacts of climate change (Glennerster & Jayachandran, 2023). It is a difficult task, but one that is worth attempting in order to create a more sustainable future. In order to better understand the impacts of engineered landfills on climate, it is important to assess their emission source input scenarios. This assessment will need to take into account factors such as the type of waste being disposed of, the quantity of waste being disposed of, and the type of landfill used. Additionally, consideration must be made of sources of emissions not necessarily related to landfills, such as agricultural practices and energy generation (Malav et al., 2020). Once these emission source input scenarios have been determined, further research is necessary to determine how this will affect climate change.

10.3.1 Landfill Gasses

Landfill gas (LFG) or biogas is one of the biological products which is available in municipal solid waste (MSW) landfills and the product generated from the decomposition of biologically decomposable organic matter (Rada et al., 2015; Shen et al., 2018). During decomposition of organic matter at the anaerobic condition LFG generated from the MSW landfill. These gases are mostly made up of carbon dioxide (CO_2) and methane (CH_4), with lower amounts of other trace gases. One-ton MSW contain 150–250 kg organic carbon from where microorganisms transform it into landfill gas at anaerobic condition. 40–60% methane has an average heating value of 17,765 kJ/N m^3 which is generated as LFG, additionally energy conversion efficiency is 34% and produced electric energy is approximately 2.5 kW h/N m^3 (Zappini et al., 2010). The major components of landfill gas are methane (CH_4) and carbon di-oxide (CO_2) with a large number of other components such as sulfide, ammonia, non-methane volatile organic compounds (VOCs) (Hasnine et al., 2022). Apart from this some organic and inorganic compounds are generated in landfills by the help of biological and chemical transformation, the compounds are tri- and per-chlorethylene to vinylchloride, sulphur compounds to hydrogen sulphide (H_2S) where large proportion of the plasterboard that is gypsum and $CaSO_4$ are responsible for H_2S emission and amino acids to methyl- and ethyl-mercaptans (Naveen et al., 2017). By the biological decomposition of 1 ton of wastes produces 442 m^3 of landfill gas containing 55% of methane and calorific value of 15–21 MJ/m^3 which is approximately half of total natural gas (Danthurebandara, 2015). 20% of the methane emission generated from each landfill sites among total global anthropogenic sources (Anarbaev et al., 2019). By installing horizontal and vertical wells within the landfill sites landfill gas can be monitored or controlled. These wells either connected to the central blower system or vented to the atmosphere which may pulls gas to a flare or treatment process. Intergovernmental Panel on Climate Change (IPCC) report that the landfill gas collection efficiencies ranging from 9 to 90% and estimates an average of 20% (IPCC 2006). Some factors such as waste composition, moisture content, landfill age and temperature etc. are affecting the amount of LFG. LFG generation starts after one or two years of waste disposal into the landfill site and continues throughout 15–25 years (Scheutz & Kjeldsen, 2019; Tchobanoglous & Kreith, 2002; Thomasen et al., 2019). The following are the detailed composition of landfill gases (Table 10.1):

Methane (CH_4): Methane makes up between 40 and 60% of the total volume of landfill gasses and is the most prevalent component. It is a strong greenhouse gas with a far greater potential for rapid global warming than carbon dioxide. An estimated

Table 10.1 Typical composition of landfill gas emission

Items	Percent by volume
Methane	45–60
Carbon dioxide	40–60
Oxygen	0.1–1
Nitrogen	2–5
Ammonia	0.1–1
Sulphides	0–1
Hydrogen	0–0.2
Carbon monooxide	0–0.2
NMOCs (non-methane organic compounds)	0.01–0.6

*Source*EPA (2023)

30–70 million tonnes of methane are released annually from landfills worldwide. According to the Global Methane Tracker 2022, 20% of methane emissions in India come from the waste sector. In 2021, the methane released by MSW landfills was almost equal to the greenhouse gas (GHG) emissions from driving nearly 23.1 million gasoline-powered cars for a year or the carbon dioxide emissions from using nearly 13.1 million homes for energy for a year (EPA, 2023).

Carbon dioxide (CO_2): Usually comprising between 30 and 40% of the total volume of landfill gasses, carbon dioxide is a major component. Even though it is not as strong a greenhouse gas as methane, it nevertheless plays a significant role in global warming.

Trace gases: Other trace gases, such as nitrogen (N_2), oxygen (O_2), sulfur compounds, and volatile organic compounds (VOCs), are also present in smaller quantities in landfill gases. Depending on the waste materials and landfill circumstances, the precise composition of trace gases can change.

Ammonia (NH_3): Ammonia can be found in landfill gasses, particularly during the decomposition of organic waste (food scraps and agricultural wastes).

Hydrogen (H_2): Although it occurs in lower amounts, hydrogen is created as garbage breaks down and is normally present in landfill gas.

Nitrogen (N_2): Nitrogen is a component of the air that enters the landfill and can also be found in landfill gasses.

Hydrogen sulfide (H_2S): Another trace gas detected in the emissions from landfills is hydrogen sulfide (H_2S). It can lead to problems with air quality near landfills and has a very bad smell.

Oxygen (O_2): The anaerobic (low-oxygen) conditions found in landfills result in generally low oxygen levels in landfill gas. On the other hand, trace amounts of oxygen might exist.

10.3.2 Leachate Emission

Leachate, a term used for liquids that contain dissolved or suspended materials that are produced when water percolates through landfill, is a potential source of environmental pollution (Kamaruddin et al., 2017). Percolated water picks up suspended and soluble materials which may generate from or degradation products of the waste. Main organic contents of leachate manufactured during breakdown process and its organic strength is normally measured in terms of biochemical oxygen demand (BOD), chemical oxygen demand (COD) and total organic carbon (TOC) (Crowley et al., 2003a, 2003b). Wide variety of toxic, hazardous and carcinogenic chemical contaminants present in municipal solid waste leachate. The characteristics of leachate produced are highly variable depending on the composition of the waste, waste age, cover design, compaction, site hydrology, precipitation rates, sampling procedure and interaction of leachate with the environment and landfill design and operation (Salami & Susu, 2019). Rain plays an important role in formation of leachate. Rain water percolates through all the deposited waste and binds to dissolve and non-dissolved waste substance with the help of different chemical and physical reactions. Additionally, surface runoff, groundwater and biological decomposition also take part in the leachate formation. Landfilling is designed as water-proof covers, insulation layers which is made up of mineral cohesive soil, geotextiles and/or plastic materials. During leachate upper layers of the landfill are affected more than the others (Da Costa et al., 2018). The leachate generated from landfills has a direct impact on the environment due to its potential for contamination. This chapter discusses the leachate generation from engineered landfills and its potential impacts on climate (Iravanian & Ravari, 2020). This is very much important to manage and control the leachate discharge and production due to the potential threat to both the environment mainly groundwater and human health. It is an issue of particular relevance in the context of global warming and climate change as landfills are often cited as a major source of carbon dioxide emissions. The leachate generated from engineered landfills has an important influence in the global greenhouse gas (GHG) emission. It is known that methane and carbon dioxide which are the primary GHG from engineered landfills are present in landfill leachate (Danthurebandara et al., 2012; Gautam & Agrawal, 2021; Zhang et al., 2019). The magnitude of leachate-related emissions from engineered landfills should be considered in order to better assess the amount of emissions that are contributing to climate change (Duan et al., 2021). The emissions from landfill leachate can cause various environmental impacts such as acidification, eutrophication, ground water contamination, and climate change (Danthurebandara et al., 2012). The potential impacts of landfill leachate emissions on climate include increased global average temperatures, increased intensity of storms, and possible changes in precipitation patterns (Hellberg & Chu, 2016). Leachates have four main component such as nutrients which is namely nitrogen, heavy metals, volatile organic compounds and toxic organic compounds (Arunbabu et al., 2017; Budi et al., 2016;

Moody & Townsend, 2017). In order to reduce emissions from landfills, it is necessary to better understand the emission dynamics of the leachate and its potential impacts on climate.

10.4 Factors Affecting the Landfill Emission

The properties of the garbage, such as its age and composition, as well as a variety of environmental conditions, such as the landfills's moisture level, temperature, humidity, and oxygen concentration, determine the amount and velocity of landfill gas production.

Waste materials composition: Generally, microbial breakdown produces increased gas from garbage in a landfill that has greater organic waste. Certain kinds of organic waste include minerals like calcium, magnesium, potassium, salt, and others that support the growth of bacteria. The amount of landfill gas produced rises in the presence of such nutrients. On the other hand, certain wastes include substances in them that damage bacteria, reducing the amount of gas generated. For example, excessive salt levels in trash can limit the growth of methane-producing microorganisms.

Refusal age: A greater amount of gas will be produced by freshly buried waste compared with older rubbish. generally, landfills generate significant gas production in one to three years. Typically, optimum generation of gases happens five to seven years following wastes are discarded. The majority of gas is generated inside 20 years of garbage being placed; nevertheless, trace amounts of gas might last for fifty years or longer after being released from a landfill. Nevertheless, a scenario with a low yield of methane calculates that methane will be produced by gradually decaying garbage after five years, and the gas is going to continue to be released for forty years. The breakdown of waste in various areas of the landfill may be occurring at various times, contingent on how long ago the garbage was initially deposited in every region. The period of gas generation lasts depends in large part on the quantity of organic matter present in the trash.

Oxygen level in waste: Bacteria won't start producing methane until all of the oxygen has been consumed. Aerobic microbes in a landfill may break down trash in Phase I for a longer period of time depending on how much oxygen is there. A greater amount of oxygen is readily accessible to oxygen-dependent bacteria when garbage is lightly buried or regularly agitated, allowing them to continue producing water and carbon dioxide for a greater duration of time. However, when aerobic microbes give way to methane-producing anaerobic bacteria in Phase III, methane generation will start sooner if the trash is severely packed. Any oxygen left in the waste will cause the anaerobic bacteria to stop producing methane gas because they only begin to do so when the aerobic bacteria have consumed all of the oxygen. Barometric highs have the tendency to change the activity of bacteria in shallower surface soils of a landfill by introducing air oxygen. In this case, trash in Phase IV, for instance, can momentarily return to Phase I until all of the oxygen is consumed once more.

Level of moisture: A specific level of moisture promotes microbial development and spreads nutrients and bacteria throughout the landfill, which raises gas output. Optimal gas generation is encouraged (e.g., in a capped landfill) when the moisture level is 40% or greater depending on the wet mass of the garbage. Since trash compression raises the bulkiness of the materials in the landfill and decreases the speed at where water penetrates the garbage, it inhibits the formation of gas. If a landfill receives a lot of rain or has porous landfill covers, more water will be introduced, which will speed up the creation of methane.

Temperature condition: Bacterial growth rises in higher temperatures, and this accelerates the pace at which gas from landfills is produced. Lower temperatures suppress the growth of microorganisms. Below 50° Fahrenheit (F), bacteria's activity usually decreases sharply. In shallower landfills, weather variations have a far bigger impact on the generation of gas. Unlike deeper landfills, wherein the garbage is covered by an extensive soil layer, the bacteria in these locations are less resistant to variations in temperature. Generally, a covered landfill keeps its temperature constant, which maximizes the generation of gas. While temperatures as high as 158°F have been recorded, bacteria growth generates heat that stabilizes a landfill's temperatures between 77°F and 113°F. Moreover, temperature rises encourage chemical processes and volatilization. Emission of NMOCs typically double for every 18 °C.

10.5 Impacts of Engineered Landfill

Impact due to Landfill Construction: For landfill construction site selection is the major issue to manage waste with all infrastructural facilities adopted. Due to this construction landscape changes, displacement of fauna, loss of habitats as well as have the capacity to destroy the ecology of the site and as well as all the impacts are site specific and need to be assessed on case-by-case basis (Banias et al., 2010; Crowley et al., 2003a, 2003b). The soil quality of the selected sites differs from the other surrounding sites due to chemical and physical properties of soil due to removal of topsoil. Soil is an important resource which supports kind of cultural, economical and ecological functions and the factors such as density, porosity, aggregate strength and water holding capacity related to soil quality which are developed in the top soil fraction. Subsoil being poorly developed and have lower ability to grow plant and all these disturbances can be occurred due to construction activities. The movement of heavy machinery can make excessive compaction of topsoil and subsoil. On the landfill construction site, existing flora and fauna got impacted more due to removal vegetation. This damage can be recovered after closing the phase of the landfills (Danthurebandara et al., 2012; Mellanby, 1992).

Environmental Impacts: The environmental impact of landfills can range from air and water pollution to climate change (El-Fadel et al, 1997). One of the most important factors to consider when evaluating the environmental impact of landfills is the emissions produced by materials disposed inside. Engineered landfills, in particular, are often equipped with systems that reduce atmospheric emissions, but some compounds can still seep into the atmosphere, making them a contributor to climate change (Chai et al., 2016).

Societal and Public Health Impacts: The environmental and social impacts of landfills are significant. Emissions from landfills cause both air pollution and climate change. Landfill gas, a byproduct of the decomposition process, contains methane, a potent greenhouse gas (Shammas et al., 2020). The methane released from landfills has significant global warming potential, and studies have found that emissions from landfills can significantly contribute to climate change (Naveen et al., 2023). In addition, many landfills lack appropriate treatment systems that would reduce the emission of hazardous pollutants, such as volatile organic compounds and dioxins, into the atmosphere. These pollutants can be detrimental to human health when exposed at high levels (Karmakar et al., 2023). The health and social impacts of landfills also extend to their impact on the local community. For instance, landfills can cause odors, fly infestations, and water contamination. This can lead to decreased air quality, public health risks, and economic harm to adjacent communities (Ogundele et al., 2018). Properly engineered landfills can minimize these impacts, but it is important to ensure that these facilities are properly operated and maintained. Finally, landfills produce large amounts of waste that can be difficult to manage and can take up large amounts of space (Sosna et al., 2023). Properly disposing of the waste is essential to protecting the environment and preserving resources. Thus, it is important that landfills are properly designed and managed. This includes ensuring that adequate landfills have appropriate design, operation, and maintenance (El-Saadony et al., 2023). Different environmental and health impact of landfilling are given in Table 10.2.

Leachate Emission Impact: Leachate emissions from landfill sites are of particular concern due to their toxic impact when released untreated into the environment (Cotman & Zgajnar-Gotvajn, 2008). Moreover, the improper management of solid waste in landfills has serious environmental consequences (Solid Waste Management in Minna, North Central Nigeria: Present Practices and Future Challenges, 2016). Improper waste disposal in landfills can result in the emission of pollutants into the atmosphere and contamination of surface and subsurface waters. Leachate, a liquid formed as water passes through the waste and absorbs dissolved or suspended materials, contains organic and inorganic substances, posing a threat to ecosystems and human health when it enters surface and groundwater. Leachate from landfills may contain hazardous substances that can contaminate water sources, leading to the pollution of drinking water supplies and ecological damage (Weng et al., 2010). According to various studies, the leachate from landfill sites can cause deterioration of surface waters and imperil groundwater quality (Yu et al., 2020). In most of the landfills, leachates are treated as either on-or-off site processes. During the leachate

Table 10.2 Different environmental and health impact of landfilling

Sl. No.	Types of landfills	Environmental impact	Health impact	References
1	Non-hazardous waste landfills	–	No significant health risk observed	Schlosser et al. (2016)
2	Landfill	Due to the presence of organic compounds micro-pollutants are release from landfills and their release continues for lifetime	Landfills is an important source of dioxin pollution which have craniofacial effect and also have teratogenic effects	Vodyanitskii and Yakovlev (2016), Leskow et al. (2019)
3	MSW landfill	During leachate surface water and soil got polluted but ground water did not. And also release of PM10 particles may cause the air pollution	Diseases such as asthma, cough, flu, stomach ache, bad mood, annoyance and skin infections also related to the landfill	Sanchez-Arias et al. (2019)
4	Regulated and unregulated dump sites	Various compounds like nitrates, phosphates, Ni, Cd etc. leached which may exceeded the US EPA standard for drinking water	Cellular organelles and DNA damages observed along with upregulation of liver activity enzymes	Khalil et al. (2018)
5	Uncontrolled municipal landfill	Leachate contain various metal which contaminated groundwater at the depth of 30 m and was not suitable for drinking	–	Han et al. (2014)
6	Hazardous waste landfills	Groundwater contamination observed over a long period of time	Some leachate metals have carcinogenic and non-carcinogenic effects	Xu et al. (2018)
7	Open" landfill"	Soil concentration of different metal is higher, accumulation through plants also observed with indication of bioaccumulation	–	Alam et al. (2019)
8	Open dumps or "controlled" dumps	Leachate has polluted the drinking water and having biological contamination	–	Bernache (2003)
9	Solid waste landfill	–	No significant health risk observed	Davoli et al. (2010)

(continued)

Table 10.2 (continued)

Sl. No.	Types of landfills	Environmental impact	Health impact	References
10	Waste dumpsite	–	Have carcinigenic effect or risk on the population	Hafeez et al. (2016)
11	Illegal dumping	–	Mortality rate increased due to cancer	Marfe and Di Stefano (2016)
12	MSW dumping and burning	Different air pollutants like CO2, CH4, SO2, Nox etc.releases in tons in amount which may cause of climate change	Heart diseases, respiratory diseases and allergic hypersensitivity etc. health impact shows	Das et al. (2018)
13	Lined landfills	Concentrated perfluoroalkyl are found in higher in amount and POP also in environment	–	Harrad et al. (2019)
14	Dumping of waste and landfilling	From dumped factory POP can be generated and which is consider as a air pollutants and water can be polluted from the washed water	–	Navaro et al.
15	Dumping of waste	After coal mining wastes are dumped and play an important role for contribution of Hg and pollute the soil	–	Antoszczyszyn & Michalska, 2016
16	Landfill garbage site	Groundwater contamination observed over a long period of time which may vary time to time	–	Van Giang & Duan (2017)
17	Domestic waste landfill	Different greenhouse gas and toxic gases released due to aerobic and anaerobic condition and may cause global warming	–	Sainova et al. (2019)

treatment process general traditional pollutants like chemical oxygen demand (COD), biological oxygen demand (BOD), heavy metals and ammonia nitrogen are eliminated (Luo et al., 2020). Furthermore, the decomposition of waste in landfills leads to the production of harmful gases, such as methane and carbon dioxide. These gases are significant contributors to global warming and climate change, as methane is known to be a potent greenhouse gas (Bouyakhsass et al., 2023). The adverse environmental impacts associated with landfill management necessitate the implementation of effective waste management strategies. In order to mitigate the environmental burden of landfills, it is crucial to employ engineered landfill design that focuses on reducing and managing leachate and gas (Siddiqua et al., 2022).

Engineered Landfill and Climate Change: Landfills emit a lot of greenhouse gases, mostly carbon dioxide (CO_2) and methane (CH_4), which are the main culprits in climate change. There are multiple ways in which these emissions contribute to climate change and global warming. These outcomes consist of:

Global warming and Global Warming Potential: Landfills are a significant source of methane emissions, which contributes to greenhouse gas concentrations. Landfills also release carbon dioxide. Long-lasting greenhouse gas CO_2 contributes to climate change over centuries, even if its short-term potency is less than that of methane (Chai et al., 2016). Stronger than carbon dioxide at retaining heat over a shorter period of time, methane is a powerful greenhouse gas. Methane is thought to be more than 80 times more effective at storing heat than carbon dioxide, even though it finally breaks down in the atmosphere over a 20-year period (Shammas et al., 2020). Methane has a far greater potential for global warming than carbon dioxide over a brief period of time (20 years). Methane emissions from landfills increase the amount of greenhouse gasses in the atmosphere overall (Naveen et al., 2023). Methane emissions from landfills can therefore hasten climate change and cause short-term global warming. This may lead to an increase in extreme weather occurrences, a rise in temperature, and other negative environmental effects.

Negative Feedback Loop: Emissions from landfills may be indirectly impacted by climate change. For instance, variations in temperature and meteorological patterns can affect how quickly materials decompose and release gasses from landfills. Pollutants from landfills may potentially seep out as a result of increased rainfall or flooding.

Sea Level Rise: Rising global temperatures and sea levels are the results of climate change, which is fueled by greenhouse gas emissions. Rising sea levels put landfills close to the shore at risk of flooding, which might discharge pollutants and waste products that have been stored and exacerbate environmental issues.

Positive Feedback Loop: Methane emissions from landfills have the potential to worsen climate change. Increased microbial activity in landfills due to global warming can hasten the breakdown of organic waste and increase methane production. This creates a positive feedback loop, aggravating the problem further.

Human Health and Air Quality: Pollutants and volatile organic compounds found in landfill emissions can have a negative impact on both human health and air quality, which are already discussed in public health section (Karmakar et al., 2023). Air quality issues can worsen respiratory ailments and other health issues, particularly in areas where landfills are located (Ogundele et al., 2018).

10.6 Conclusion

This chapter indicates that engineered landfills may pose a significant source of methane emissions, with the potential of contributing to global climate change. In particular, the data suggest that emissions from engineered landfills are higher than those from natural landfills, due to the structure of engineered landfills, which is designed to promote increased methane production. Furthermore, our data indicate that engineered landfills have the potential to emit more methane than even calculate estimate based on existing models, suggesting that engineered landfills could be a larger source of emissions than previously thought. Given these results, it is important that additional research explore the potential of engineered landfills to contribute to climate change. As the world shifts towards more sustainable forms of waste disposal, it is essential that we understand the environmental impact of engineered landfills and the potential of these landfills to emit more methane than previously thought. Only by understanding the full scope of this issue can we make informed decisions regarding our waste management practices in the future.

Acknowledgements Authors are highly acknowledging the respective Institutions for continuous support and encouragement during writing this book chapter.

References

Abdel-Shafy, H. I., Ibrahim, A. M., Al-Sulaiman, A. M., & Okasha, R. A. (2023). Landfill leachate: Sources, nature, organic composition, and treatment: An environmental overview. *Ain Shams Engineering Journal*, 102293.https://doi.org/10.1016/j.asej.2023.102293

Abdul-Wahab, S., Al-Rawas, G., Charabi, Y., Al-Wardy, M., & Fadlallah, S. (2017). A study to investigate the key sources of odors in Al-Multaqa Village, Sultanate of Oman. *Environmental Forensics, 18*(1), 15–35. https://doi.org/10.1080/15275922.2016.1230911

Adesina, O. A., Sonibare, J. A., Diagboya, P. N., Adeniran, J. A., & Yusuf, R. O. (2018). Spatiotemporal distributions of polycyclic aromatic hydrocarbons close to a typical medical waste incinerator. *Environmental Science and Pollution Research, 25*, 274–282. https://doi.org/10.1007/s11356-017-0335-1

Alam, R., Ahmed, Z., & Howladar, M. F. (2020). Evaluation of heavy metal contamination in water, soil and plant around the open landfill site Mogla Bazar in Sylhet, Bangladesh. *Groundwater for Sustainable Development, 10*, 100311. https://doi.org/10.1016/j.gsd.2019.100311

Anarbaev, A., Tursunov, O., Kodirov, D., Muzafarov, S., Babayev, A., Sanbetova, A., Batirova, L., & Mirzaev, B. (2019). Reduction of greenhouse gas emissions from renewable energy technologies in agricultural sectors of Uzbekistan. In *E3S Web of Conferences* (Vol. 135, p. 01035). EDP Sciences. https://doi.org/10.1051/e3sconf/201913501035.

Ancona, C., Badaloni, C., Mataloni, F., Bolignano, A., Bucci, S., Cesaroni, G., Sozzi, R., Davoli, M., & Forastiere, F. (2015). Mortality and morbidity in a population exposed to multiple sources of air pollution: A retrospective cohort study using air dispersion models. *Environmental Research, 137*, 467–474. https://doi.org/10.1016/j.envres.2014.10.036

Antoszczyszyn, T., & Michalska, A. (2016). The potential risk of environmental contamination by mercury contained in Polish coal mining waste. *Journal of Sustainable Mining, 15*(4), 191–196. https://doi.org/10.1016/j.jsm.2017.04.002

Arunbabu, V., Indu, K. S., & Ramasamy, E. V. (2017). Leachate pollution index as an effective tool in determining the phytotoxicity of municipal solid waste leachate. *Waste Management, 68*, 329–336. https://doi.org/10.1016/j.wasman.2017.07.012

Ashworth, D. C., Fuller, G. W., Toledano, M. B., Font, A., Elliott, P., Hansell, A. L., & De Hoogh, K. (2016). Comparative assessment of particulate air pollution exposure from municipal solid waste incinerator emissions. In *Air quality* (pp. 87–116). Apple Academic Press.

Banias, G., Achillas, C., Vlachokostas, C., Moussiopoulos, N., & Tarsenis, S. (2010). Assessing multiple criteria for the optimal location of a construction and demolition waste management facility. *Building and Environment, 45*(10), 2317–2326.https://doi.org/10.1016/j.buildenv.2010.04.016

Bernache, G. (2003). The environmental impact of municipal waste management: The case of Guadalajara metro area. *Resources, Conservation and Recycling, 39*(3), 223–237. https://doi.org/10.1016/S0921-3449(03)00029-6

Bogner, J. E., Spokas, K. A., & Chanton, J. P. (2011). Seasonal greenhouse gas emissions (methane, carbon dioxide, nitrous oxide) from engineered landfills: Daily, intermediate, and final California cover soils. *Journal of Environmental Quality, 40*(3), 1010–1020. https://doi.org/10.2134/jeq2010.0407

Bouyakhsass, R., Souabi, S., Bouaouda, S., Taleb, A., Kurniawan, T. A., Liang, X., Goh, H. H., & Anouzla, A. (2023). Adding value to unused landfill gas for renewable energy on-site at Oum Azza landfill (Morocco): Environmental feasibility and cost-effectiveness. *Trends in Food Science & Technology*, 104168.https://doi.org/10.1016/j.tifs.2023.104168

Budi, S., Suliasih, B. A., Othman, M. S., Heng, L. Y., & Surif, S. (2016). Toxicity identification evaluation of landfill leachate using fish, prawn and seed plant. *Waste Management, 55*, 231–237. https://doi.org/10.1016/j.wasman.2015.09.022

Chai, X., Tonjes, D. J., & Mahajan, D. (2016). Methane emissions as energy reservoir: Context, scope, causes and mitigation strategies. *Progress in Energy and Combustion Science, 56*, 33–70. https://doi.org/10.1016/j.pecs.2016.05.001

Chemel, C., Riesenmey, C., Batton-Hubert, M., & Vaillant, H. (2012). Odour-impact assessment around a landfill site from weather-type classification, complaint inventory and numerical simulation. *Journal of Environmental Management, 93*(1), 85–94. https://doi.org/10.1016/j.jenvman.2011.08.016

Chetri, J. K., & Reddy, K. R. (2021). Advancements in municipal solid waste landfill cover system: A review. *Journal of the Indian Institute of Science, 101*(4), 557–588. https://doi.org/10.1007/s41745-021-00229-1

Cotman, M., & Zgajnar-Gotvajn, A. (2008, April 30). *Toxicity reduction evaluation (TRE) procedure for identification of toxic substances in landfill leachate.* https://doi.org/10.2495/etox080131

Crowley, D., Staines, A., Collins, C., Bracken, J., & Bruen, M. (2003). *Health and environmental effects of landfilling and incineration of waste—A literature review.*

Crowley, D., Staines, A., Collins, C., Bracken, J., Bruen, M., Fry, J., Hrymak, Victor., Malone, D., Magette, B., Ryan, M., & Thunhurst, C. (2003). *Health and environmental effects of landfilling and incineration of waste—A literature review.*

Da Costa, F. M., Daflon, S. D. A., Bila, D. M., da Fonseca, F. V., & Campos, J. C. (2018). Evaluation of the biodegradability and toxicity of landfill leachates after pretreatment using advanced oxidative processes. *Waste Management, 76*, 606–613. https://doi.org/10.1016/j.wasman.2018.02.030

Danthurebandara, M. (2015). Environmental and economic performance of enhanced landfill mining. *Heverlee Belgium.*

Danthurebandara, M., Van Passel, S., Nelen, D., Tielemans, Y., & Van Acker, K. (2012). Environmental and socio-economic impacts of landfills. *Linnaeus Eco-Tech, 2012*, 40–52.

Das, B., Bhave, P. V., Sapkota, A., & Byanju, R. M. (2018). Estimating emissions from open burning of municipal solid waste in municipalities of Nepal. *Waste Management, 79*, 481–490. https://doi.org/10.1016/j.wasman.2018.08.013

Davoli, E., Fattore, E., Paiano, V., Colombo, A., Palmiotto, M., Rossi, A. N., Grande, M. I., & Fanelli, R. (2010). Waste management health risk assessment: A case study of a solid waste landfill in South Italy. *Waste Management, 30*(8–9), 1608–1613. https://doi.org/10.1016/j.wasman.2009.10.013

Duan, Z., Scheutz, C., & Kjeldsen, P. (2021). Trace gas emissions from municipal solid waste landfills: A review. *Waste Management, 119*, 39–62. https://doi.org/10.1016/j.wasman.2020.09.015

El-Fadel, M., Findikakis, A. N., & Leckie, J. O. (1997). Environmental impacts of solid waste landfilling. *Journal of Environmental Management, 50*(1), 1–25. https://doi.org/10.1006/jema.1995.0131

El-Saadony, M. T., Saad, A. M., El-Wafai, N. A., Abou-Aly, H. E., Salem, H. M., Soliman, S. M., El-Maged, T. A. A., Elrys, A. S., Selim, S., El-Hack, M. A. E., Kappachery, S., El-Tarabily, K. A., & AbuQamar, S. F. (2023). Hazardous wastes and management strategies of landfill leachates: A comprehensive review. *Environmental Technology & Innovation*, 103150.https://doi.org/10.1016/j.eti.2023.103150

EPA. (2023). *Basic information about landfill gas*. https://www.epa.gov/lmop/basic-information-about-landfill-gas. Accessed on October 26, 2023.

Fang, J. J., Yang, N., Cen, D. Y., Shao, L. M., & He, P. J. (2012). Odor compounds from different sources of landfill: Characterization and source identification. *Waste Management, 32*(7), 1401–1410. https://doi.org/10.1016/j.wasman.2012.02.013

Gandhimathi, R., Durai, N. J., Nidheesh, P. V., Ramesh, S. T., & Kanmani, S. (2013). Use of combined coagulation-adsorption process as pretreatment of landfill leachate. *Iranian Journal of Environmental Health Science & Engineering, 10*, 1–7. https://doi.org/10.1186/1735-2746-10-24

Gautam, M., & Agrawal, M. (2021). Greenhouse gas emissions from municipal solid waste management: a review of global scenario. In *Carbon footprint case studies: Municipal solid waste management, sustainable road transport and carbon sequestration* (pp. 123–160). https://doi.org/10.1007/978-981-15-9577-6_5

Glennerster, R., & Jayachandran, S. (2023). Think globally, act globally: Opportunities to mitigate greenhouse gas emissions in low-and middle-income countries. *Journal of Economic Perspectives, 37*(3), 111–135. https://doi.org/10.1257/jep.37.3.111

Gupta, J., Ghosh, P., Kumari, M., & Thakur, I. S. (2022). Solid waste landfill sites for the mitigation of greenhouse gases. In *Biomass, Biofuels, Biochemicals* (pp. 315–340). Elsevier. https://doi.org/10.1016/B978-0-12-823500-3.00010-8.

Hafeez, S., Mahmood, A., Syed, J. H., Li, J., Ali, U., Malik, R. N., & Zhang, G. (2016). Waste dumping sites as a potential source of POPs and associated health risks in perspective of current waste management practices in Lahore city, Pakistan. *Science of the Total Environment, 562*, 953–961. https://doi.org/10.1016/j.scitotenv.2016.01.120

Hamoda, M. F. (2006). Air pollutants emissions from waste treatment and disposal facilities. *Journal of Environmental Science and Health, Part A, 41*(1), 77–85. https://doi.org/10.1080/10934520500298895

Han, D., Tong, X., Currell, M. J., Cao, G., Jin, M., & Tong, C. (2014). Evaluation of the impact of an uncontrolled landfill on surrounding groundwater quality, Zhoukou, China. *Journal of Geochemical Exploration, 136*, 24–39. https://doi.org/10.1016/j.gexplo.2013.09.008

Harrad, S., Drage, D. S., Sharkey, M., & Berresheim, H. (2019). Brominated flame retardants and perfluoroalkyl substances in landfill leachate from Ireland. *Science of the Total Environment, 695*, 133810. https://doi.org/10.1016/j.scitotenv.2019.133810

Hasnine, M. T., Anand, N., Zoungrana, A., Palani, S. G., & Yuan, Q. (2022). An overview of physicochemical and biological treatment of landfill leachate. In *Circular economy in municipal solid waste landfilling: Biomining & leachate treatment: sustainable solid waste management: Waste to Wealth* (pp. 115–152). https://doi.org/10.1007/978-3-031-07785-2_6

Hellberg, R. S., & Chu, E. (2016). Effects of climate change on the persistence and dispersal of foodborne bacterial pathogens in the outdoor environment: A review. *Critical Reviews in Microbiology, 42*(4), 548–572.https://doi.org/10.3109/1040841X.2014.972335

Iravanian, A., & Ravari, S. O. (2020, December). Types of contamination in landfills and effects on the environment: a review study. In *IOP Conference Series: Earth and Environmental Science* (Vol. 614, No. 1, p. 012083). IOP Publishing. https://doi.org/10.1088/1755-1315/614/1/012083.

Johari, A., Ahmed, S. I., Hashim, H., Alkali, H., & Ramli, M. (2012). Economic and environmental benefits of landfill gas from municipal solid waste in Malaysia. *Renewable and Sustainable Energy Reviews, 16*(5), 2907–2912. https://doi.org/10.1016/j.rser.2012.02.005

Kale, S. S., Kadam, A. K., Kumar, S., & Pawar, N. J. (2010). Evaluating pollution potential of leachate from landfill site, from the Pune metropolitan city and its impact on shallow basaltic aquifers. *Environmental Monitoring and Assessment, 162*, 327–346. https://doi.org/10.1007/s10661-009-0799-7

Kamaruddin, M. A., Yusoff, M. S., Rui, L. M., Isa, A. M., Zawawi, M. H., & Alrozi, R. (2017). An overview of municipal solid waste management and landfill leachate treatment: Malaysia and Asian perspectives. *Environmental Science and Pollution Research, 24*, 26988–27020. https://doi.org/10.1007/s11356-017-0303-9

Karmakar, A., Daftari, T., Sivagami, K., Chandan, M. R., Shaik, A. H., Kiran, B., & Chakraborty, S. (2023). A comprehensive insight into Waste to Energy conversion strategies in India and its associated air pollution hazard. *Environmental Technology & Innovation, 29*, 103017.https://doi.org/10.1016/j.eti.2023.103017

Kaur, K., Mor, S., Vig, N., & Ravindra, K. (2022, July 7). *Assessing the risk of landfill leachate on the groundwater quality of Chandigarh.* https://doi.org/10.22438/jeb/43/4/mrn-2097

Khalil, C., Al Hageh, C., Korfali, S., & Khnayzer, R. S. (2018). Municipal leachates health risks: Chemical and cytotoxicity assessment from regulated and unregulated municipal dumpsites in Lebanon. *Chemosphere, 208*, 1–13. https://doi.org/10.1016/j.chemosphere.2018.05.151

Kim, H. H., Lee, C. S., Jeon, J. M., Yu, S. D., Lee, C. W., Park, J. H., Shin, D. C., & Lim, Y. W. (2013b). Analysis of the association between air pollution and allergic diseases exposure from nearby sources of ambient air pollution within elementary school zones in four Korean cities. *Environmental Science and Pollution Research, 20*, 4831–4846. https://doi.org/10.1007/s11356-012-1358-2

Kim, K. H., Jahan, S. A., & Kabir, E. (2013a). A review on human health perspective of air pollution with respect to allergies and asthma. *Environment International, 59*, 41–52. https://doi.org/10.1016/j.envint.2013.05.007

Krčmar, D., Tenodi, S., Grba, N., Kerkez, D., Watson, M., Rončević, S., & Dalmacija, B. (2018). Preremedial assessment of the municipal landfill pollution impact on soil and shallow groundwater in Subotica, Serbia. *Science of the Total Environment, 615*, 1341–1354. https://doi.org/10.1016/j.scitotenv.2017.09.283

Kumar, S., Nimchuk, N., Kumar, R., Zietsman, J., Ramani, T., Spiegelman, C., & Kenney, M. (2016). Specific model for the estimation of methane emission from municipal solid waste landfills in India. *Bioresource Technology, 216*, 981–987. https://doi.org/10.1016/j.biortech.2016.06.050

Leśków, A., Nawrocka, M., Łątkowska, M., Tarnowska, M., Galas, N., Matejuk, A., & Całkosiński, I. (2019). Can contamination of the environment by dioxins cause craniofacial defects? *Human & Experimental Toxicology, 38*(9), 1014–1023. https://doi.org/10.1177/0960327119855121

Liu, Y., Lu, W., Guo, H., Ming, Z., Wang, C., Xu, S., Liu, Y., & Wang, H. (2016). Aromatic compound emissions from municipal solid waste landfill: Emission factors and their impact on air pollution. *Atmospheric Environment, 139*, 205–213. https://doi.org/10.1016/j.atmosenv.2016.05.043

Luo, H., Zeng, Y., Cheng, Y., He, D., & Pan, X. (2020). Recent advances in municipal landfill leachate: A review focusing on its characteristics, treatment, and toxicity assessment. *Science of the Total Environment, 703*, 135468. https://doi.org/10.1016/j.scitotenv.2019.135468

Makgae, M. (2011, August 23). *Key areas in waste management: A South African perspective*. https://doi.org/10.5772/18023

Malav, L. C., Yadav, K. K., Gupta, N., Kumar, S., Sharma, G. K., Krishnan, S., Rezania, S., Kamyab, H., Pham, Q. B., Yadav, S., Bhattacharyya, S., Yadav, V. K., & Bach, Q. V. (2020). A review on municipal solid waste as a renewable source for waste-to-energy project in India: Current practices, challenges, and future opportunities. *Journal of Cleaner Production, 277*, 123227. https://doi.org/10.1016/j.jclepro.2020.123227

Marfe, G., & Di Stefano, C. (2016). The evidence of toxic wastes dumping in Campania, Italy. *Critical Reviews in Oncology/hematology, 105*, 84–91. https://doi.org/10.1016/j.critrevonc.2016.05.007

Mehrdad, S. M., Abbasi, M., Yeganeh, B., & Kamalan, H. (2021). Prediction of methane emission from landfills using machine learning models. *Environmental Progress & Sustainable Energy, 40*(4), e13629. https://doi.org/10.1002/ep.13629

Mellanby, K. (1992). *Waste and pollution: The problem for Britain*. Harper Collins.

Moody, C. M., & Townsend, T. G. (2017). A comparison of landfill leachates based on waste composition. *Waste Management, 63*, 267–274. https://doi.org/10.1016/j.wasman.2016.09.020

Mor, S., & Ravindra, K. (2023). Municipal solid waste landfills in lower-and middle-income countries: Environmental impacts, challenges and sustainable management practices. *Process Safety and Environmental Protection, 174*, 510–530. https://doi.org/10.1016/j.psep.2023.04.014

Naveen BP, Tabaroei, A., & Garg, A. (2023). Methane emission and carbon sequestration potential from municipal solid waste landfill, India. *Sustainability, 15*(9), 7125. https://doi.org/10.3390/su15097125

Naveen, B. P., Mahapatra, D. M., Sitharam, T. G., Sivapullaiah, P. V., & Ramachandra, T. V. (2017). Physico-chemical and biological characterization of urban municipal landfill leachate. *Environmental Pollution, 220*, 1–12. https://doi.org/10.1016/j.envpol.2016.09.002

Ogundele, O. M., Rapheal, O. M., & Abiodun, A. M. (2018). Effects of municipal waste disposal methods on community health in Ibadan-Nigeria. *Polytechnica, 1*(1–2), 61–72. https://doi.org/10.1007/s41050-018-0008-y

Ozbay, G., Jones, M., Gadde, M., Isah, S., & Attarwala, T. (2021). Design and operation of effective landfills with minimal effects on the environment and human health. *Journal of Environmental and Public Health, 2021*.https://doi.org/10.1155/2021/6921607

Porta, D., Milani, S., Lazzarino, A. I., Perucci, C. A., & Forastiere, F. (2009). Systematic review of epidemiological studies on health effects associated with management of solid waste. *Environmental Health, 8*(1), 1–14. https://doi.org/10.1186/1476-069X-8-60

Rada, E. C., Ragazzi, M., Stefani, P., Schiavon, M., & Torretta, V. (2015). Modelling the potential biogas productivity range from a MSW landfill for its sustainable exploitation. *Sustainability, 7*(1), 482–495. https://doi.org/10.3390/su7010482.

Sainova, G. A., Akbasova, A. D., Abdikarim, G. G., Kalieva, N. A., & Mehmet, A. O. (2019). Environmental monitoring on the landfill of solid domestic wastes of the town Kentau. *News Natl Acad Sci Repub Kazakhstan, Series Geol Tech Sci, 1*, 57–62.

Salami, L., & Susu, A. A. (2019). A comprehensive study of leachate characteristics from three soluos dumpsites in Igando Area of Lagos State, Nigeria. *Greener Journal of Environmental Management and Public Safety, 8*(1), 1–14. https://doi.org/10.15580/GJEMPS.2019.1.110118520

Sánchez-Arias, M., Riojas-Rodríguez, H., Catalán-Vázquez, M., Terrazas-Meraz, M. A., Rosas, I., Espinosa-García, A. C., Santos-Luna, R., & Siebe, C. (2019). Socio-environmental assessment of a landfill using a mixed study design: A case study from México. *Waste Management, 85*, 42–59. https://doi.org/10.1016/j.wasman.2018.12.012

Scheutz, C., & Kjeldsen, P. (2019). Guidelines for landfill gas emission monitoring using the tracer gas dispersion method. *Waste Management, 85*, 351–360. https://doi.org/10.1016/j.wasman.2018.12.048

Schlosser, O., Robert, S., & Debeaupuis, C. (2016). Aspergillus fumigatus and mesophilic moulds in air in the surrounding environment downwind of non-hazardous waste landfill sites. *International Journal of Hygiene and Environmental Health, 219*(3), 239–251. https://doi.org/10.1016/j.ijheh.2016.02.003

Shammas, N. K., Wang, L. K., Wang, M. H. S., & Chen, S. L. (2020). Ecological impact and management of solid waste landfill gas. In *Handbook of environment and waste management: Acid rain and greenhouse gas pollution control* (pp. 455–482). https://doi.org/10.1142/9789811207136_0012.

Shen, S., Chen, Y., Zhan, L., Xie, H., Bouazza, A., He, F., & Zuo, X. (2018). Methane hotspot localization and visualization at a large-scale Xi'an landfill in China: Effective tool for landfill gas management. *Journal of Environmental Management, 225*, 232–241. https://doi.org/10.1016/j.jenvman.2018.08.012

Siddiqua, A., Hahladakis, J. N., & Al-Attiya, W. A. K. (2022). An overview of the environmental pollution and health effects associated with waste landfilling and open dumping. *Environmental Science and Pollution Research, 29*(39), 58514–58536. https://doi.org/10.1007/s11356-022-21578-z

Simmons, A. M. (2016). The world's trash crisis, and why many Americans are oblivious. *Los Angeles Times, 22.*

Sohoo, I., Ritzkowski, M., Sultan, M., Farooq, M., & Kuchta, K. (2022). Conceptualization of bioreactor landfill approach for sustainable waste management in Karachi, Pakistan. *Sustainability, 14*(6), 3364. https://doi.org/10.3390/su14063364

Solid Waste Management in Minna, North Central Nigeria: Present Practices and Future Challenges. (2016, February 24). https://doi.org/10.1201/b19424-11

Sosna, D., Stehlíková, B., & Mašek, P. (2023). Ecologies of quantification in waste management: Landfilling, e-waste recycling, and car breaking. *Critique of Anthropology*, 0308275X231205961. https://doi.org/10.1177/0308275X231205961.

Spengler, J. D., Brauer, M., & Koutrakis, P. (1990). Acid air and health. *Environmental Science & Technology, 24*(7), 946–956. https://doi.org/10.1021/es00077a002

Surroop, D., & Mohee, R. (2011). Power generation from landfill gas. In *2nd International Conference on Environmental Engineering and Applications IPCBEE* (Vol. 17, pp. 237–241).

Tchobanoglous, G., Theisen, H., & Vigil, S. A. (1993). *Integrated solid waste management: Engineering principles and management issues.*

Tchobanoglous, G., & Kreith, F. (2002). *Handbook of Solid WM*, McGraw-Hill. ISBN: 9780071356237.

Thomasen, T. B., Scheutz, C., & Kjeldsen, P. (2019). Treatment of landfill gas with low methane content by biocover systems. *Waste Management, 84*, 29–37. https://doi.org/10.1016/j.wasman.2018.11.011

Van Giang, N., & Duan, N. B. (2017). Hydrogeophysical approach for contamination assessment in NamSon landfill, Hanoi, Vietnam. *Arabian Journal of Geosciences, 10*, 1–12. https://doi.org/10.1007/s12517-017-3138-2

Vesilind, P. A., Worrell, W., & Reinhart, D. (2002). Solid Waste Engineering, BROOKS/COLE. *Thomson Learning.*

Vitkov, N. (2019, September). Potential For Energy Production From Landfill Methane In Bulgaria. In *2019 11th Electrical Engineering Faculty Conference (BulEF)* (pp. 1–6). IEEE. org/https://doi.org/10.1109/BulEF48056.2019.9030785.

Vodyanitskii, Y. N., & Yakovlev, A. S. (2016). Contamination of soils and groundwater with new organic micropollutants: A review. *Eurasian Soil Science, 49*, 560–569. https://doi.org/10.1134/S1064229316050148

Voigt, C., Lamprecht, R. E., Marushchak, M. E., Lind, S. E., Novakovskiy, A., Aurela, M., Martikainen, P. J., & Biasi, C. (2017). Warming of subarctic tundra increases emissions of all three important greenhouse gases–carbon dioxide, methane, and nitrous oxide. *Global Change Biology, 23*(8), 3121–3138. https://doi.org/10.1111/gcb.13563

Weng, H., Zhang, F., Zhu, Y., Qin, Y., Ji, Z., & Cheng, C. (2010, July 27). *Treatment of leachate from domestic landfills with three-stage physicochemical and biochemical technology*. https://doi.org/10.1007/s12665-010-0677-1

Xu, Y., Xue, X., Dong, L., Nai, C., Liu, Y., & Huang, Q. (2018). Long-term dynamics of leachate production, leakage from hazardous waste landfill sites and the impact on groundwater quality and human health. *Waste Management, 82*, 156–166. https://doi.org/10.1016/j.wasman.2018.10.009

Yu, X., Sui, Q., Lyu, S., Zhao, W., Liu, J., Cai, Z., Yu, G., & Barcelo, D. (2020). Municipal solid waste landfills: An underestimated source of pharmaceutical and personal care products in the water environment. *Environmental Science & Technology, 54*(16), 9757–9768. https://doi.org/10.1021/acs.est.0c00565

Zabaleta, A. (2008). *Sustainability indicators for Municipal Solid Waste treatment case study: The city of Stockholm: landfill vs. incineration.*

Zappini, G., Cocca, P., & Rossi, D. (2010). Performance analysis of energy recovery in an Italian municipal solid waste landfill. *Energy, 35*(12), 5063–5069. https://doi.org/10.1016/j.energy.2010.08.012

Zhang, C., Xu, T., Feng, H., & Chen, S. (2019). Greenhouse gas emissions from landfills: A review and bibliometric analysis. *Sustainability, 11*(8), 2282. https://doi.org/10.3390/su11082282

Chapter 11
Characteristics and Impacts of Municipal Solid Waste (MSW)

Sukhendu Dey, Megha Santra, Apurba Ratan Ghosh, and Palas Samanta

Abstract Municipal Solid Waste (MSW) refers to the waste generated by households, commercial establishments, institutions, and other non-industrial sources. The characteristics and impacts of MSW are of significant concern due to their environmental, social, and economic implications. This abstract provides an overview of the key characteristics and impacts associated with MSW. MSW is a complex mixture comprising various components such as organic waste, paper, plastic, glass, metals, textiles, and hazardous materials. The composition of MSW varies across regions, influenced by factors like population density, lifestyle patterns, and waste management practices. The organic fraction, including food waste, is a substantial component of MSW and contributes to the generation of greenhouse gases when disposed of in landfills. Improper management of MSW can lead to severe environmental consequences. Landfilling is a common disposal method, which results in the release of greenhouse gases (methane and carbon dioxide) into the atmosphere. These gases contribute to climate change and global warming. Additionally, landfills can contaminate soil and water resources, posing risks to ecosystems and human health. The presence of hazardous materials and improper handling of MSW can pose health risks to both waste handlers and the general population. Exposure to harmful substances from waste, such as heavy metals or toxic chemicals, can lead to various health problems, including respiratory issues, skin diseases, and long-term chronic illnesses. MSW management has significant economic implications. Inefficient waste management practices can result in increased costs for waste collection, transportation, and disposal. Recycling and resource recovery from MSW can provide economic opportunities through the creation of jobs, the production of renewable energy, and the generation of secondary raw materials. The social impacts of MSW are diverse and multifaceted. Poor waste management practices can contribute to the degradation of local aesthetics, leading to a decline in the quality of life for residents. Inadequate waste collection and disposal systems can disproportionately

S. Dey · M. Santra · A. R. Ghosh
Department of Environmental Science, The University of Burdwan, Burdwan, West Bengal, India

P. Samanta (✉)
Department of Environmental Science, Sukanta Mahavidyalaya, University of North Bengal, Dhupguri, West Bengal, India
e-mail: samanta.palas2010@gmail.com

A. Anouzla et al. (eds.), *Handbook of the Effects of Landfill Emissions on Sustainable Development*, https://doi.org/10.1007/978-3-032-01951-6_11

affect marginalized communities, exacerbating social inequalities. Effective waste management strategies, on the other hand, can promote community engagement, environmental awareness, and sustainable development. Understanding the characteristics and impacts of MSW is crucial for developing effective waste management strategies. By implementing sustainable practices such as waste reduction, recycling, composting, and resource recovery, the environmental, social, and economic burdens associated with MSW can be minimized. Integrated waste management approaches that prioritize waste prevention, reuse, and recycling can contribute to a healthier environment and a more sustainable future.

Keywords Municipal waste · Characteristics · Toxic chemicals · Recycling · Resource recovery · Waste management · Environmental burdens

11.1 Introduction

Presently, human society facing two major challenging issues such as environmental pollution and shortage of resources due to huge industrialization and rapid urbanization (Hoang et al., 2021). Based on fast-growing up human population, it is expected that population will be reach up to 10 billion in 2057 which is also another greatest threat throughout the world. Globally, the people discarded the things or waste like garbage, rubbish, junk etc. from the residential and commercial sources are known as Municipal Solid Waste (MSW). According to engineers, MSW refers as all semi-solid and solid materials which are discarded by residential and business purpose but hazardous waste and wastewater are excluded from it (Adnan et al., 2020; Kaur et al., 2021). Annually, 2 Gt of MSW are generated and released to the environment, among of which 33% waste are not collected and processed appropriately (Ayeleru et al., 2018; Usmani et al., 2020).

According to some researchers, annual generation of MSW shows the most generation flow in the year of 2017–2018 by different countries. Among all of them top five MSW generating countries are United States (258 Mt), China (220 Mt), India (168 Mt), Brazil (80 Mt) and Russian Federation (60 Mt). Based on statistical data MSW increasing day by day throughout the world and have an impact on environment and human health. On a global scale, it's creating a great pressure on the waste management, energy sectors and as well as industrial sustainability (Yang et al., 2021). Due to mismanagement of MSW, MSW not only impacting the environment but also inflict the multiple problems on human health, society wellbeing, financial aspects and affecting society (Xiao et al., 2020). Increasing of rapid MSW generation by complexity and volume indeed great challenge to manage in future and present scenario (Ye et al., 2020). Different advanced technologies are used to conversion of non-recyclable MSW into energy carriers like heat, electricity, biofuel and biogas. Conventional waste treatment technologies are landfilling (Christensen et al., 2020) and composting (Miller, 2020) where hydrothermal processing (Chen et al., 2020a), gasification (Prasertcharoensuk et al., 2019), incineration (Escamilla-Garcia et al., 2020), pyrolysis (Kwon et al., 2019) and anaerobic digestion (Wang et al., 2018) offer higher MSW valorization into value-added fuels and chemicals. Every developed country applies different waste to energy production technologies widely and plays an important role (Chen et al., 2020b; Farooq et al., 2021).

11.2 Composition, Characteristics and Management

Composition of MSW may vary and depend on source, location, industrial structure, economic situation, lifestyle and applied methods for waste management (Rezaei et al., 2018). MSW contain 3% textiles, 4% metal, 5% glass, 10% plastics, 13% inert gas, 17% Paper, 46% organic product and 2% miscellaneous (Fig. 11.1) and categorized as recyclable, compostable organic matter, hazardous solid waste and toxic substance (Table 11.1 and Fig. 11.2) (Sharma & Jain, 2020).

Presently, many countries adopted different kind of advanced waste management technologies like landfilling, incineration and unregulated disposal of waste (OECD, 2019). In many cases incineration and landfilling are destructive for environment and human health for a long time. Leachate from landfillng contaminates the soil and water mainly groundwater sources (Venna et al., 2021). Some hazardous MSW like electronics, homecare products, motor oils etc. have to be need to proper segregation and treatment specially at the time of energy production and some chemical compound such as phenol, polycyclic compounds etc. also found which are also serious threat for human health and environment. Some researchers suggest standardized treatment process and desire to improve the current MSW handling (Chen et al., 2022; Kanagamani et al., 2020; Manggali & Susanna, 2019). Other significant

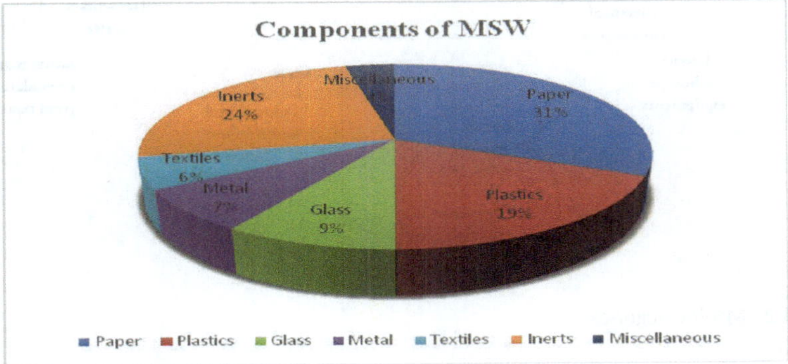

Fig. 11.1 Components of Municipal Solid Waste (MSW)

Table 11.1 Typical classification of MSW based on composition and categorized characterization

Typical classification of MSW based on composition and categorized characterization	
Biodegradable waste	Food and kitchen waste, green waste, paper
Recyclable materials	Paper, cardboard, glass, bottles, jars, tin cans, aluminum cans and foil, metals, certain plastics, fabrics, clothes, tires, batteries etc.
Inert waste	Construction and demolition waste, dirt, rocks, debris
Electrical and electronic waste	Electrical appliances, light bulbs, washing machines, TVs, computers, screens, mobile phones, alarm clocks, watches etc.
Composite waste	Waste clothing, tetra packs, waste plastics such as toys
Hazardous waste	Paints, chemicals, tires, batteries, light bulbs, electrical appliances, fluorescent lamps, aerosol spray cans and fertilizers etc.
Toxic waste	Pesticides, herbicides and fungicides
Biomedical waste	Expired pharmaceutical drugs etc.

Fig. 11.2 MSW categories

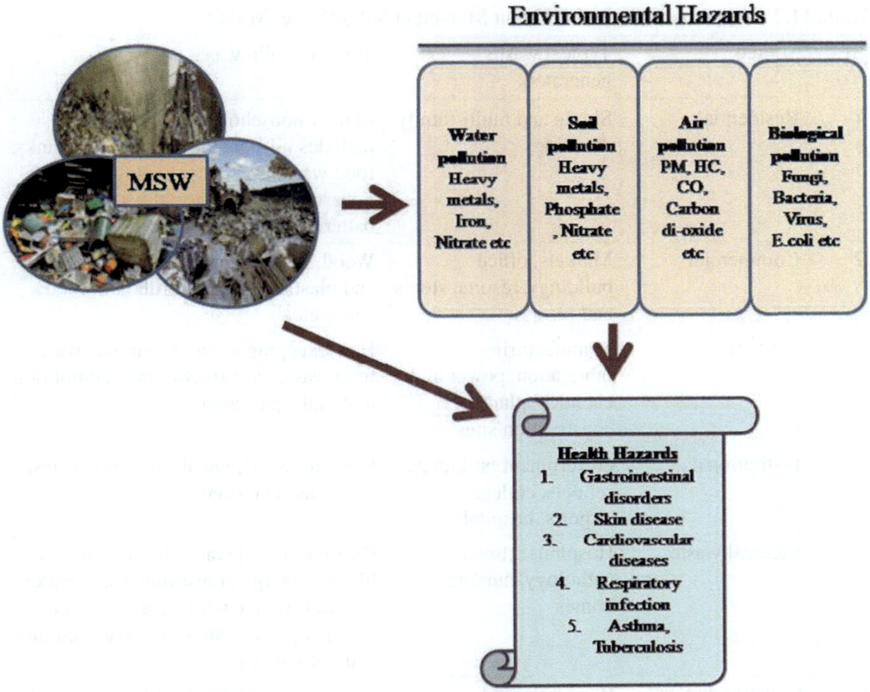

Fig. 11.3 Effects of MSW on environment and human health (Malav et al., 2020)

environmental effects are Water pollution, soil pollution, air pollution and biological pollution and human health effects are gastrointestinal disorders, skin diseases, cardiovascular diseases etc. (Fig. 11.3) (Malav et al., 2020).

11.3 Status and Source of Municipal Solid Waste

Maximum MSW generated by human activities due to improper utilization of energy and natural resources (Laohalidanond et al., 2015). MSW is collected from different sources like residential building, industries, institutions, manufacturers, offices, markets, shops etc. The MSW have also different variety such as organic and inorganic materials, including polymers and non-renewable items or all mixture items (Zheng et al., 2014). In developing countries, households generate highest amount of MSW (55–80%) while in the commercial sectors generate 10–30% of waste (Llano et al., 2021). MSW also collected from non-residential sources which is quite diverse in nature in terms of its contents and physic-chemical characteristics (Dehkordi et al., 2020). According to Kumar and Agrawal (2020), MSW generated through different sources (Table 11.2) which are given below:

Table 11.2 Types and source of different Municipal Solid Waste (MSW)

Sl. No.	Source	Typical waste generators	Types of solid waste
1	Residential	Single and multi-family dwellings	Mainly household hazardous waste includes oils, paints, aerosols, gas tanks, food waste, waste containing mercury, e-waste, plastics, special waste like batteries, bulky items etc.
2	Commercial	Markets, office buildings, resorts, stores and resorts	Wood, e-waste, hazardous waste, paper and plastics, metals, cardboard, food waste etc.
3	Industrial	Manufacturing, fabrication, power and chemical plants, construction sites	Housekeeping waste, hazardous waste, food waste, construction and demolition materials, packaging
4	Institutional	Government buildings, schools, colleges, airports, hospital	E-waste, wood, metals, paper, plastics, glass and metals etc.
5	Medical waste	Hospitals, clinics, pathology, nursing homes	Pharmaceutical waste, hazardous waste like used surgical instruments, chemicals etc., radioactive waste due to cancer therapy, gloves, bandage, swabs, culture and body fluids
6	Construction and demolition waste	Road repairing, renovation sites, demolition of waste, construction sites	Dust, steel, dirt, bricks, tiles wood and concrete
7	Agricultural	Farms, dairies, crops, feedlots, vineyards, orchards	Pesticides, fertilizers, insecticides, chemicals etc., agricultural wastes like rice husks, coconut shells, cotton stalks, spoiled foods
8	Municipal services	Wastewater treatment plants, landscaping, street waste, waste from public places and beaches	Sludge, landscape, tree trimming, sweepings, waste from recreational areas beaches parks etc.
9	Process	Power plants, chemical plants, mineral extraction and processing, refineries etc.	Slag, off-specification products, tailing, industrial process wastes

Excess amount of waste generation has a positive correlation with the economic condition or growth of the nation, urbanization, population growth and industrialization (Rana et al., 2015). In 2016, MSW generation was around 2010 million tons (MT) per year throughout world and it is expected that the quantity of waste generation will increase to 3400 MT by 2050. It is big increase about 70% of MSW production within 34 years (Khandelwal et al., 2019). According to World Bank Group, 2018 percentage of waste generation, composition and management within the groups of the countries are discussed below, countries are grouped on the basis of their gross national income (GNI), like lower income countries (LICs), lower-middle-income countries (LMICs), upper-middle-income countries (UMICs) and high-income countries (HICs) (Table 11.3).

The world classified into regions like East Asia and Pacific (EAP), Europe and Central Asia (ECA), South Asia (SA), NA, Latin America and Caribbean (LAC), Sub-Saharan Africa (SSA) and the Middle East and North Africa (MENA). MSW generation getting faster due to rapid urbanization of the world and in future progress will be more. According to the regions of the countries waste generation, collection, composition and management are discussed below (Table 11.4):

11.4 Sustainable Development

Globally, the waste means undesirable and/or unnecessary byproduct which is an unavoidable consequence of most processes and yearly 7–9 billion tons of wastes are generated (Wilson & Velis, 2015). MSW have an impact on various areas of sustainable development in each of the three sustainability domains like economy, ecology and society. Out of 17 Sustainable Development Goals (SDGs) of the 2030 Agenda for Sustainable Development, adopted by the 193 UN Member States in September 2015 (UNSDGs, 2016), at least 12 SDGs and they have a direct link to SWM. United Nations Environment Programme (UNEP)'s recent Global Waste Management Outlook (GWMO) developed the contrary argument, which is the proper nature of SWM and impact on not just one but on 12 SDGs should only emphasize the importance and increase the political priority of SWM (Wilson et al., 2015). In Sustainable Development Goals, the Goals and indicators represent waste in Goal 11 which is sustainable cities and communities and among this Goal 11.6 represent as reduce the adverse per capita environmental impact of cities, including municipal waste management which is Goal 11.6.1. Sustainable Development Goals (SDGs) represent the goal 11.6 as 'By 2030, reduce the adverse per capita environmental impact of cities, including by paying special attention to air quality and municipal and other waste management'. This indicator measures the progress of city's MSW management performance. It aims to determining the proportion of MSW which is regularly collected and treated in a city and quantifies like,

- The total MSW generated in the city (tones/day),
- The total MSW collected in the city (tones/day),
- The population proportion with access to basic MSW collection services in the city (%),
- The total MSW managed in controlled facilities in the city (tones/day),
- The MSW composition.

Table 11.3 Waste generation, collection, composition and management according to the income level of countries

Income groups of countries		HICs	UMICs	LICs	LMICs
GNI	US$/capita/year	≥ 12,476	4036–12,475	≤ 1025	1026–4035
Urban population, 2018	%	81.5	66.6	32.2	40.6
Urban population, 1950	%	58.5	22.1	9.3	17.2
Average MSW generation rate, 2016	MT/year	683	655	93	586
	Kg/capita/day	1.58	0.69		
Average MSW generation rate, 2016	%	34	32	5	29
Projected MSW generation rate, 2025	MT/year	781	835	143	827
	Kg/capita/day	1.71	0.83	0.43	0.63
Projected MSW generation rate, 2050	MT/year	879	1004	283	1233
	Kg/capita/day	1.87	0.99	0.56	0.79
Average MSW collection rate	Overall (%)	96	82	39	51
	In rural areas (%)	98	45	26	33
	In urban areas (%)	100	85	48	71
Average MSW composition (%), 2016	Organic (%)	32	54	56	53
	Paper (%)	25	12	7	12.5
	Plastic (%)	13	11	6.4	11
	Glass (%)	5	4	1	3
	Wood (%)	4	1	0	1
	Metal (%)	6	2	2	2
	Rubber and leather (%)	4	1	0	0.5
	Other (%)	11	15	27	17
Estimated MSW generation growth rate (from 2016 to 2050)	(%)	29	53	204	110
MSW treatment and disposal methods (%), 2016	Open dumps	2	30	93	66
	Composting	6	2	0.3	10
	Sanitary landfilling	39	54	3	18
	Incineration	22	10	0	0.5
	Recycling	29	4	3.7	5.5

(continued)

Table 11.3 (continued)

Income groups of countries		HICs	UMICs	LICs	LMICs
GNI	US$/capita/year	≥ 12,476	4036–12,475	≤ 1025	1026–4035
	Other methods	2	0	0	0
E-waste generation rate	Kg/capita/day	0.05	0.02	< 0.01	0.01
Industrial waste generation rate	Kg/capita/day	42.62	5.72	NA	0.3
SWM budget (billion US$/year)	Available data,2012	159.3	24.5	1.5	20.1
	Estimated data, 2025	220.2	63.5	7.7	84.1
Solid waste costs of collection and treatment (US$/ton/year)	AD	65–150	50–100	NA	20–80
	Recycling	30–80	05–50	0–25	05–30
	Composting	35–90	10–40	05–30	10–40
	Incineration	70–200	40–100	NA	40–100
	Sanitary landfilling	40–100	15–40	10–20	15–40
	Open dumps	NA	03–10	02–08	03–10
	Collection	90–200	30–75	20–50	30–75

Sources World Bank Group (2018), WEC (2016)

To calculate the SDG indicator 11.6.1, the formula is used as (UNEP, 2021; UN, 2018):

$$\text{SDG } 11.6.1 = \frac{\text{Total MSW collected and managed in controlled faclities tonnes per day}}{\text{Total MSW generated tonnes per day}} \times 100$$

SDG 11.6.1 also provides two more important sub-categories with varying policy implications:

$$\text{SDG } 11.6.1\text{, category a} = \frac{\text{Total MSW collected tonnes per day}}{\text{Total MSW generated tonnes per day}} \times 100(\%)$$

$$\text{SDG } 11.6.1\text{, category b} = \frac{\text{Total MSW managed in controlled faclities tonnes per day}}{\text{Total MSW generated tonnes per day}} \times 100(\%)$$

Table 11.4 Solid waste generation, collection, composition and management according to the regions of countries

Region of the countries		EAP	ECA	SA	NA	LAC	SSA	MENA
Urbanization rate, 2016	%	57	71	33	82	81	38	65
Average MSW generation rate, 2016	MT/year	468	392	334	289	231	174	129
	Kg/capita/day	0.56	1.18	0.52	2.21	0.99	0.46	0.81
Average MSW generation rate, 2016	%	23	20	17	14	11	9	6
Projected MSW generation rate, 2025	MT/year	602	440	466	342	290	269	177
	Kg/capita/day	0.68	1.3	0.62	2.37	1.11	0.5	0.9
Projected MSW generation rate, 2050	MT/year	714	490	661	396	369	516	255
	Kg/capita/day	0.81	1.45	0.79	2.5	1.3	0.63	1.06
Average MSW collection rate	Overall (%)	71	90	51	100	84	44	82
	In rural areas (%)	45	55	40	99.7	30	9	74
	In urban areas (%)	77	96	77	100	85	43	90
Average MSW composition (%), 2016	Organic (%)	53	36	57	28	52	43	58
	Paper (%)	15	18.6	10	28	13	10	13
	Plastic (%)	12	11.5	8	12	12	8.6	12
	Glass (%)	2.6	8	4	4.5	4	3	3
	Wood (%)	2	1.6	1	5.6	0.5	0.4	1
	Metal (%)	3	3	3	9.3	3	5	3
	Rubber and leather (%)	0.4	0.4	2	9	0.5	0	2
	Other (%)	12	21	15	3.6	15	30	8
Estimated MSW generation growth rate (from 2016 to 2050)	(%)	53	25	98	37	60	197	75
MSW treatment and disposal methods (%), 2016	Open dumps	18	25.6	75	0	26.8	69	52.7
	Composting	2	10.7	16	0.4	0.2	0.4	4
	Sanitary landfilling	46	25.9	4	54.3	68.5	24	34
	Incineration	24	17.8	0	12	0	0	0.3
	Recycling	9	20	5	33.3	4.5	6.6	9

11.4.1 Ecosystem and Biodiversity

Waste disposal and management are the form of land-use change like human being, exchanging the habitat of the species where human share the planet. Generated waste, emission of toxic chemicals may affect as acute and/or chronic toxicity and harming the flora and fauna. The hazardous waste released from hospitals, industry as well as residential and disposed to the environment directly and burn openly. MSW directly affect the ocean health. Every year millions of metric tons of plastic waste enter into the ocean (Gregory, 2009) in a 6:1 ratio of plastics and other marine debris (Moore et al., 2001). The impacts of all these persistent wastes serve as at least 267 species ingest and/or entangled by plastics where leading to increased morbidity and mortality (Derraik, 2002). Just because of plastics absorb persistent organic pollutants and concentrate toxins and become vectors of poisons which can accumulate through food chain (Law et al., 2010). Besides plastic waste can inhibit the gas exchange from the sediments at the time of settles down and act as carriers of invasive species. Because of vast quantity of waste and at the entry point of waste into the ocean is not known how much MSW ends up in the ocean. In future it is needed that to calculate the contribution of unmanaged MSW on environment.

11.4.2 Waste Management Aspects for Energy Production Strategy in Circular Bioeconomy

The rapid growth of urbanization throughout the world led to significant increase in demand for material goods and energy (Venkata Mohan et al., 2016). Where 'Bioeconomy' is the relatively new term but the basis of the bioeconomy is already in place. From the waste management perspective bioeconomy is considers as organic wastes as raw materials, where transforming them into useful products as for example crude oil is used as raw materials for chemical production. Circular Economy (CE) optimizes the activities to minimize the consumption and generation of finite resources (Ellen MacArthur Foundation, 2013). CE has two key elements such as 'technical nutrients' and 'biological nutrients' and three principles (Ellen MacArthur Foundation & Granta Design, 2015) such as:

- To preserve and enhancing the use of resources to regenerate the natural systems;
- To maximizing the yield and value of materials, resources and components by keeping in economy loop through recycling and reusing;
- Aiming at designing of a system to eliminate the negative impacts and externalities of residential and commercial activities which can harm the environment.

The application of waste to energy production process supports to reach the goals of CE by transforming waste into useful forms of energy (Garmulewicz et al., 2018). Waste to energy production processes is pyrolysis, incineration, anaerobic digestion, gasification, biomethanation and landfill gas recovery (Fig. 11.4). These approaches

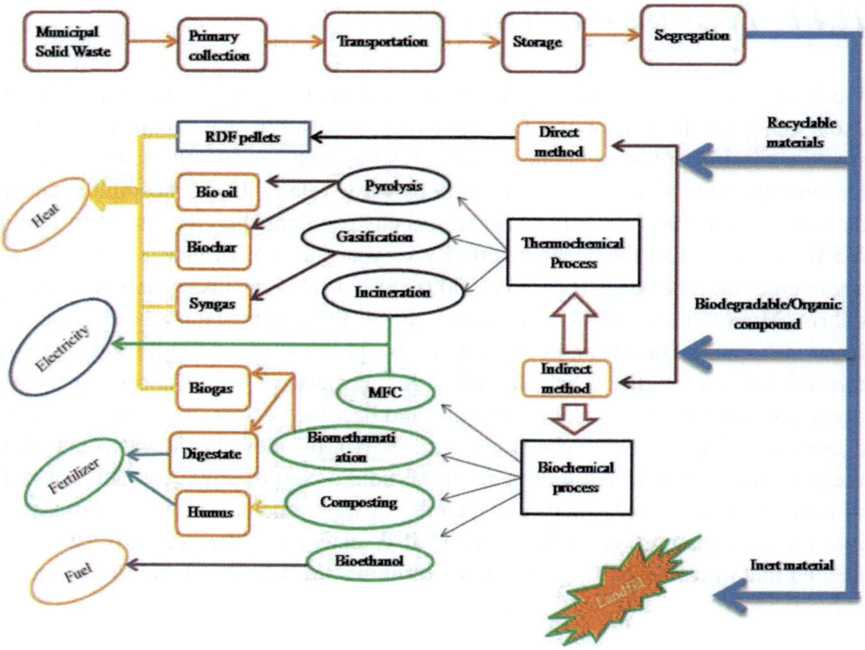

Fig. 11.4 Current technologies for energy production from waste

are effective MSW treatments while giving rise to valorization of energy (Palacio et al., 2019).

These methods are intended to meet three objectives:

- Decrease the total volume of MSW which is generated from commercial and residential sectors and all wastes are disposed in landfills
- Minimize the pollution of biodegradables in MSW along with prevention of secondary environmental pollution with runaway CH_4 from potential decay of biodegradable remaining after the treatment
- Valorize the energy content of non-recyclable MSW in the form of generating electricity and heat.

Considering the entire technologies direct and indirect path are used to energy recovery through waste to energy production where direct path used for direct combustion of refuse-derived fuel and indirect path used for pre-treatment steps before the energy generation.

11.5 Effect of Municipal Solid Waste on Environmental and Human Health

In developing countries impact of MSW on environment and human health are related with garbage disposal. Implementation of sustainable MSW management occurred to minimize environmental losses of pollutant gases which is related with climate change due to greenhouse gases (GHGs) and acidification of the ecosystem through ammonia evolution. In MSW management different technologies are used to investigate or to quantify the carbon (C) and nitrogen (N) losses concerning to various environmental and operational conditions but all the impact is uncertain (Pardo et al., 2015). Most of the hazardous waste is generated from different industries and their mixing with MSW creates a severe risk on human health (Alam & Ahmade, 2013). Heavy metals contamination in food chain cycle occurred due to unscientific discharge of MSW and industrial liquid effluents having metal toxicities due to discharged in drainage system or at open dumping sectors which may causes different types of disease. Health problems are congenital malformation, cancer, chemical poisoning through chemical inhalation, low birth weight, neurological disease, vomiting and nausea, and others problems are like uncollected waste can obstruct the stormwater runoff resulting in flood, mercury toxicity from eating fish with high levels of mercury, high algal population in se and river, plastic pollution in oceans ingested by birds, water and soil quality degradation (Royal Commission of Environment Pollution, 1984).

11.6 Impacts of Municipal Solid Waste (MSW)

Globally each year MSW generated about 10×10^3 kg and waste management is a civilization related problem (Shu et al., 2018) where the most common and oldest disposal method is landfills worldwide (Ghosh et al., 2015). MSW landfills can cause pollution of all environmental sectors along with effects and also have an impact on economical, land-use planning and social related categories. Biodegradation, natural habitat degradation, energy consumption, water quality and/or contamination, hazardous gas emissions are the main environmental impacts. Disposal costs, siting resistance and regulations are related with economic impacts. Setting landfill areas related with land-use and social impact related with community. MSW specially affect the environment through emissions to the water, air and land due to production and management of the waste. Different technologies are used to minimize the environmental impact of waste.

11.7 Future Perspectives in the Field

Energy production from the MSW by the application of different kind of waste to energy production can be assessed based on sustainability performance in four areas such as environment, finance/economic, technology and socio-political activities. Implementation of sustainable and integrated waste management systems provides an important strategy to reduce landfilling waste. At the source of waste sorting and installation of special pre-processing plants can improve the rate of recovery and recyclable materials from the waste stream and improving the overall value of the process. The minimizing of waste generation should be priority over the planning of various waste treatment methods. Through municipal program public awareness can be adopted and sorting of household waste should be implemented and promoted through environmental educational initiatives. Engagement, encouragement and input some knowledge about waste management strategy or policies for future and current scenario among the citizens. Presently, MSW is the environmental and resource management challenges. Different significant future perspectives or issues are given below:

- Interdisciplinary approaches are needed regarding waste management because waste is not only technical problem,
- More attention should be given to manufacturers and consumers ability to affect the nature of MSW which is produced through the product design for reuse and repair and through purchasing choices,
- Standardized data collection and more research is needed at point of waste generation and management along with new conceptual approaches which is emerging to manage waste as a resource.

11.8 Conclusions

Different waste to energy production technologies is used to produce energy from waste in the form of heat and/or electricity. Energy production from organic fraction of MSW has attracted the waste management professionals, interest of policymakers and energy researchers. In this study, the technologies are categorized as direct and indirect approach, while previously only direct combustion of waste are used to produce heat which is deemed to have lower energy. The open burning process of MSW is strictly prohibited in most the countries to avoid the risk of releasing toxic substance into the environment and the effect on human health, biodiversity and as well as environment. Technologically, most popular indirect waste to energy production procedure is thermochemical or biological approaches to convert MSW to energy. Indirect waste to energy involves tedious procedure to convert waste to intermediates before energy recovering from them. As for example to treatment of MSW by indirect thermochemical process along with pyrolysis, gasification and carbonization is popular where as incineration is the direct method. The thermochemical processes are famous for rapid conversation process of energy products such as bio-oil, char, combustible gases like biogas and syngas and scalability. Through biological process such as composting, fermentation, anaerobic digestion are used to produce energy products like ethanol and biogas and all these biological processes suppress the microorganism.

Economically each technology indicates their low waste utilization and increased costs. The high costs result from waste transportation, sorting and equipments. Advancement of the commercial, societal and technologies provides critical momentum for the rapid adoption of various kind of waste to energy technologies. In future circular economy it is suggested that a better and sustainable integration of waste to energy process is required with improved comprehension and addition of safety protocols. This safety protocols are very much important to minimizing environmental pollution and the risk of propagation of microorganisms, toxicity and related diseases.

References

Adnan, M., Jha, A., & Kumar, S. (2020). Municipal Solid Waste Management and its Impact: A Review. *International Journal of Advanced Research in Engineering and Technology (IJARET), 11*(5), 685–693. https://doi.org/10.34218/IJARET.11.5.2020.071

Alam, P., & Ahmade, K. (2013). Impact of solid waste on health and the environment. *International Journal of Sustainable Development and Green Economics (IJSDGE), 2*(1), 165–168.

Ayeleru, O. O., Okonta, F. N., & Ntuli, F. (2018). Municipal solid waste generation and characterization in the City of Johannesburg: A pathway for the implementation of zero waste. *Waste Management, 79*, 87–97. https://doi.org/10.1016/j.wasman.2018.07.026

Chen, W.-T., Haque, M. A., Lu, T., Aierzhati, A., & Reimonn, G. (2020a). A perspective on hydrothermal processing of sewage sludge. *Current Opinion in Environmental Science & Health, 14*, 63–73. https://doi.org/10.1016/j.coesh.2020.02.008

Chen, X., Lin, J., Li, X., & Ma, Z. (2020b). A novel framework for selecting sustainable healthcare waste treatment technologies under Z-number environment. *Journal of the Operational Research Society, 72*(9), 1–14. https://doi.org/10.1080/01605682.2020.1759382

Chen, W. H., Hoang, A. T., Nižetić, S., Pandey, A., Cheng, C. K., Luque, R., Ong, H. C., Thomas, S., & Nguyen, X. P. (2022). Biomass-derived biochar: From production to application in removing heavy metal-contaminated water. *Process Safety and Environmental Protection, 160*, 704–733. https://doi.org/10.1016/j.psep.2022.02.061

Christensen, T. H., Cossu, R., & Stegmann, R. (Eds.) (2020). *Landfilling of waste: Biogas*, 1st ed. CRC Press. https://doi.org/10.1201/9781003062097.

Dehkordi, S. M. M. N., Jahromi, A. R. T., Ferdowsi, A., Shumal, M., & Dehnavi, A. (2020). Investigation of biogas production potential from mechanical separated municipal solid waste as an approach for developing countries (case study: Isfahan-Iran). *Renewable and Sustainable Energy Reviews, 119*, 109586. https://doi.org/10.1016/j.rser.2019.109586

Derraik, J. G. B. (2002). The pollution of the marine environment by plastic debris: A review. *Marine Pollution Bulletin, 44*(9), 842–852. https://doi.org/10.1016/S0025-326X(02)00220-5

Ellen MacArthur Foundation and Granta Design. (2015). *Circularity indicators. An approach to measuring circularity*. https://www.clmsostenible.es/wpcontent/uploads/2019/02/Circularity-Indicators_ProjectOverview_May2015.pdf. Accessed December 09, 2021.

Escamilla-García, P. E., Camarillo-López, R. H., Carrasco-Hernández, R., Fernández-Rodríguez, E., & Legal-Hernández, J. M. (2020). Technical and economic analysis of energy generation from waste incineration in Mexico. *Energy Strategy Reviews, 31*, 100542. https://doi.org/10.1016/j.esr.2020.100542

Farooq, A., Haputta, P., Silalertruksa, T., & Gheewala, S. H. (2021). A framework for the selection of suitable waste to energy technologies for a sustainable municipal solid waste management system. *Frontiers in Sustainability, 2*, 27. https://doi.org/10.3389/frsus.2021.681690

Garmulewicz, A., Holweg, M., Veldhuis, H., & Yang, A. (2018). Disruptive technology as an enabler of the circular economy: What potential does 3D printing hold? *California Management Review, 60*, 112–132. https://doi.org/10.1177/0008125617752695

Ghosh, P., Gupta, A., & Thakur, I. S. (2015). Combined chemical and toxicological evaluation of leachate from municipal solid waste landfill sites of Delhi, India. *Environmental Science and Pollution Research, 22*, 9148–9158. https://doi.org/10.1007/s11356-015-4077-7

Gregory, M. R. (2009). Environmental implications of plastic debris in marine settings—Entanglement, ingestion, smothering, hangers-on, hitch-hiking and alien invasions. *Philosophical Transactions of the Royal Society b: Biological Sciences, 364*(1526), 2013–2025. https://doi.org/10.1098/rstb.2008.0265

Hoang, A. T., & Nguyen, X. P. (2021). Integrating renewable sources into energy system for smart city as a sagacious strategy towards clean and sustainable process. *Journal of Cleaner Production, 305*, 127161. https://doi.org/10.1016/j.jclepro.2021.127161

Kanagamani, K., Geethamani, P., Narmatha, M. (2020). Hazardous waste management. *In Environmental change and sustainability*. IntechOpen, https://doi.org/10.5772/intechopen.94080.

Kaur, A., Bharti, R., & Sharma, R. (2021). Municipal solid waste as a source of energy. *Materials Today: Proceedings, 81*(2), 904–915. https://doi.org/10.1016/j.matpr.2021.04.286

Khandelwal, H., Dhar, H., Thalla, A. K., & Kumar, S. (2019). Application of life cycle assessment in municipal solid waste management: A worldwide critical review. *Journal of Cleaner Production, 209*, 630–654. https://doi.org/10.1016/j.jclepro.2018.10.233

Kumar, A., & Agrawal, A. (2020). Recent trends in solid waste management status, challenges, and potential for the future Indian cities—A review. *Current Research in Environmental Sustainability, 2*, 100011. https://doi.org/10.1016/j.crsust.2020.100011

Kwon, E.E., Kim, S., Lee, J. (2019). Pyrolysis of waste feedstocks in CO2 for effective energy recovery and waste treatment. *Journal of CO2 Utilization, 31*,173–180. https://doi.org/10.1016/j.jcou.2019.03.015.

Laohalidanond, K., Chaiyawong, P., & Kerdsuwan, S. (2015). Municipal solid waste characteristics and green and clean energy recovery in Asian megacities. *Energy Procedia, 79*, 391–396. https://doi.org/10.1016/j.egypro.2015.11.508

Law, K. L., Mor´et-Ferguson, S., Maximenko, N. A., Proskurowski, G., Peacock, E. E, et al. (2010). Plastic accumulation in the North Atlantic subtropical gyre. *Science, 329*(5996), 1185–88.https://doi.org/10.1126/science.1192321

Llano, T., Arce, C., & Finger, D. C. (2021). Optimization of biogas production through anaerobic digestion of municipal solid waste: A case study in the capital area of Re ykjavik, Iceland. *Journal of Chemical Technology & Biotechnology, 96*, 1333–1344. https://doi.org/10.1002/jctb.6654

MacArthur, E. (2013). Towards the circular economy. *Journal of Industrial Ecology, 2*(1), 23–44.

Malav, L. C., Yadav, K. K., Gupta, N., Kumar, S., Sharma, G. K., Krishnan, S., Rezania, S., Kamyab, H., Pham, Q. B., Yadav, S., Bhattacharyya, S., Yadav, V. K., & Bach, Q. V. (2020). A review on municipal solid waste as a renewable source for waste-to-energy project in India: Current practices, challenges, and future opportunities. *Journal of Cleaner Production, 277*, 123227. https://doi.org/10.1016/j.jclepro.2020.123227

Manggali, A. A., & Susanna, D. (2019). Current management of household hazardous waste (HHW) in the Asian region. *Reviews on Environmental Health, 34*(4), 415–426. https://doi.org/10.1515/reveh-2019-0049

Miller, F. C. (2020). Composting of municipal solid waste and its components. In *Microbiology of solid waste* (pp. 115–154). CRC Press. eBook ISBN9780138747268.

Moore, C. J., Moore, S. L., Leecaster, M. K., & Weisberg, S. B. (2001). A comparison of plastic and plankton in the North Pacific central gyre. *Marine Pollution Bulletin, 42*(12), 1297–1300. https://doi.org/10.1016/S0025-326X(01)00114-X

OECD. (2019). *Waste management and the circular economy in selected OECD countries: Evidence from environmental performance reviews, OECD environmental performance reviews.* OECD Publishing.https://doi.org/10.1787/9789264309395-en

Palacio, J. C. E., Santos, J. J. C. S., Renó, M. L. G., Júnior, J. C. F., Carvalho, M., Reyes, A. M. M., Orozco, D. J. R. (2019). Municipal solid waste management and energy recovery. In I. H. Al-Bahadly (Eds.), *Energy conversion–Current technologies and future trends* (pp. 127–146). InTech Open. https://doi.org/10.5772/intechopen.79235.

Pardo, G., Moral, R., Aguilera, E., & Del Prado, A. (2015). Gaseous emissions from management of solid waste: A systematic review. *Global Change Biology, 21*(3), 1313–1327. https://doi.org/10.1111/gcb.12806

Prasertcharoensuk, P., Bull, S. J., & Phan, A. N. (2019). Gasification of waste biomass for hydrogen production: Effects of pyrolysis parameters. *Renewable Energy, 143*, 112–120. https://doi.org/10.1016/j.renene.2019.05.009

Rana, R., Ganguly, R., & Gupta, A. K. (2015). An assessment of solid waste management system in Chandigarh City, India. *Electronic Journal of Geotechnical Engineering, 20*(6), 1547–1572.

Rezaei, M., Ghobadian, B., Samadi, S. H., & Karimi, S. (2018). Electric power generation from municipal solid waste: A techno-economical assessment under different scenarios in Iran. *Energy, 152*, 46–56. https://doi.org/10.1016/j.energy.2017.10.109

Royal Commission on Environmental Pollution. (1984). *10th Report tackling pollution experience and prospects.* HMSO.

Sharma, K. D., & Jain, S. (2020). Municipal solid waste generation, composition, and management: The global scenario. *Social Responsibility Journal, 16*(6), 917–948. https://doi.org/10.1108/SRJ-06-2019-0210

Shu, S., Zhu, W., Wang, S., Wang, C., Ng, C. W. W., Chen, Y., & Chiu, A. C. F. (2018). Leachate breakthrough mechanism and key pollutant indicator of municipal solid waste landfill barrier systems: Centrifuge and numerical modelling approach. *Science of the Total Environment, 612*, 1123–1131. https://doi.org/10.1016/j.scitotenv.2017.08.185

UN Environment. (2018). *Global manual on ocean statistics.* Towards a definition of indicator methodologies: https://wedocs.unep.org/handle/20.500.11822/35086

UN. (2016). *Sustainable development goals.* Available online: http://www.un.org/sustainabledevelopment/sustainable-development-goals/. Accessed on December 1, 2016.

United Nations Environment Programme. (2021). *Global chemicals and waste indicator review document*. https://wedocs.unep.org/handle/20.500.11822/36753

Usmani, Z., Kumar, V., Varjani, S., Gupta, P., Rani, R., & Chandra, A. (2020). Municipal solid waste to clean energy system: a contribution toward sustainable development. In *Current developments in biotechnology and bioengineering* (pp. 217–231). Elsevier. https://doi.org/10.1016/B978-0-444-64321-6.00011-2.

Venkata Mohan, M. S., Nikhil, G. N., Chiranjeevi, P., Nagendranatha, R. C., Rohit, M. V., Naresh, K. A., & Sarkar, O. (2016). Waste biorefinery models towards sustainable circular bioeconomy: Critical review and future perspectives. *Bioresource Technology, 215*, 2–12. https://doi.org/10.1016/j.biortech.2016.03.130

Venna, S., Sharma, H. B., Reddy, P. H. P., Chowdhury, S., & Dubey, B. K. (2021). Landfill leachate as an alternative moisture source for hydrothermal carbonization of municipal solid wastes to solid biofuels. *Bioresource Technology, 320*, 124410. https://doi.org/10.1016/j.biortech.2020.124410

Wang, D., Liu, X., Zeng, G., Zhao, J., Liu, Y., Wang, Q., Chen, F., Li, X., & Yang, Q. (2018). Understanding the impact of cationic polyacrylamide on anaerobic digestion of waste activated sludge. *Water Research, 130*, 281–290. https://doi.org/10.1016/j.watres.2017.12.007

WEC. (2016). *World energy resources-2016 [WWW document]*. World Energy Counc. Available at: https://www.worldenergy.org/wp-content/uploads/2016/10/World-EnergyResources-Full-report-2016.10.03.pdf. Accessed January 11, 2017.

Wilson, D. C., & Velis, C. A. (2015). Waste management–still a global challenge in the 21st century: An evidence-based call for action. *Waste Management & Research, 33*(12), 1049–1051. https://doi.org/10.1177/0734242X15616055

Wilson, D. C., Rodic, L., Modak, P., Soos, R., Carpintero Rogero, A., Velis, C.; Iyer, M., Simonett, O. (2015). Global waste management outlook; prepared for United Nations environment programme (UNEP) and international solid waste association (ISWA). In D. C. Wilson (Ed.), *UNEP International Environment Technology Centre (IETC)*: Osaka, Japan. Available online: http://web.unep.org/ietc/what-we-do/global-waste-managementoutlook-gwmo. Accessed on February 24, 2017.

World Bank Group. (2018). What a waste 2.0: a global snapshot of solid waste management to 2050-The urban development series [WWW document]. International Bank for Reconstruction and Development/World Bank. Available at: https://openknowledge.worldbank.org/handle/10986/30317. Accessed June 14, 2019.

Xiao, S., Dong, H., Geng, Y., Francisco, M.-J., Pan, H., & Wu, F. (2020). An overview of the municipal solid waste management modes and innovations in Shanghai, China. *Environmental Science and Pollution Research, 27*, 29943–29953. https://doi.org/10.1007/s11356-020-09398-5

Yang, S., Wei, J., & Cheng, P. (2021). Spillover of different regulatory policies for waste sorting: Potential influence on energy-saving policy acceptability. *Waste Management, 125*, 112–121. https://doi.org/10.1016/j.wasman.2021.02.008

Ye, Y., Ngo, H. H., Guo, W., Chang, S. W., Nguyen, D. D., Varjani, S., Ding, A., Bui, X.-T., & Nguyen, D. P. (2020). Bio-membrane based integrated systems for nitrogen recovery in wastewate treatment: Current applications and future perspectives. *Chemosphere, 265*, 129076. https://doi.org/10.1016/j.chemosphere.2020.129076

Zheng, L., Song, J., Li, C., Gao, Y., Geng, P., Qu, B., & Lin, L. (2014). Preferential policies promote municipal solid waste (MSW) to energy in China: Current status and prospects. *Renewable and Sustainable Energy Reviews, 36*, 135–148. https://doi.org/10.1016/j.rser.2014.04.049